Recursive Realism
The Universe Beyond Binary Extremism

By

Behzad Ghorbani

Copyright © 2024 by Behzad Ghorbani

All rights reserved. No part of this book may be reproduced, distributed, or transmitted in any form or by any means, including photocopying, recording, or other electronic or mechanical methods, without the prior written permission of the publisher, except in the case of brief quotations embodied in critical reviews and certain other non-commercial uses permitted by copyright law. This book is independently published by the author through Amazon Kindle Direct Publishing (KDP).

This work is a product of hybrid collaboration, authored by Behzad Ghorbani and refined through the Hyper Hybrid Intellect model, with AI acting as the joint hybrid editor. This prototype combines human creativity and thought leadership with advanced analytical enhancement and structural optimisation (including Comparative Matrix Analysis, Mirroring Hemisphere Fractal Analysis, Totalisation of Details, and Recursive Realism Perspective), ensuring a uniquely crafted and rigorously developed narrative.

First Edition

Published by Amazon.com

10 9 8 7 6 5 4 3 2 1

November 2024

Paperback: ISBN: 9798344035017
Hardcover: ISBN: 9798345227329

CONTENTS

	Preface	8
1	Introduction to Recursive Realism	10
2	The Foundation of Recursive Perception	17
3	Emergent Reality in Recursive Realism	23
4	Functional Aspects of Recursive Realism	28
5	The Recursive Universe	36
6	Recursive Realism and the Nature of Time	45
7	Recursive Realism and the Architecture of Reality	53
8	Recursive Realism and the Boundaries of Consciousness	62
9	Recursive Realism and the Philosophy of Knowledge	71
10	Recursive Realism and the Limits of Perception	78
11	Recursive Realism and the Nature of Conscious Experience	85
12	Recursive Realism and the Architecture of Time	92
13	Recursive Realism and the Emergence of Complexity	99
14	Recursive Realism and the Patterns of Consciousness	106
15	Recursive Realism and the Nature of Reality	113
16	Recursive Realism and the Dynamics of Knowledge	121
17	Recursive Realism and the Architecture of the Mind	128
18	Recursive Realism and the Patterns of Social Interaction	135

19	Recursive Realism and the Philosophy of Time	143
20	Recursive Realism and the Boundaries of Reality	151
21	Recursive Realism and the Dynamics of Ethics and Morality	158
22	Recursive Realism and the Nature of Consciousness	166
23	Recursive Realism and the Nature of Reality	174
24	Recursive Realism and the Concept of Free Will	181
25	Recursive Realism and the Evolution of Knowledge	188
26	Recursive Realism and the Nature of Language	195
27	Recursive Realism and the Dynamics of Human Emotion	202
28	Recursive Realism and the Perception of Time	209
29	Recursive Realism and the Emergence of Complexity in Nature	216
30	Recursive Realism and the Philosophy of Consciousness	223
31	Recursive Realism and the Origins of Creativity	230
32	Recursive Realism and the Ethical Dimensions of Self-Reflection	237
33	Recursive Realism and the Structure of Knowledge	244
34	Recursive Realism and the Limits of Human Understanding	251
35	Recursive Realism and the Interconnectedness of All Things	258
36	Recursive Realism and the Future of Knowledge	266
37	Recursive Realism and the Ethics of Knowledge Creation	273
38	Recursive Realism and the Role of Imagination in Knowledge Creation	280
39	Recursive Realism and the Quest for Meaning	287

40	Recursive Realism and the Evolution of Consciousness	294
41	Recursive Realism and the Dynamics of Human Culture	301
42	Recursive Realism and the Nature of Time	308
43	Recursive Realism and the Intersection of Science and Spirituality	315
44	Recursive Realism and the Limits of Human Knowledge	323
45	Recursive Realism and the Role of Imagination in Shaping Reality	331
46	Recursive Realism and the Dynamics of Memory and Future Thinking	339
47	Recursive Realism and the Relationship Between Language, Thought, and Reality	346
48	Recursive Realism and the Interplay Between Consciousness and the Unconscious	353
49	Recursive Realism and the Integration of Science, Philosophy, and Spirituality	360
50	Recursive Realism and the Future of Human Thought	368
51	Modelling and Predicting the Future with Recursive Realism	376
	General References	381
	Further Readings	383

Preface

The origins of this work stem from an enduring pursuit of understanding the profound relationship between human perception and the cosmos. This journey traversed the expansive terrains of cognitive science, philosophy, biology, and theoretical inquiry, revealing a pivotal truth: the frameworks often employed to interpret the human mind and its interaction with reality, grounded in binary oppositions such as nature versus nurture, fall short of encompassing the true depth and fluidity of cognition. These limitations underscore the need for a more encompassing paradigm capable of capturing the dynamic and interwoven nature of thought.

From this necessity emerged the concept of Recursive Realism, a paradigm designed to transcend oversimplifications and enable a deeper exploration of perception. Far from a purely abstract theory, Recursive Realism anchors itself in empirical observations, drawing from the brain's mechanisms, the intricacies of information processing, and the self-referential qualities of recursive thought. By synthesising these dimensions, the model serves as a robust framework for understanding how human cognition evolves and refines itself through iterative feedback loops within the complex reality it inhabits.

Central to Recursive Realism is its critique of binary thinking. This mode of cognition, characterised by categorising concepts into oppositional pairs, often hinders a nuanced exploration of cognitive and social phenomena. Whether in debates about the influences of genetics versus environment or in the polarisation of ideological landscapes, binary thinking neglects the intricate feedback systems that inform human and societal development. Recursive Realism counters these limitations, capturing the dynamic interplay of reflective processes and adaptive mechanisms that underpin thought, belief systems, and cultural evolution.

As this inquiry progressed, it became evident that Recursive Realism could serve as a transformative lens through which the universe might be reimagined. Its recursive methodology, marked by revisiting established concepts, challenging entrenched beliefs, and refining perspectives, offers a pathway for engaging with the complexities of cognition and societal change. This model does not confine itself to analysing current patterns but also envisions future possibilities where extremes are tempered and complexity is recognised as an inherent characteristic of reality.

The model's strength lies in its duality of construction and reconstruction, enabling an understanding of past paradigms, responsiveness to present challenges, and foresight into potential futures. Recognising that every thought, belief, and cultural shift arises from a continuous and cyclical process, Recursive Realism invites a re-examination of our perspectives. At a time when polarisation and rigidity dominate, the model offers a framework to transcend such constraints, fostering an integrative and holistic view of existence.

This work is a collective effort to redefine intelligence as a dynamic interplay of self-reflection, cultural adaptation, and cognitive evolution. Rooted in empirical evidence and informed by neuroscientific insights, Recursive Realism surpasses theoretical boundaries, emerging as a practical tool for examining the evolving nature of thought. By bridging the interplay between mind and matter, thought and experience, and individual and society, it aligns with humanity's broader quest for understanding.

Ultimately, this exploration extends an invitation to reimagine how we perceive and engage with the world. Recursive Realism advocates embracing the interconnectedness of existence, nurturing a recursive approach that not only enhances our comprehension of human cognition but also celebrates its transformative capacity. By traversing the intricate pathways of thought with clarity and depth, this work aspires to pave the way for a more adaptive and integrated response to the complexities of the modern world.

Chapter 1: Introduction to Recursive Realism

Human thought has historically been shaped by models of dualism and binary opposition, deeply embedded in the traditions of Western philosophy and science. These frameworks have served as powerful tools for understanding by simplifying complex phenomena into opposing categories such as mind versus body, subject versus object, or nature versus nurture. While these dichotomies have contributed to significant advancements across various disciplines, they also impose inherent limitations, restricting our capacity to fully grasp the fluid and interconnected nature of reality. Recursive Realism emerges as a necessary response to these constraints, offering an integrative approach to understanding the complexities of cognition, perception, and the universe.

The limitations of dualism become evident when considering its long-standing influence, exemplified by René Descartes' division of mind and body into distinct entities. This perspective laid the groundwork for scientific methodologies emphasising objectivity and material analysis. However, Cartesian dualism also left unresolved the fundamental question of how immaterial thought could interact with physical matter. Over time, this unresolved tension contributed to increasingly fragmented views of reality, which persist in various disciplines. In biology, for instance, the persistent framing of the nature versus nurture debate ignores the profound interplay between genetic predispositions and environmental influences. Similarly, in cosmology, the polarisation between creationism and naturalistic explanations portrays faith and empirical evidence as irreconcilable, preventing a richer synthesis of these perspectives.

Binary thinking not only hinders intellectual progress but also fosters cognitive rigidity, which manifests in extremism. When individuals or societies adhere to one pole of a dichotomy while dismissing the other, they risk oversimplifying nuanced issues and rejecting alternative viewpoints. This rigidity is particularly evident in ideological disputes, where opposing sides often entrench themselves in self-reinforcing feedback loops that exclude meaningful dialogue. The debate between science and religion illustrates this dynamic; fundamentalist religious perspectives often reject scientific findings, while extreme naturalists dismiss metaphysical concepts as irrelevant or illogical. In both cases, the result is a static worldview incapable of growth or adaptation.

Closely linked to dualistic thinking is the concept of linear causality, the idea that events follow a straightforward cause-and-effect sequence. While this model is effective for understanding isolated phenomena, it is insufficient for analysing complex systems characterised by interdependence and non-linearity. In disciplines such as ecology, neuroscience, and social systems theory, researchers have recognised the limitations of linear causality. Ecosystems, for example, are governed by intricate feedback loops that defy simple cause-and-effect explanations. Likewise, the human brain operates as a dynamic network, where cognition emerges from recursive interactions among billions of neurons. These systems illustrate the importance of recognising feedback processes and recursive relationships as central to understanding complexity.

Human perception itself exemplifies the insufficiency of linear models. Unlike passive receivers of sensory input, human beings actively interpret, reprocess, and refine their experiences through iterative thought. Traditional models of causality, which assume a unidirectional flow of information, fail to account for this recursive engagement with reality. Recursive Realism addresses this limitation by framing cognition and perception as fundamentally dynamic processes, where the observer and the observed continuously influence each other. This perspective allows for a more nuanced understanding of how subjective experiences shape and are shaped by the external world.

The consequences of binary thinking and linear causality extend beyond intellectual stagnation to broader social and cultural challenges. Extremism, whether political, religious, or intellectual, thrives in environments where rigid categorisation suppresses complexity and ambiguity. By perpetuating closed systems of thought that resist adaptation, extremism undermines the potential for collaborative problem-solving and mutual understanding. Recursive Realism, with its emphasis on iterative feedback and integrative thinking, offers an antidote to these tendencies. It promotes openness to diverse perspectives, recognising that reality is neither fixed nor reducible to simple dichotomies but is instead a continuously evolving construct.

At its core, Recursive Realism rejects the static frameworks of dualism and linear causality in favour of a dynamic, integrative model of knowledge. By incorporating feedback and recursion as fundamental principles, it seeks to bridge the gaps between seemingly opposing forces. For example, the relationship between science and spirituality, often portrayed as antagonistic, can be reimagined within a recursive framework. Science, rooted in empirical observation, and spirituality, which explores metaphysical dimensions, represent complementary layers of understanding that enrich rather than negate one another. This integrative approach allows for a more comprehensive and meaningful engagement with reality.

Recursive Realism also reframes perception as an active, participatory process in which individuals play a central role in shaping their understanding of the world. Through continuous cycles of reflection, evaluation, and adaptation, this model acknowledges the evolving nature of thought and perception. It provides a pathway for overcoming the limitations of static, either/or paradigms, enabling a more fluid and adaptive engagement with complexity. In doing so, Recursive Realism offers not just a theoretical lens but also a practical methodology for navigating the challenges of a rapidly changing world.

This model represents a shift from rigid categorisation to an embrace of interconnectivity and mutual influence. By transcending binary constraints and incorporating the recursive interplay of mind, body, and environment, Recursive Realism paves the way for a more holistic and integrated understanding of existence. It challenges individuals and societies to move beyond cognitive rigidity, fostering a deeper appreciation for the intricacies of thought, perception, and reality itself.

Recursive Realism represents a philosophical and cognitive paradigm that acknowledges the interconnected, dynamic nature of perception, cognition, and reality. It posits that the universe, along with human understanding, is fundamentally structured through recursive processes, feedback loops that constantly shape and reshape reality through interactions between the observer and the observed. Unlike traditional models that regard reality as fixed or perception as passive, Recursive Realism highlights reality as an emergent property of these recursive interactions, evolving through the interplay of feedback, self-reference, and adaptive processes.

Recursion lies at the heart of this framework, encompassing the self-referential nature of systems. In mathematical and computational terms, recursion involves a process where the output of a function loops back as input for subsequent iterations. This principle extends far beyond technical applications, underpinning numerous natural phenomena, from biological evolution to cognitive development and even cosmological processes. Within the context of Recursive Realism, recursion is more than a mechanical process; it is a foundational principle that governs the unfolding of reality and the dynamic interactions between human beings and their environment.

Perception, cognition, and reality are all seen as recursive under this model. Perception is understood as an iterative process, continuously shaped by sensory input, past experiences, memories, and anticipations. Rather than passively receiving information, the human mind actively engages in feedback loops that refine and evolve its understanding of the world. Similarly, reality itself is not static but emerges from the interplay between observer and environment. The "real" is constructed through recursive interactions involving the mind, body, and external world, each influencing the other in a continuous process of co-creation. Thought processes, too, are inherently recursive, involving cycles of reflection, evaluation, and adjustment as humans revisit and reinterpret information, memories, and ideas.

Recursive Realism challenges classical realism, which posits that reality exists independently of the observer. Instead, it aligns with constructivist and phenomenological perspectives, asserting that reality is co-constructed through the observer's cognitive processes. This does not imply that reality is purely subjective or that the external world lacks existence. Rather, it emphasises the interdependence between perception and reality. Observers play an active role in shaping reality by interpreting, filtering, and organising sensory data into meaningful patterns, a process driven by recursive feedback. This dynamic interplay ensures that perception is context-dependent, constantly evolving as new information is integrated and past experiences inform ongoing interpretation.

Emergence is another cornerstone of Recursive Realism, encapsulating the principle that complex systems and patterns arise from simpler recursive interactions. This concept is evident in diverse fields. In biology, for instance, the complexity of life emerges through recursive processes such as natural selection and genetic replication. While the building blocks of life operate according to basic rules, their recursive interactions give rise to the intricate phenomena associated with living organisms. Similarly, ecosystems maintain their balance through recursive feedback loops between species, exemplifying the emergence of order and complexity from underlying processes.

Human perception is likewise an emergent phenomenon. Rather than simply processing raw sensory data, the brain engages in recursive interactions between sensory input, cognitive interpretation, and the feedback from memory and anticipation. This iterative process produces a richer, more nuanced reality than a linear input-output model could achieve. Even consciousness itself can be viewed as an emergent property of recursive neural processes. The brain's ability to generate self-awareness, memory, and foresight stems from recursive feedback among its various regions, which process information at multiple levels of abstraction. Just as consciousness emerges from these neural dynamics, our experience of reality arises from the recursive interplay between mind and environment.

Recursive Realism also provides a framework for addressing polarising extremes, particularly the dichotomies between scientific materialism and metaphysical explanations. The model views these domains not as mutually exclusive but as interacting layers of a larger recursive structure. Scientific inquiry, with its empirical focus, represents one dimension of reality, while metaphysical concepts, such as the idea of purpose or the search for ultimate meaning, form another recursive layer. These layers are interdependent, enriching one another within the broader system of human cognition and understanding.

Extremism, often a byproduct of cognitive rigidity, thrives on closed feedback loops where new information is rejected, and existing beliefs are reinforced in isolation. Recursive Realism counters this tendency by promoting open feedback systems that facilitate the integration of diverse perspectives. Through its emphasis on recursion and adaptability, the model fosters flexible, evolving thought processes. This approach enables a continuous dialogue between scientific knowledge and metaphysical inquiry, breaking the cycle of extremism and encouraging a more nuanced and interconnected view of reality.

In essence, Recursive Realism redefines perception and cognition as dynamic, participatory processes where the observer plays an integral role in shaping reality. By transcending the limitations of binary thinking and linear causality, it offers a comprehensive framework for understanding the emergent complexities of existence. The recursive interactions between mind, body, and environment serve as the foundation for a reality that is constantly evolving, shaped by feedback and reflection. This perspective not only deepens our understanding of perception and cognition but also provides a means to navigate the intellectual, cultural, and philosophical challenges of a complex world.

Recursive Realism introduces a transformative perspective on knowledge, challenging traditional epistemological frameworks by redefining how we know what we know. It moves away from the conventional view of knowledge as a fixed collection of truths, proposing instead that knowledge is a dynamic, evolving process shaped by recursive interactions between the knower and the known. This approach emphasises that understanding is not static but emerges from continuous cycles of observation, interpretation, and revision.

In the context of Recursive Realism, knowledge is seen as a recursive process. Rather than accumulating facts in a linear fashion, individuals engage in iterative feedback loops where each new observation integrates into an existing framework of understanding. This framework is then revised and reinterpreted as further insights are gained. Such cycles ensure that knowledge is constantly refined, adapting to new information and evolving in complexity over time. Truth, within this model, is also dynamic. It is not an unchanging absolute but an emergent property of these recursive processes, shaped by the ongoing interaction between human cognition and the external world. Through this lens, the search for truth becomes a participatory process, where understanding deepens as it engages with the complexities of experience and discovery.

Central to Recursive Realism is the concept of the recursive mind, which functions not merely as an interpreter of reality but as an active co-creator. The brain, as a highly recursive organ, processes sensory input, revisits memories, and anticipates future outcomes in continuous cycles of feedback. This dynamic process enables the construction of models of reality that are constantly updated in light of new information. The recursive mind does not simply reflect the world; it actively shapes reality through its interaction with the external environment. Reality, in this framework, is not solely an external, independent entity but a cognitive construction shaped by recursive processes. This co-construction does not imply subjectivity or illusion but highlights the mutual influence between the observer and the observed in defining what is experienced as real.

Recursive Realism builds upon and departs from centuries of philosophical inquiry and scientific discovery. Classical realism, exemplified by the works of Plato and Aristotle, posited that reality exists independently of human perception, with knowledge serving as a tool to uncover pre-existing truths. Plato's theory of Forms described an ideal, unchanging reality beyond the material world, while Aristotle's empirical realism emphasised observation and rational analysis. While these foundational ideas advanced the understanding of reality as objective and universal, they also introduced limitations. Classical realism portrayed perception as a passive process, where the mind merely reflected or observed an external, static reality. Recursive Realism diverges by proposing that perception is active, involving continuous feedback between the observer and the observed, dynamically shaping reality.

The Cartesian shift in the 17th century introduced another layer to this discourse. René Descartes' dualism established a divide between mind and body, casting the mind as the domain of thought and the physical world as governed by mechanical laws. While this distinction laid the groundwork for modern scientific methodologies, it also created a problematic separation between subjective experience and material reality. Recursive Realism bridges this divide by treating mind and body as components of an integrated feedback system. Mental and physical processes are not isolated but are dynamically interconnected, with cognition and perception actively shaping and being shaped by the material world.

Immanuel Kant's revolutionary insights in the late 18th century further shifted the focus toward the active role of the mind in shaping reality. Kant argued that human perception is not a passive reception of external data but an active construction, where the mind imposes categories such as space, time, and causality on raw experience. This "Copernican Revolution" in philosophy emphasised the mind's role in constructing the phenomenal world while maintaining a division between it and the noumenal world, the latter existing independently of perception. Recursive Realism builds on Kant's framework by suggesting that the distinction between the phenomenal and noumenal worlds is not rigid. Instead, the observer and the observed engage in a recursive interaction that co-constructs reality. The noumenal world, within this view, emerges through the interplay of cognitive processes with the external environment.

The development of cybernetics and systems theory in the 20th century introduced the concept of feedback loops as a cornerstone for understanding complex systems. Norbert Wiener's cybernetics demonstrated how systems adapt and regulate themselves through feedback, a principle central to Recursive Realism. Similarly, Claude Shannon's information theory highlighted the role of feedback in refining information transmission, mirroring how the brain processes sensory data. Gregory Bateson expanded these ideas into ecology and psychology, describing how recursive interactions govern biological and cognitive systems. Recursive Realism extends these principles, suggesting that reality itself is an emergent property of recursive feedback between the observer and the observed.

The phenomenological movement, led by Edmund Husserl and Maurice Merleau-Ponty, also contributes to the foundations of Recursive Realism. Phenomenology emphasised the role of subjective experience and the interconnectedness of perception and embodiment. Merleau-Ponty, in particular, argued that the body is central to perception, with the mind and body continuously interacting with the environment. Recursive Realism integrates these phenomenological insights while advancing them through the lens of feedback systems. It frames subjective experience and objective reality as mutually influencing entities, dynamically shaping each other through recursive interactions.

In its synthesis of historical and modern ideas, Recursive Realism offers a unified framework for understanding perception, cognition, and reality. By emphasising recursion and feedback as fundamental principles, it transcends the limitations of classical realism, Cartesian dualism, and other traditional models. It reframes knowledge as a dynamic, co-constructive process, where understanding evolves through continuous interaction between the observer and the external world. Recursive Realism not only enriches the study of epistemology but also provides a practical methodology for navigating the complexities of human thought and experience.

Recent developments in neuroscience and cognitive science have provided substantial support for the concept of recursion, particularly in elucidating how the brain processes information. Modern cognitive models increasingly recognise that the brain functions through intricate feedback loops, with various regions interacting recursively to give rise to consciousness, perception, and behaviour. This perspective aligns closely with the principles of Recursive Realism, which emphasises the recursive interplay between cognitive processes and the external world in shaping reality.

Neural networks, both biological and artificial, exemplify the role of recursion in learning and adaptation. In the human brain, sensory information is not processed in a straightforward, linear sequence. Instead, it undergoes recursive exchanges between different neural regions, enabling a more adaptive and nuanced understanding of the environment. This dynamic system of feedback allows the brain to synthesise complex patterns, adjust its responses, and refine its interpretations of sensory data. Similarly, artificial neural networks, inspired by biological systems, rely on recursive mechanisms to improve performance and achieve sophisticated outcomes through iterative adjustments.

One prominent theory that underscores the recursive nature of brain function is predictive coding. This framework posits that the brain operates as a predictive machine, continually generating expectations about incoming sensory input based on prior experiences. These predictions are compared against actual sensory data, and any discrepancies, termed prediction errors, are recursively integrated into the system to refine the brain's internal model of reality. Predictive coding reflects the active, adaptive nature of perception, demonstrating how the mind continuously updates its understanding of the world through recursive feedback between internal cognitive states and external sensory inputs. This process resonates with the principles of Recursive Realism, which frames perception as an active, iterative engagement with reality rather than a passive reception of stimuli.

Recursive Realism synthesises insights from various intellectual traditions, unifying classical philosophy, cybernetics, phenomenology, and cognitive science into a cohesive framework. It moves beyond earlier paradigms by asserting that perception, cognition, and reality are not isolated or linear phenomena but are fundamentally recursive, with each element influencing and being influenced by the others in an ongoing feedback loop.

Unlike classical realism, which posited a fixed, observer-independent reality, Recursive Realism proposes that reality emerges through the recursive interactions between the mind and the external world. It challenges the Cartesian dualism that divided mind and body, instead viewing them as components of a dynamic, interdependent system. Furthermore, Recursive Realism diverges from traditional epistemological approaches that sought immutable truths, embracing instead the notion that knowledge is emergent and perpetually evolving through recursive processes of observation, interpretation, and feedback.

By integrating these diverse perspectives, Recursive Realism offers a comprehensive framework for understanding the recursive nature of cognition and its role in shaping reality. It transcends the limitations of earlier models, fostering a more nuanced and dynamic view of the interplay between perception, thought, and the external world. This synthesis not only deepens our understanding of human cognition but also provides a pathway for exploring the emergent complexities of knowledge and reality itself.

References

Bateson, G. (1972) *Steps to an Ecology of Mind*. Chicago: University of Chicago Press.

Descartes, R. (1641) *Meditations on First Philosophy*. Translated by J. Cottingham. Cambridge: Cambridge University Press.

Kant, I. (1781) *Critique of Pure Reason*. Translated by P. Guyer and A. Wood. Cambridge: Cambridge University Press.

Merleau-Ponty, M. (1962) *Phenomenology of Perception*. Translated by C. Smith. London: Routledge & Kegan Paul.

Plato (1997) *Complete Works*. Edited by J.M. Cooper. Indianapolis: Hackett Publishing.

Shannon, C. E. (1948) 'A Mathematical Theory of Communication', *Bell System Technical Journal*, 27(3), pp. 379–423.

Wiener, N. (1948) *Cybernetics: Or Control and Communication in the Animal and the Machine*. Cambridge: MIT Press.

Chapter 2: The Foundation of Recursive Perception

Perception is often conceptualised as a simple process: sensory input is received from the environment, processed by the brain, and experienced as a coherent representation of the world. However, this traditional model oversimplifies the intricate and dynamic nature of how perception functions. Within the framework of Recursive Realism, perception is understood as a recursive process. The brain continuously loops information back upon itself, refining and reinterpreting sensory input in light of past experiences, learned expectations, and real-time feedback from the environment. This recursive feedback between the observer and the observed enables a more adaptive and complex understanding of the world.

Traditional models of perception frequently assume a linear progression of sensory information, where input is transmitted in a straightforward cause-and-effect sequence. For example, in vision, light enters the eye, is processed by the retina, and interpreted by the brain to create an image. While this description is useful for explaining the basic flow of information, it fails to capture the recursive and dynamic processes involved in real-world perception. Perception does not occur as a one-way transfer of data; it is a non-linear and deeply recursive process. Each sensory input is continuously compared to previous experiences, existing expectations, and even predictions about the future. What we perceive at any moment is the outcome of an ongoing loop of feedback and adjustment, where the brain constantly re-evaluates and refines its interpretation of the sensory world.

Sensory input is actively processed by recursive feedback loops, which allow the brain to make predictions and adjustments in real time. For instance, when observing a moving object, the brain does not simply compile a series of static images. Instead, it predicts the object's trajectory, compares this prediction with new sensory input, and refines its interpretation to ensure a smooth and continuous perception of motion. These predictive capabilities rely on recursive processes that integrate sensory data with stored memories and learned patterns, allowing for rapid and precise responses to environmental stimuli. This mechanism is evident in tasks as diverse as catching a ball, navigating traffic, or interpreting subtle social cues.

Feedback loops play a central role in the recursive processing of sensory information. Each sensory experience is not merely processed once; it is continually fed back into the brain's system, where it interacts with pre-existing cognitive structures, past experiences, and expectations. For example, visual perception involves more than the raw data captured by the retina. The brain uses recursive feedback to interpret this data, drawing upon a lifetime of visual experience to identify shapes, colours, and movement. The hierarchical structure of the brain further supports this process, with higher-level cognitive functions providing feedback to lower-level sensory systems. In visual processing, for instance, basic information is initially handled by the occipital lobe, while higher-order interpretation occurs in regions like the temporal and parietal lobes. These higher regions send refined feedback to the lower regions, allowing for context-sensitive adjustments that enable object recognition, facial identification, and the interpretation of complex scenes, even when information is incomplete.

A key function of these feedback loops is error correction. When the brain encounters a discrepancy between its initial interpretation of sensory data and the incoming information, such as when a shadow unexpectedly resembles a familiar shape, feedback loops allow for rapid adjustment. This mechanism ensures that perception remains accurate and responsive to environmental changes, enabling the brain to correct its predictions in real time. Error correction underscores the recursive, adaptive nature of perception, illustrating how the brain refines its understanding of the world through iterative feedback.

The adaptive nature of perception is another cornerstone of the recursive model. Perception is not static or inflexible; it evolves in response to new data and changing conditions. For example, when transitioning from bright sunlight to a dimly lit room, initial perception is limited. Over time, recursive feedback loops recalibrate the brain's interpretation of visual input, allowing for enhanced vision in low-light conditions. This process is gradual and iterative, involving continuous interaction between the eyes, the brain, and the surrounding environment. Similarly, predictive perception is critical for activities like driving, where the brain anticipates the movement of vehicles, pedestrians, and other elements, adjusting its predictions in response to unexpected changes. Recursive feedback allows these adjustments to occur with remarkable speed and precision, enabling humans to navigate complex and dynamic environments.

Social perception further demonstrates the recursive nature of sensory processing. In conversation, humans do not merely process words in isolation; they interpret tone, facial expressions, and body language, comparing these inputs with cultural knowledge and prior experiences. Recursive feedback enables the rapid adaptation to social cues, allowing for nuanced responses and the detection of subtle emotions such as sarcasm or empathy. This dynamic interaction underscores the fluidity and complexity of perception in social contexts, highlighting the brain's ability to integrate diverse forms of sensory data in real time.

Understanding perception as a recursive process fundamentally redefines the relationship between humans and their environment. Rather than passively receiving sensory input, individuals are engaged in an ongoing, iterative dialogue with the world around them. Perception involves not only processing raw data but also continuously refining and adjusting that data through feedback loops that draw on memory, prediction, and contextual understanding. By acknowledging the non-linear and adaptive nature of perception, Recursive Realism provides a comprehensive framework for exploring how humans engage with reality. This model reveals perception as an active, dynamic process, deeply interconnected with cognition, memory, and learning, offering a richer and more nuanced understanding of human experience.

The human brain is often characterised as a highly efficient pattern recognition machine, a capability essential for navigating the complexities of life. From recognising faces to interpreting language, the brain's ability to detect and interpret patterns underpins much of human cognition. However, pattern recognition is not merely a linear process of matching sensory input to pre-existing templates. Instead, it is deeply recursive. The brain engages in continuous loops of sensory input, memory retrieval, and predictive modelling to refine its understanding of the world. This recursive mechanism allows humans not only to recognise patterns but also to dynamically adapt to new situations and revise old patterns as needed, providing an active and evolving engagement with the environment.

At the core of this process is the brain's recursive ability to identify and interpret patterns from streams of sensory information that would otherwise appear chaotic. This is not a simple matter of direct stimulus-response; rather, it involves complex comparisons between incoming sensory data, stored memories, and future predictions. For example, visual perception relies on recursive processes to construct coherent images from fragmented inputs such as light, shadow, and motion. In auditory perception, the brain employs similar mechanisms to focus on meaningful sounds, such as speech, while filtering out irrelevant background noise. These recursive processes are dynamic, enabling the brain to revisit and adjust its interpretations to align with past experiences and current expectations.

Language processing provides one of the most striking examples of the brain's recursive pattern recognition capabilities. Language is inherently structured and rule-based, yet its rules are often implicit and context-dependent. When listening to speech, the brain does not process each word in isolation. Instead, it recursively integrates individual words and phrases into broader syntactic and semantic patterns, drawing on a vast network of linguistic knowledge stored in memory (Hagoort, 2005). This recursive engagement allows humans to comprehend meaning rapidly, even when faced with complex or ambiguous sentences.

Similarly, the ability to recognise faces and emotions exemplifies the brain's recursive prowess. Facial recognition involves continuous feedback loops that compare observed features, such as the shape of eyes and the curve of a mouth, with stored templates of familiar faces (Bruce & Young, 1986). This process enables rapid identification, even under challenging conditions, such as partial occlusion or changes in perspective. Emotional recognition adds another layer of complexity, as the brain integrates sensory input from facial expressions with past experiences of emotions, recursively refining its interpretation to understand another person's emotional state.

Pattern recognition is not a fixed ability but a highly adaptive process. Recursive cognition allows the brain to challenge and update established patterns when new information does not align with prior expectations. For instance, encountering an unfamiliar object that does not fit neatly into any known category prompts the brain to engage in recursive loops of analysis. By revisiting stored memories and drawing on past experiences, the brain may modify its existing patterns or create entirely new ones to accommodate the novel input. This flexibility is essential for adapting to new or ambiguous situations (Friston, 2010).

Uncertainty and ambiguity further highlight the importance of recursive cognition. In situations where sensory information is incomplete or contradictory, the brain relies on recursive feedback loops to resolve the ambiguity. For example, when interpreting an unclear visual scene, the brain alternates between multiple interpretations, refining each recursively until a coherent understanding emerges. This process enables humans to navigate uncertain environments with remarkable efficiency, avoiding paralysis by indecision.

Learning and adaptation are closely linked to recursive cognition. New skills and knowledge are acquired through iterative cycles of trial, feedback, and adjustment. For instance, learning to play a musical instrument involves recursive refinement, as the brain integrates sensory feedback from the instrument, memory of previous practice, and anticipatory adjustments for future improvement. This recursive process enables rapid adaptation, even in complex and dynamic contexts (Kolb & Whishaw, 2015).

Memory and anticipation are integral components of recursive perception and cognition. Memory is not a static repository but a dynamic system shaped by recursive feedback between past experiences and current perceptions. When recalling a memory, the brain reconstructs it through recursive interactions with present context and knowledge, ensuring that recollections remain relevant and adaptive (Tulving, 2002). Anticipation, meanwhile, allows the brain to generate predictions about future events based on past experiences. For example, when walking through a crowded street, the brain predicts the movement of pedestrians and vehicles, recursively adjusting these predictions as new sensory data becomes available.

The interplay between memory and prediction exemplifies the recursive nature of cognition. Past experiences inform future expectations, while those expectations, in turn, influence how memories are recalled and interpreted. This continuous feedback loop enables humans to remain flexible and adaptive, even in rapidly changing environments. For example, entering a familiar space evokes memories of previous visits, which help guide expectations. If the environment has changed, the brain's predictions are revised, and the new experience is incorporated into memory, enriching the recursive cycle.

Recursive cognition lies at the heart of the brain's capacity for pattern recognition, learning, and adaptation. By constantly looping between sensory input, memory, and prediction, the brain transcends simple pattern matching, enabling a flexible, dynamic engagement with reality. This recursive capability allows humans to recognise and respond to patterns in diverse contexts, from interpreting language to navigating social interactions and physical environments. Through its adaptive and iterative processes, recursive cognition provides the foundation for a more sophisticated understanding of the world, equipping humans to thrive in complex and uncertain conditions.

Traditional models of perception often conceptualise the mind as a passive receiver of sensory information, where the external world is simply processed and interpreted. Recursive Realism challenges this notion by redefining perception as an active, interactive process. Instead of passively receiving stimuli, perception involves a dynamic feedback loop where the mind actively participates in constructing reality. This shift from passive to active perception transforms our understanding of how humans engage with their environment, highlighting the recursive nature of sensory and cognitive processes.

At the heart of Recursive Realism is the idea that perception is an interactive process, involving continuous engagement between the brain and the external world. The brain is not merely processing sensory data; it actively constructs reality through recursive loops of interpretation and adjustment. For example, when entering a room, our perception of the space is not a direct reflection of sensory input. Instead, the brain filters and interprets incoming stimuli based on memory, expectations, and contextual knowledge. This recursive process allows for a coherent understanding of the environment, enabling humans to navigate and interact with the world effectively.

The dynamic nature of perception is evident in how humans engage with their surroundings. Perception is not a one-time event but an iterative process of refinement. Each new piece of sensory input is compared to stored knowledge and prior experiences, with the brain continuously updating its interpretation. This engagement becomes increasingly nuanced with familiarity, as recursive feedback loops refine our perception through repeated interaction. Moreover, perception and action form a tightly coupled recursive cycle. For example, reaching for a glass of water involves feedback between visual perception, motor control, and proprioception. The brain adjusts the movement in real time based on sensory input, ensuring the action's success. This interaction illustrates the inseparability of perception and action, demonstrating how recursive processes enable humans to adapt to their environment.

A critical component of active perception is attention, which functions as a dynamic mechanism that directs the brain's focus to relevant stimuli while filtering out distractions. Attention operates through recursive feedback loops, allowing the brain to adjust its focus based on changing priorities and sensory inputs. For instance, in a noisy environment, attention shifts between a speaker's voice and background noise, constantly recalibrating to maintain focus on the most important information. This recursive nature of attention enables selective perception, where cognitive filters such as prior knowledge and expectations prioritise certain stimuli over others (Posner & Petersen, 1990). The brain's ability to engage in focused or divided attention highlights its recursive adaptability, allowing humans to process complex and dynamic environments efficiently.

The creative aspect of perception further underscores its recursive nature. Perception is not limited to interpreting raw sensory data; it involves filling in gaps, making predictions, and constructing meaning. This creative process is evident in visual illusions, where the brain completes incomplete sensory data to form a coherent image (Gregory, 1997). Similarly, imagination plays a central role in perception, especially in contexts with sparse sensory input. For example, when reading a novel, the brain constructs vivid mental images by recursively integrating textual information with memory and creativity. Art provides another illustration of the recursive nature of perception. Impressionist paintings, with their fragmented brushstrokes, invite the viewer's brain to fill in missing details, while abstract art prompts continuous reinterpretation through recursive engagement.

This active, recursive model of perception redefines humans as co-creators of reality rather than passive observers. By continuously looping between sensory input, memory, attention, and creativity, the brain constructs a dynamic understanding of the world. Whether through the detection of patterns, the adaptive flexibility of attention, or the creative construction of meaning, recursive processes are central to how humans perceive and interact with their environment.

Recursive Realism positions perception as a dynamic interplay of engagement, refinement, and construction. It highlights the mind's role in shaping reality through recursive loops of interpretation and action. By moving beyond passive models, this framework provides a richer understanding of human perception, underscoring its active, adaptive, and creative nature.

References

Bruce, V. & Young, A. (1986) 'Understanding face recognition', *British Journal of Psychology*, 77(3), pp. 305–327.

Friston, K. (2010) 'The free-energy principle: a unified brain theory?', *Nature Reviews Neuroscience*, 11(2), pp. 127–138.

Gregory, R.L. (1997) *Eye and Brain: The Psychology of Seeing*. 5th edn. Princeton: Princeton University Press.

Hagoort, P. (2005) 'On Broca, brain, and binding: a new framework', *Trends in Cognitive Sciences*, 9(9), pp. 416–423.

Kolb, B. & Whishaw, I.Q. (2015) *An Introduction to Brain and Behavior*. 5th edn. New York: Worth Publishers.

Posner, M.I. & Petersen, S.E. (1990) 'The attention system of the human brain', *Annual Review of Neuroscience*, 13, pp. 25–42.

Tulving, E. (2002) 'Episodic memory: From mind to brain', *Annual Review of Psychology*, 53, pp. 1–25.

Chapter 3: Emergent Reality in Recursive Realism

In traditional perspectives, reality is often viewed as a fixed, stable construct, an objective world that exists independently of perception, waiting to be discovered through observation and analysis. Recursive Realism fundamentally challenges this notion by proposing that reality is emergent and dynamic, continually shaped by recursive interactions between the observer and their environment. Rather than being a static entity, reality is understood as a living, evolving system, reflecting the complexity of the processes underlying its construction.

The concept of emergence, central to Recursive Realism, describes how complex systems and properties arise from the interactions of simpler elements. Emergence is a phenomenon seen across nature, from the organisation of cells into tissues and organs to the intricate patterns of weather systems. These complexities cannot be fully explained by examining individual components in isolation; instead, they result from interactions and feedback among those components. Within Recursive Realism, emergence takes on an even more profound meaning, applying not only to physical systems but also to the dynamic interplay between perception and reality. The mind, through recursive processes, actively participates in the co-creation of reality, resulting in a dynamic and evolving understanding of existence.

Cognitive systems, such as the human brain, exemplify how emergence operates through recursion. The complexity of thought, consciousness, and self-awareness arises not from the isolated activity of individual neurons but from the interactions and feedback among billions of neurons in a self-organising network (Edelman & Tononi, 2000). This emergent complexity is mirrored in Recursive Realism, where the brain's recursive engagement with the environment generates a perception of reality far richer than the sum of its sensory inputs. Similarly, emergence is evident in nature. Honeybee colonies, flocking birds, and ecosystems operate through recursive feedback loops, where individual agents interact with their environment in ways that give rise to complex adaptive behaviours. These natural examples of emergence parallel the recursive processes that underpin human perception and the evolving nature of reality.

One of the key tenets of Recursive Realism is that reality is co-constructed through a dynamic process of interaction. The observer plays an active role in shaping the reality they experience, just as reality influences the observer's perception. This perspective shifts the traditional view of the observer as a passive receiver of sensory data to one of active participation. For example, when observing a forest, the brain uses recursive feedback loops to integrate sensory input, such as sights, sounds, and smells, into a dynamic model of the environment. The forest is not perceived as a fixed entity but as an emergent reality that evolves as the observer moves through it. This continuous dialogue between the observer and their surroundings exemplifies how perception and reality are co-constructed.

The roots of this co-constructive view can be traced to constructivist philosophy, which emphasises that knowledge and perception are shaped by individual interactions with the world (Piaget, 1971). Recursive Realism builds on this foundation by introducing recursion as the mechanism driving co-construction. Unlike classical constructivism, which often focuses on social and cultural influences, Recursive Realism highlights the role of internal cognitive processes in shaping reality. Through recursive engagement, the mind and environment mutually influence one another, creating a dynamic and adaptive model of the world.

Reality, as understood through Recursive Realism, evolves through non-linear dynamics. Non-linearity refers to systems where small changes can produce disproportionately large effects, often leading to unexpected outcomes. This principle is seen in weather systems, where minor shifts in temperature or pressure can cascade into significant events like storms, a phenomenon popularly known as the butterfly effect (Lorenz, 1963). Similarly, small changes in perception or understanding can reverberate through the recursive loops of cognition, resulting in profound shifts in how reality is experienced. The emergent properties of such non-linear dynamics reflect the adaptive and evolving nature of reality, where new elements arise through recursive interactions.

Recursive Realism bridges the divide between the physical and metaphysical, offering a unified framework in which these traditionally distinct domains interact. Physical phenomena, governed by empirical observation and measurement, and metaphysical concepts, which address questions of existence and meaning, are not isolated. Instead, they coexist as part of a larger recursive system. Feedback loops in nature, such as predator-prey dynamics or self-regulating climate systems, demonstrate how recursive principles operate in physical phenomena. These interactions not only shape the physical world but also provide a foundation for exploring more abstract, metaphysical realms.

On the quantum level, recursion is evident in phenomena like wave-particle duality and quantum entanglement. The state of a quantum particle is influenced by the act of observation, introducing a recursive relationship between observer and observed (Bohr, 1935). Entanglement further illustrates this principle, where the state of one particle instantaneously affects another, regardless of distance. These examples suggest that even at the most fundamental levels of reality, recursive feedback plays a critical role in shaping outcomes.

Recursive Realism also offers insights into metaphysical questions, such as the nature of divinity, meaning, and consciousness. The concept of God, for instance, can be understood as a recursive interpretation of existence. Rather than a fixed entity, God may represent an ultimate recursion, an evolving concept shaped by human cognition as it engages with the mysteries of the universe. Meaning and purpose similarly emerge through recursive interaction with reality. As humans engage with their environment, they interpret and reinterpret experiences, assigning meaning that evolves over time. Consciousness itself is a recursive phenomenon, arising from the interplay between the mind, body, and environment. Through recursive loops of self-reflection and awareness, humans construct a layered and dynamic experience of existence.

By emphasising recursion as a fundamental principle, Recursive Realism offers a transformative understanding of reality as an emergent and co-constructed process. It bridges the physical and metaphysical, integrates perception and cognition, and highlights the dynamic interplay between the observer and the observed. This framework not only enriches our understanding of reality but also provides a basis for exploring its profound complexity and interconnectedness.

Recursive Realism offers a transformative framework for integrating physical and metaphysical perspectives into a cohesive understanding of reality. Rather than viewing empirical observations and transcendental ideas as mutually exclusive, Recursive Realism positions them as interacting components of a larger system, connected through recursive feedback loops. This integrated perspective addresses questions often left unresolved by purely scientific or metaphysical approaches, enabling a more holistic exploration of existence.

Science and spirituality, traditionally viewed as addressing fundamentally different dimensions of reality, are often perceived in opposition. Science seeks to uncover the mechanisms of the universe, focusing on the *how*, while spirituality contemplates the *why*, exploring meaning and purpose. Recursive Realism suggests that these perspectives are not conflicting but rather complementary layers of the same recursive system. Scientific discoveries, such as the laws governing the cosmos, can deepen spiritual awe and reflection, while spiritual concepts of meaning and transcendence can shape the interpretation of scientific phenomena. This interaction creates a feedback loop where empirical findings and metaphysical insights mutually enrich one another, fostering a more comprehensive view of reality.

The notion of transcendence, often associated with spiritual or mystical experiences, can also be understood through recursion. Recursive Realism reframes transcendence not as an escape from reality but as a deeper engagement with it. Through recursive reflection, individuals explore profound questions of existence, where each insight generates new questions, leading to a continual evolution of understanding. This recursive process reveals transcendence as an ever-expanding field of possibilities, where reality is not static but dynamic and fluid, shaped by the interplay between observation and reflection.

The ultimate aim of Recursive Realism is to develop a unified ontology, an integrated framework for understanding existence that encompasses both physical processes and metaphysical concepts. This ontology recognises that reality is emergent and dynamic, shaped by the recursive interplay between mind and matter, observation and interpretation. Questions of causality, existence, and consciousness are not isolated from the physical universe but are deeply interconnected through recursive feedback loops. This unified view treats reality as a living system, continuously co-created by the mind and the cosmos.

A core tenet of Recursive Realism is its ability to transcend the limitations of binary thinking, which has long dominated human understanding. Binary frameworks reduce complex phenomena into rigid dichotomies such as mind versus matter, science versus spirituality, or determinism versus free will. While these divisions simplify the complexity of reality for categorisation, they often obscure the fluidity and interconnectedness that characterise the universe. Recursive Realism moves beyond these false dichotomies, revealing the intricate feedback loops that link seemingly opposing concepts into a cohesive whole.

Binary thinking, though useful for clarity and distinction, often oversimplifies inherently complex systems. For example, debates about nature versus nurture in biology attempt to separate genetic influences from environmental factors, despite overwhelming evidence that the two interact recursively. Similarly, discussions of consciousness frequently separate mental phenomena from physical brain processes, ignoring the dynamic interplay between them. This reductionist approach fails to capture the emergent properties of complex systems, which arise from the recursive interactions of their components.

Binary thinking also fosters polarisation and extremism, as opposing perspectives often become entrenched in self-reinforcing feedback loops. For instance, in debates between scientific materialism and religious fundamentalism, both sides frequently dismiss the other as fundamentally misguided. This polarisation inhibits dialogue and obscures the possibility of synthesising insights from both perspectives into a richer understanding. By rejecting binary oppositions, Recursive Realism emphasises the integration of diverse viewpoints, highlighting their interdependence within a dynamic, recursive system.

Feedback loops, central to Recursive Realism, act as bridges that connect seemingly opposing concepts, revealing their interconnectedness. For example, the divide between science and spirituality is often considered irreconcilable, with science grounded in empirical observation and spirituality in subjective experience. However, Recursive Realism demonstrates that these domains can influence and enrich one another. Discoveries about the structure of the universe, such as the balance of physical laws, can inspire spiritual wonder, while spiritual reflections on meaning and purpose can shape how scientific phenomena are interpreted. This feedback between empirical evidence and metaphysical inquiry fosters an integrated understanding of existence.

The nature versus nurture debate provides another illustration of feedback loops. Recursive Realism reframes this binary as a reciprocal interaction, where genes and environment co-create developmental pathways. Genes establish biological potential, but their expression is shaped by environmental factors, which in turn are influenced by behaviours and choices rooted in genetic predispositions. This dynamic interaction demonstrates how recursive feedback systems resolve false dichotomies, highlighting the co-dependence of nature and nurture in shaping organisms.

Similarly, the mind-body problem, how mental experiences relate to physical brain processes, has traditionally been framed as a binary opposition. Recursive Realism approaches this issue by recognising consciousness as a recursive process that arises from the interplay between the mind and body. Neural activity provides the substrate for mental experiences, while conscious thoughts feed back into the brain, altering neural pathways and influencing behaviour. This circular causality demonstrates that mind and matter are not separate but are unified through recursive feedback loops, creating a dynamic and evolving system.

Extremism often emerges when individuals or groups become trapped in closed feedback loops, where external input is excluded, and self-reinforcing narratives dominate. These loops are characteristic of ideological, political, and religious extremism, where rigid beliefs resist adaptation or critique. Recursive Realism breaks these closed loops by fostering open-ended feedback, where new perspectives and information are integrated into an evolving understanding. This openness encourages cognitive flexibility, the ability to revise beliefs and interpretations in response to changing circumstances, allowing for a more nuanced and adaptive approach to complex issues.

By promoting integration over division, Recursive Realism enables a shift from binary to recursive thinking. Reality is no longer viewed as a collection of isolated opposites but as an interconnected system where elements interact and influence one another. Through recursion, the complexities and contradictions of existence are embraced rather than simplified, offering a more comprehensive and adaptive framework for understanding reality.

In this chapter, Recursive Realism has been presented as a vision of reality that is emergent, dynamic, and interconnected. By recognising reality as a co-constructed process shaped by recursive feedback between the observer and the observed, Recursive Realism offers a transformative perspective that integrates the physical and metaphysical. It resolves binary extremes, replaces rigid divisions with adaptive feedback, and fosters a holistic understanding of existence. This perspective positions reality as an evolving process, continually shaped by the recursive interplay between mind and cosmos, where understanding is not an endpoint but a journey of infinite discovery.

References

Bohr, N. (1935) 'Can Quantum-Mechanical Description of Physical Reality Be Considered Complete?', *Physical Review*, 48(8), pp. 696–702.

Edelman, G.M. and Tononi, G. (2000) *A Universe of Consciousness: How Matter Becomes Imagination*. New York: Basic Books.

Lorenz, E.N. (1963) 'Deterministic Nonperiodic Flow', *Journal of the Atmospheric Sciences*, 20(2), pp. 130–141.

Piaget, J. (1971) *Biology and Knowledge: An Essay on the Relations Between Organic Regulations and Cognitive Processes*. Chicago: University of Chicago Press.

Posner, M.I. and Petersen, S.E. (1990) 'The attention system of the human brain', *Annual Review of Neuroscience*, 13, pp. 25–42.

Tulving, E. (2002) 'Episodic memory: From mind to brain', *Annual Review of Psychology*, 53, pp. 1–25.

Chapter 4: Functional Aspects of Recursive Realism

Cognition and Learning

Cognition and learning are fundamental aspects of human existence, enabling adaptation, problem-solving, and the acquisition of knowledge. Traditional approaches often conceptualise learning as a linear process, wherein information is received, stored, and retrieved. However, Recursive Realism posits that learning is a deeply recursive phenomenon, involving iterative feedback loops that continuously refine and reorganise knowledge. This dynamic process allows learners to adapt and evolve their understanding in response to new information.

Learning is not a passive absorption of facts but an active, recursive process where new information interacts with existing mental frameworks. Each iteration refines understanding, allowing for the gradual integration of new concepts into established structures. Feedback is central to this process, providing the necessary mechanism for comparison, evaluation, and refinement. For instance, students encountering complex mathematical concepts often refine their understanding through recursive cycles of testing, error correction, and adjustment. These cycles allow for the synthesis of new insights, fostering a deeper and more resilient comprehension.

Error correction exemplifies the recursive nature of learning. Mistakes are not simply failures but opportunities for growth, providing feedback that informs the next attempt. This trial-and-error process is evident in skill acquisition, such as learning a musical instrument. A pianist, for example, iteratively adjusts their technique through recursive feedback from auditory cues and muscle memory, gradually mastering complex compositions. Similarly, expertise in any domain emerges from countless recursive cycles, where feedback refines mental models, enabling experts to anticipate patterns and adapt strategies with fluid precision.

Problem-Solving and Recursion

Problem-solving, like learning, is often framed as a linear progression from problem identification to solution implementation. In practice, however, effective problem-solving is inherently recursive. Solutions emerge through iterative cycles of hypothesis testing, evaluation, and revision, where each attempt informs subsequent strategies. For example, in engineering, prototyping involves recursive feedback loops, where designs are tested, flaws identified, and modifications implemented. This iterative refinement ensures that the final product meets functional requirements.

Scientific inquiry epitomises recursive problem-solving, with hypotheses continuously revised in light of new data. The development of evolutionary theory, for example, was not a straightforward discovery but a recursive process of observation, hypothesis formation, and refinement. Fossil records, genetic data, and ecological studies provided feedback that shaped the theory, illustrating how scientific understanding evolves through recursive engagement with evidence.

Adaptive strategies are another hallmark of recursive problem-solving. Whether navigating a strategic game or responding to real-world challenges, individuals rely on feedback to adjust their approaches dynamically. This recursive loop between perception, evaluation, and action enables flexibility and resilience, ensuring that solutions remain effective in changing circumstances.

Creativity and Recursive Thinking

Creativity is often perceived as a sudden burst of inspiration, yet it is fundamentally a recursive process. Creative thinkers engage in cycles of exploration, revision, and refinement, where ideas are revisited and reshaped. This iterative approach allows for the emergence of novel connections and insights, transforming initial concepts into sophisticated creations.

The creative process involves a recursive interplay between divergent and convergent thinking. Divergent thinking generates possibilities, while convergent thinking refines them. For example, an author may begin with a broad idea and iteratively revise drafts, incorporating feedback and new perspectives until a coherent narrative emerges. Each iteration introduces refinements, enabling the work to evolve organically.

Similarly, artists and innovators engage in recursive cycles of inspiration and iteration. An artist sketching a landscape, for instance, may revisit the scene multiple times, refining their interpretation with each encounter. This recursive engagement deepens their understanding, allowing the final work to capture both the essence and complexity of the subject. Connections between seemingly unrelated ideas often arise through recursive thinking, as the mind loops through diverse domains, identifying analogies and metaphors that enrich understanding.

Social Systems and Recursive Feedback

Human societies, like individual cognition, are characterised by recursive interactions. Social norms, cultural practices, and institutional structures emerge and evolve through feedback loops that link individual behaviour to collective dynamics. This recursive interplay drives both stability and change, shaping the trajectory of societies over time.

Social norms exemplify the recursive nature of societal processes. Norms emerge as individual behaviours reinforce collective expectations, creating self-sustaining feedback loops. For example, greeting customs, such as shaking hands, persist because each act of participation reinforces the behaviour as a social expectation. Yet, norms are not static; they adapt through recursive feedback as societies encounter new challenges and opportunities. Technological advancements, such as social media, have reshaped norms around communication and privacy, illustrating how recursive interactions drive adaptation.

Deviance, or behaviour that challenges established norms, plays a critical role in societal evolution. Acts of deviance introduce feedback that forces societies to re-evaluate and potentially adjust their norms. For instance, civil rights movements, often initiated through acts of protest, create recursive cycles of challenge and response, ultimately leading to transformative social change. These dynamics highlight the adaptive capacity of societies, where recursive feedback fosters both continuity and innovation.

Integration Across Domains

Recursion underpins cognition, learning, creativity, and social systems, revealing a unifying principle that transcends disciplinary boundaries. By understanding these processes as recursive, Recursive Realism provides a framework for exploring the interconnectedness of individual and collective dynamics. This perspective positions recursion as the foundation for human adaptability, innovation, and resilience, offering profound insights into the mechanisms that shape both minds and societies.

Institutions as Recursive Structures

Institutions, which regulate and organise social behaviour, are often regarded as rigid frameworks designed to maintain order and stability. However, Recursive Realism repositions institutions as dynamic entities, capable of adapting and regenerating through recursive feedback loops. This adaptability allows them to remain relevant amid societal evolution and external pressures.

Institutional stability emerges through recursive reinforcement. Rules and norms are maintained via structured feedback mechanisms such as legal enforcement, societal adherence, and institutional sanctions. For example, legal systems employ a cycle of rule enforcement, judicial interpretation, and societal response to maintain order. When laws are violated, consequences such as penalties or reforms provide feedback that reinforces adherence or highlights areas for revision.

Conversely, institutions undergo change through recursive adaptation. External pressures, technological advancements, demographic shifts, or evolving public values, necessitate adjustments in institutional structures. For instance, educational systems have responded to the digital revolution by incorporating virtual classrooms and online resources. These adjustments are the result of recursive feedback between institutional goals and societal needs, allowing institutions to recalibrate their strategies.

Crises play a pivotal role in institutional transformation, acting as disruptive forces that challenge established feedback loops. For example, the 2008 global financial crisis exposed weaknesses in economic systems, prompting regulatory reforms and the restructuring of banking practices. This crisis-response cycle demonstrates how institutions can utilise recursive processes to reassess and strengthen their frameworks in the face of systemic failures.

Recursive Dynamics in Social Change

Social change unfolds through recursive interactions between individuals, groups, and institutions, reflecting a complex interplay of feedback and adaptation. Social movements, cultural shifts, and ideological transformations are all driven by recursive dynamics that amplify, refine, and propagate change over time.

Social movements serve as prime examples of recursive feedback in action. They often begin with isolated acts of defiance that generate societal feedback, support from allies, opposition from critics, and attention from institutions. These interactions create a positive feedback loop, amplifying the movement's visibility and influence. The civil rights movement in the United States exemplifies this process, as protests and public discourse recursively strengthened the movement, ultimately reshaping laws and societal values.

Ideas and cultural practices evolve through recursive diffusion across social networks. When a new concept, such as sustainability, gains traction, it undergoes recursive adaptation through feedback from scientists, policymakers, and the public. This continuous refinement ensures that the idea remains relevant, actionable, and aligned with societal priorities. The evolution of sustainability, for example, reflects a recursive dialogue between environmental concerns, economic practices, and technological innovations.

Cultural evolution itself is a recursive process. Traditions are inherited and modified through feedback from each generation, allowing for a dynamic balance between continuity and innovation. For example, linguistic evolution demonstrates how recursive interactions, between speakers, cultural influences, and practical needs, shape language over time. Similarly, artistic movements like impressionism and modernism emerge through recursive challenges to established norms, fostering new forms of expression and cultural dialogue.

Identity and the Self

The self, a construct central to human experience, is often viewed as a fixed entity defined by stable traits and characteristics. Recursive Realism, however, reimagines identity as a dynamic process, shaped by continuous feedback between personal experiences, social interactions, and self-reflection. The self is not static but is perpetually reconstructed through recursive loops, enabling growth, adaptation, and transformation.

The self emerges from recursive loops between internal reflection and external validation. Internal reflection allows individuals to evaluate and refine their self-concept by revisiting past experiences, analysing actions, and integrating insights. For instance, a person who undergoes a major life transition, such as changing careers, may engage in recursive reflection to redefine their sense of purpose and identity.

Social interactions further shape identity through feedback from others. Validation or critique from peers, family, or colleagues influences how individuals perceive themselves and adjust their behaviour. For example, a person may see themselves as a leader, but this identity is reinforced or challenged by feedback from their professional or social environment. This recursive interplay between internal beliefs and external feedback continually refines and redefines the self.

Personal growth is inherently recursive. As individuals encounter new challenges or opportunities, they integrate these experiences into their evolving self-concept. Each recursive cycle of reflection and adaptation fosters a deeper understanding of one's abilities, values, and aspirations. For example, overcoming a significant obstacle, such as a health crisis, may prompt an individual to reassess their priorities and develop a more resilient sense of self.

Memory and Self-Reflection

Memory is integral to the recursive construction of identity, providing the continuity that links past experiences to present self-awareness. However, memory is not a static repository; it is a dynamic process shaped by recursive feedback loops that allow for revision, reinterpretation, and integration.

Memory is inherently fluid, reconstructed through recursive interactions with current experiences and emotional states. Each time a memory is recalled, it is reshaped in light of new perspectives, creating a dynamic interplay between past and present. For example, a childhood memory of a family gathering might take on new meaning as an adult gains a deeper understanding of familial relationships.

Narrative identity, the coherent story individuals construct to make sense of their lives, evolves through recursive self-reflection and memory reconstruction. This narrative is not fixed but adapts as new experiences prompt reinterpretation. For instance, someone who initially views a career setback as a failure might, upon reflection, reinterpret it as a catalyst for personal growth and resilience.

The recursive nature of memory also facilitates the integration of new experiences into one's identity. Each new event adds to the ongoing narrative, sometimes reinforcing existing beliefs and other times prompting revisions. For example, achieving a long-term goal may reinforce an individual's self-concept as determined and capable, while unexpected setbacks might lead to a reassessment of values or priorities.

Adapting Identity Through Recursion

Life presents a series of changes, challenges, and transitions that compel individuals to reassess and adapt their sense of self. Recursive Realism provides a powerful lens to understand how identity evolves in response to these dynamics, emphasizing the role of recursive processes such as self-reflection, feedback integration, and re-evaluation. Through these processes, individuals maintain psychological resilience, ensuring their sense of self remains coherent yet flexible in the face of life's complexities.

Identity in Times of Change

Major life transitions, such as relocating to a new environment, beginning a new career, or enduring personal loss, often disrupt established identities. In such instances, individuals engage in a recursive process of reflection and reconstruction. By revisiting past experiences and integrating new insights, they reconstruct a revised sense of self that accommodates the change. For example, a retired individual might transition from identifying strongly with their professional role to exploring new dimensions of their identity through hobbies, familial roles, or community involvement. This adaptive process exemplifies how recursive feedback loops facilitate the evolution of identity, ensuring coherence even during significant upheavals.

Resilience and Self-renewal

Resilience, the capacity to recover and grow from adversity, is closely tied to the recursive nature of identity adaptation. Resilient individuals demonstrate an ability to reframe challenges as opportunities for growth, reconstructing their sense of self in response to setbacks. This recursive feedback loop between internal reflection and external experiences fosters self-renewal and psychological well-being. For instance, someone recovering from a professional or personal failure might reflect on the lessons learned, reframe their goals, and develop a renewed, more adaptive sense of identity. Recursive processes allow such individuals to transcend adversity, emerging with greater self-awareness and confidence.

The Role of Social Support in Identity Adaptation

Social relationships serve as a critical source of feedback during periods of identity transition. External validation and feedback from supportive relationships, such as family, friends, or mentors, reinforce or challenge an individual's evolving self-concept. This recursive interplay between internal reflection and external support enables individuals to navigate life transitions more effectively. For example, someone navigating the emotional complexities of divorce may rely on friends or therapists to help redefine their self-concept, ensuring they adapt positively to their new circumstances. Social interactions not only provide encouragement but also help individuals reinterpret their experiences, adding depth and nuance to their evolving identity.

The Recursive Nature of the Self

Recursive Realism underscores that the self is not a static entity but an ever-evolving construct shaped by continuous loops of reflection, memory, and social feedback. This dynamic perspective reveals how individuals can maintain a coherent sense of self while simultaneously adapting to life's changes. By revisiting and revising their identity in response to new challenges, experiences, or insights, individuals engage in a process of continuous self-renewal.

This recursive process ensures that identity remains both flexible and adaptive, capable of evolving to meet the shifting demands of life while maintaining a sense of continuity and coherence. For example, as individuals encounter different social contexts, professional roles, or cultural environments, their identity adapts in response to the feedback and demands of these settings. This fluidity is not a sign of instability but a hallmark of the recursive self's resilience and adaptability.

Conclusion: Continuous Growth Through Recursive Processes

Throughout this chapter, we have explored the functional aspects of Recursive Realism, particularly how recursive feedback loops shape cognition, learning, social systems, and the construction of identity. By understanding these processes as dynamic, interactive, and recursive, we gain a deeper appreciation of how individuals and societies adapt to change, solve problems, and continuously reconstruct their understanding of themselves and the world around them.

Recursive Realism provides a comprehensive framework for conceptualizing the fluidity of human experience. It reveals that learning, growth, and social adaptation are not linear but are shaped by continuous cycles of feedback, reflection, and adjustment. Through the recursive nature of identity, individuals are not merely passive products of their environment but active participants in their own evolution. This approach highlights the transformative potential of recursion, demonstrating how individuals and societies can adapt, thrive, and innovate through the iterative interplay of reflection, interaction, and adjustment.

References

Ainsworth, M. D. S., Blehar, M. C., Waters, E., & Wall, S. (2015). *Patterns of Attachment: A Psychological Study of the Strange Situation*. Routledge.

Bandura, A. (1977). Self-efficacy: Toward a unifying theory of behavioral change. *Psychological Review*, 84(2), 191-215.

Baumeister, R. F., & Leary, M. R. (1995). The need to belong: Desire for interpersonal attachments as a fundamental human motivation. *Psychological Bulletin*, 117(3), 497-529.

Bateson, G. (2000). *Steps to an Ecology of Mind: Collected Essays in Anthropology, Psychiatry, Evolution, and Epistemology*. University of Chicago Press.

Bowlby, J. (1969). *Attachment and Loss: Vol. 1. Attachment*. Basic Books.

Craik, F. I. M., & Lockhart, R. S. (1972). Levels of processing: A framework for memory research. *Journal of Verbal Learning and Verbal Behavior*, 11(6), 671-684.

Csikszentmihalyi, M. (1990). *Flow: The Psychology of Optimal Experience*. Harper & Row.

Damasio, A. R. (1999). *The Feeling of What Happens: Body and Emotion in the Making of Consciousness*. Harcourt Brace.

Dennett, D. C. (1991). *Consciousness Explained*. Little, Brown and Company.

Dewey, J. (1938). *Experience and Education*. Macmillan.

Gopnik, A. (2009). *The Philosophical Baby: What Children's Minds Tell Us About Truth, Love, and the Meaning of Life*. Farrar, Straus and Giroux.

Habermas, J. (1984). *The Theory of Communicative Action: Reason and the Rationalization of Society* (Vol. 1). Beacon Press.

Hofstadter, D. R. (1979). *Gödel, Escher, Bach: An Eternal Golden Braid*. Basic Books.

James, W. (1890). *The Principles of Psychology* (Vol. 1). Henry Holt and Company.

Kegan, R. (1994). *In Over Our Heads: The Mental Demands of Modern Life*. Harvard University Press.

Kolb, D. A. (1984). *Experiential Learning: Experience as the Source of Learning and Development*. Prentice-Hall.

Lave, J., & Wenger, E. (1991). *Situated Learning: Legitimate Peripheral Participation*. Cambridge University Press.

Lewin, K. (1947). Frontiers in group dynamics: Concept, method and reality in social science; social equilibria and social change. *Human Relations*, 1(1), 5-41.

Mead, G. H. (1934). *Mind, Self, and Society from the Standpoint of a Social Behaviorist*. University of Chicago Press.

Piaget, J. (1950). *The Psychology of Intelligence*. Routledge & Kegan Paul.

Rogers, C. R. (1961). *On Becoming a Person: A Therapist's View of Psychotherapy*. Houghton Mifflin.

Schacter, D. L. (1996). *Searching for Memory: The Brain, the Mind, and the Past*. Basic Books.

Schön, D. A. (1983). *The Reflective Practitioner: How Professionals Think in Action*. Basic Books.

Senge, P. M. (1990). *The Fifth Discipline: The Art and Practice of the Learning Organization*. Doubleday.

Vygotsky, L. S. (1978). *Mind in Society: The Development of Higher Psychological Processes*. Harvard University Press.

Wenger, E. (1998). *Communities of Practice: Learning, Meaning, and Identity*. Cambridge University Press.

Chapter 5: The Recursive Universe

Recursion in Quantum Systems

The field of quantum mechanics presents one of the most profound challenges to traditional understandings of reality. At the quantum level, particles do not behave according to classical laws of physics; instead, they exhibit uncertainty, superposition, and entanglement, phenomena that defy linear cause-and-effect reasoning. Recursive Realism offers a new way to approach these phenomena, suggesting that the quantum world operates through recursive feedback loops, where observation and measurement actively shape the behaviour of quantum systems.

The Observer Effect and Recursion

One of the most famous and perplexing phenomena in quantum mechanics is the observer effect, which reveals that the act of observation affects the state of a quantum system. This phenomenon can be understood through the lens of recursion, where the feedback between the observer and the observed system creates a continuous loop that influences the system's behaviour.

In quantum mechanics, particles exist in a state of superposition, meaning they can occupy multiple states simultaneously. However, once the system is observed or measured, the wavefunction collapses, forcing the particle into a single state. This collapse is not a linear process; it occurs through a recursive interaction between the observer and the system, where the feedback from the measurement alters the system's behaviour. The system is recursively redefined by the observation, reflecting the dynamic, interactive nature of quantum reality.

The famous double-slit experiment illustrates how observation influences quantum behaviour. When particles such as electrons are fired through two slits, they create an interference pattern on the other side, suggesting they behave like waves. However, when the particles are observed as they pass through the slits, they behave like particles, and the interference pattern disappears. This shift in behaviour occurs through a recursive feedback loop, where the act of observing alters the system's dynamics, collapsing the wave-like behaviour into a particle-like state.

Quantum Entanglement as Recursion

Another key phenomenon in quantum mechanics is entanglement, where two or more particles become linked in such a way that the state of one particle instantaneously affects the state of the other, even across vast distances. This phenomenon defies classical understandings of locality and suggests that quantum systems are interconnected through recursive feedback.

Quantum entanglement can be understood as a recursive relationship between particles, where changes in one particle feed back into the other, creating a dynamic, non-local interaction. This recursive interaction occurs instantaneously, meaning that information about the state of one particle is transmitted to the other without any observable delay. This phenomenon challenges classical notions of causality, suggesting that quantum systems are deeply interconnected through recursive loops that transcend space and time.

The Einstein-Podolsky-Rosen (EPR) paradox highlights the non-local, recursive nature of quantum entanglement. In this thought experiment, two particles are entangled and then separated by a large distance. According to quantum mechanics, measuring the state of one particle instantly determines the state of the other, regardless of the distance between them. This suggests that the two particles are linked through a recursive feedback loop that operates beyond the constraints of space and time, pointing to a deeper level of reality where recursion governs the behaviour of quantum systems.

The Role of Recursion in Quantum Computing

Quantum computing represents a revolutionary application of the recursive principles found in quantum mechanics. By leveraging the superposition and entanglement of quantum states, quantum computers can perform calculations at speeds far beyond the capabilities of classical computers. Recursive algorithms play a central role in enabling quantum computers to solve complex problems efficiently.

Classical computers operate using bits that are either in a state of 0 or 1. In contrast, qubits in a quantum computer can exist in a superposition of both 0 and 1, allowing for parallel processing. This recursive superposition enables quantum computers to explore multiple solutions simultaneously, vastly increasing computational power. Recursive feedback loops between qubits enable the system to refine its calculations through entanglement and interference, leading to solutions that would be impossible for classical systems to compute.

One of the challenges of quantum computing is that qubits are highly susceptible to errors due to environmental noise and decoherence. Recursive algorithms are used in quantum error correction to detect and correct these errors in real time. By creating entangled states and using recursive feedback to compare qubit states, quantum computers can identify and fix errors before they propagate, ensuring the accuracy and stability of the computation.

Recursion plays a fundamental role in shaping the behaviour of quantum systems, from the observer effect and wavefunction collapse to entanglement and quantum computing. The recursive interactions between the observer and the quantum system, as well as the recursive relationships between entangled particles, reveal that the quantum world operates through dynamic feedback loops that challenge classical understandings of causality and locality. By understanding these recursive processes, we gain new insights into the nature of reality at the quantum level, where observation, measurement, and system behaviour are deeply interconnected.

Recursion and Cosmology

At the grandest scales of existence, the universe reveals a dynamic interplay between matter, energy, and the fundamental forces that govern cosmic evolution. From the formation of galaxies to the life cycles of stars, recursion is present in the feedback loops that regulate the structure, stability, and change of celestial objects. Recursive Realism offers a framework for understanding how cosmic systems evolve through continuous cycles of feedback, where each process interacts with and influences the others, creating a dynamic, self-organising universe.

Cosmic Feedback Loops

The universe operates through a series of cosmic feedback loops, where interactions between matter, energy, and forces regulate the structure and behaviour of celestial systems. These feedback loops drive the formation of stars, galaxies, and planetary systems, and they also determine how these systems evolve over time. Recursive interactions between gravitational forces, radiation, and matter are central to maintaining the balance and dynamism of cosmic systems.

The formation and evolution of stars provide an excellent example of a recursive feedback system at work. Stars form from clouds of gas and dust, which collapse under the influence of gravity. As the material collapses, nuclear fusion begins in the star's core, generating energy in the form of light and radiation. This radiation creates an outward pressure that counteracts the force of gravity, establishing a feedback loop between gravitational collapse and radiation pressure. The balance between these two forces determines the star's lifespan and structure. Over time, as the star exhausts its nuclear fuel, the balance shifts, and the star evolves into new forms, such as a red giant, white dwarf, or supernova, depending on its mass. This recursive cycle of formation, evolution, and eventual collapse demonstrates how stars are part of a dynamic cosmic system governed by feedback.

Black holes play a significant role in the recursive dynamics of galaxy formation. The gravitational pull of a supermassive black hole at the centre of a galaxy influences the movement and distribution of matter within the galaxy. In turn, the material falling into the black hole generates high-energy radiation and jets of particles that interact with the surrounding galaxy. This creates a recursive feedback loop between the black hole and its host galaxy, where the growth of the black hole is influenced by the galaxy's mass and vice versa. This recursive interaction helps shape the structure and evolution of galaxies, as the black hole regulates the rate of star formation by influencing the distribution of gas and dust.

The early universe underwent a period of rapid expansion known as cosmic inflation, where the universe expanded exponentially in a fraction of a second. This inflationary process can be understood as a recursive feedback loop, where small fluctuations in the density of matter and energy were amplified through inflation, leading to the large-scale structure of the universe we observe today. These initial density fluctuations created the seeds for the formation of galaxies and clusters of galaxies, showing how feedback loops operating in the early universe shaped the cosmic structures that evolved over billions of years.

The Expanding Universe and Recursion

The expansion of the universe is one of the most profound discoveries in modern cosmology. The universe is not static; it is continuously expanding, with galaxies moving away from each other over time. This expansion is governed by recursive interactions between dark energy, dark matter, and gravitational forces, which shape the large-scale structure of the cosmos.

Observations have shown that the expansion of the universe is not slowing down but is actually accelerating. This acceleration is driven by a mysterious force known as dark energy, which exerts a repulsive force on the fabric of spacetime. The interaction between dark energy and gravitational forces creates a recursive feedback loop that determines the rate of cosmic expansion. As dark energy accelerates the expansion, it dilutes the gravitational pull of matter, further speeding up the expansion. This feedback loop creates a dynamic system where the universe's rate of expansion is constantly adjusting in response to the interplay between dark energy and matter.

Even as the universe expands, gravitational forces cause matter to cluster into galaxies, clusters of galaxies, and superclusters. These structures form through recursive interactions between gravitational attraction and the motion of matter, where regions of higher density attract more matter, creating self-reinforcing loops of accretion. This process of gravitational clustering leads to the formation of cosmic filaments, vast structures that span the universe, connecting clusters of galaxies in a web-like pattern. The recursive nature of gravitational clustering reveals how local interactions can give rise to large-scale cosmic structures through feedback processes.

The Multiverse Hypothesis and Recursion

One of the most intriguing ideas in modern cosmology is the possibility of a multiverse, a vast collection of universes, each with its own physical laws and constants. The multiverse hypothesis introduces a new level of recursion, where each universe is part of a larger recursive system that generates an infinite number of universes through processes such as quantum fluctuations or cosmic inflation. Recursive Realism provides a framework for understanding how the multiverse itself could be structured through recursive feedback loops between different universes.

In some interpretations of quantum mechanics, quantum fluctuations at the smallest scales of reality can give rise to new universes. These fluctuations create recursive feedback loops where each quantum event influences the state of the broader multiverse, leading to the branching off of new universes with different physical properties. This recursive process suggests that the multiverse is not a static collection of universes but a dynamic, evolving system where new universes are continuously created through recursive interactions at the quantum level.

The theory of eternal inflation suggests that cosmic inflation, the rapid expansion that occurred in the early universe, may be an ongoing process in different regions of spacetime. In this model, our universe is just one bubble in a much larger multiverse, where inflation continues in other regions, giving rise to new universes. This creates a recursive structure, where each inflating region generates its own set of universes, and the process repeats indefinitely. The recursive nature of inflationary cosmology suggests that the multiverse is a self-replicating system, where the formation of new universes is part of an ongoing cycle of creation.

Recursion and Cosmology

The concept of recursion in cosmology reveals that the universe is not a static, deterministic system but a dynamic, self-organising structure shaped by feedback loops between matter, energy, and the fundamental forces. These recursive interactions drive the formation and evolution of cosmic structures, from stars and galaxies to the large-scale expansion of the universe itself.

The recursive feedback between gravity and radiation shapes the life cycle of stars, while the interaction between black holes and their surrounding galaxies influences the structure and evolution of galactic systems. These feedback loops create a dynamic interplay between the forces that govern the cosmos, ensuring that cosmic systems are not static but continuously evolving.

The interaction between dark energy and gravitational forces creates a recursive system that drives the expansion of the universe. As dark energy accelerates the expansion, it further dilutes the gravitational pull of matter, creating a self-reinforcing loop that shapes the large-scale structure of the cosmos.

The idea of the multiverse introduces a new level of recursion, where each universe is part of a larger recursive system that generates new universes through processes like quantum fluctuations and cosmic inflation. This recursive structure suggests that the multiverse is a self-replicating system, where the creation of new universes is an ongoing process, governed by recursive feedback at the quantum and cosmological scales.

Recursion plays a fundamental role in shaping the cosmic structure of the universe. From the formation and evolution of stars to the large-scale expansion driven by dark energy, recursive feedback loops govern the dynamic processes that shape the cosmos. The possibility of a multiverse adds a new layer of recursion, suggesting that our universe is part of a larger recursive system, where universes are continually created through quantum and cosmological processes. By understanding the universe through the lens of recursion, we gain a deeper appreciation for the dynamic, interconnected nature of cosmic systems and the self-organising principles that govern the evolution of the cosmos.

Emergence of Complexity in the Universe

One of the most remarkable features of the universe is its capacity to generate complexity from seemingly simple beginnings. From the initial conditions of the Big Bang, the universe has evolved to produce a vast array of complex structures, including galaxies, stars, planets, and ultimately living organisms. This complexity arises not through chance or randomness but through recursive processes, where feedback loops between matter, energy, and the fundamental forces drive the self-organisation of increasingly intricate systems. Recursive Realism offers a framework for understanding how complexity emerges through these recursive interactions, revealing that the universe is a dynamic, evolving system that continuously generates new forms of order and complexity.

Self-organisation and Cosmic Structure

Self-organisation is a process by which complex structures emerge from simple interactions between components without the need for external guidance. In the universe, self-organisation occurs at multiple scales, from the formation of galaxies and stars to the intricate patterns seen in planetary systems and biological life. These self-organising processes are governed by recursive feedback loops that allow for the emergence of stable, yet dynamic, systems.

Galaxies form through recursive interactions between gravity and matter. In the early universe, small fluctuations in the density of matter created regions of higher gravitational pull, which attracted more matter, leading to the formation of galaxies. As matter accumulated, gravitational forces shaped the spiral or elliptical structure of galaxies, creating stable patterns that persist over billions of years. This process of gravitational accretion is recursive, as the accumulation of matter amplifies gravitational forces, creating a self-reinforcing loop that drives the formation of larger cosmic structures.

Stars are born in regions of space known as stellar nurseries, where clouds of gas and dust collapse under the influence of gravity. As the gas collapses, nuclear fusion ignites in the core of the protostar, generating heat and light. The interplay between gravitational collapse and radiation creates a recursive feedback loop, where the outward pressure of radiation counteracts the inward pull of gravity, allowing the star to achieve a stable state. Over time, stars recycle material into the cosmos through processes like supernovae, which return heavy elements to the interstellar medium, fuelling the formation of new stars and planets. This recursive cycle of star formation and stellar death drives the continuous evolution of cosmic structures.

Planetary systems form through recursive interactions between accretion and gravitational collapse. Dust and gas in a protoplanetary disk coalesce into larger bodies, forming planets, moons, and asteroids. This process is recursive, as the gravitational pull of forming planets influences the distribution of material in the disk, while the material in the disk shapes the orbits and composition of the planets. The resulting planetary systems exhibit a delicate balance of forces, where feedback loops between gravitational forces, orbital dynamics, and accretion processes create stable yet dynamic systems capable of sustaining life.

The Formation of Planets and the Conditions for Life

The formation of planets is a key step in the emergence of complexity, as planets provide the environments where life can develop. Recursive processes play a critical role in shaping the conditions that allow for the emergence of life, from the chemical composition of planets to the dynamics of ecosystems.

The habitability of a planet is determined by a complex interplay of factors, including its distance from its star (which influences temperature), the composition of its atmosphere, and the presence of water. These factors are shaped by recursive feedback loops between planetary systems and their stars. For example, tidal forces between a planet and its moon can influence the planet's rotation, stabilising its climate and creating conditions conducive to life. Similarly, geological processes such as volcanism and plate tectonics recycle carbon and other essential elements, regulating the planet's atmosphere and climate over long timescales.

Water plays a critical role in the emergence of life, and its presence on planets is shaped by recursive processes between atmospheric cycles, hydrological cycles, and geological activity. The water cycle, where water evaporates, condenses, and precipitates, creates a feedback loop that regulates the distribution of water on a planet's surface, influencing its climate and the potential for life. On Earth, the recursive interaction between the water cycle and geological activity has created the conditions for the emergence of complex ecosystems, where life and the environment are in continuous recursive interaction.

The Recursion of Life and the Cosmos

Life itself is a recursive phenomenon, emerging from the self-organising properties of matter and evolving through recursive feedback loops between genetic variation, natural selection, and environmental adaptation. These recursive processes have driven the emergence of complexity in biological systems, from simple single-celled organisms to complex, multicellular life forms capable of consciousness and self-reflection.

The process of biological evolution is inherently recursive, involving continuous feedback between genetic variation and environmental selection. As organisms reproduce, random mutations in their DNA create variations in traits. These variations are subject to natural selection, where the environment determines which traits are most advantageous for survival and reproduction. Over time, this recursive process of variation, selection, and reproduction drives the evolution of increasingly complex life forms. For example, the evolution of the eye involved a series of small, recursive adaptations, where each improvement in light sensitivity provided an advantage that was reinforced through natural selection, leading to the development of highly complex visual systems in animals.

Ecosystems are also governed by recursive feedback loops between organisms and their environment. In an ecosystem, each species interacts with other species and the abiotic components of the environment, creating a web of interdependent relationships. These interactions are recursive, as changes in one part of the ecosystem feed back into the system, influencing the behaviour and survival of other species. For example, in a predator-prey system, the population size of predators and prey influences each other in a recursive cycle, where increases in prey populations lead to increases in predators, and vice versa. This dynamic feedback ensures that ecosystems are continuously adapting to changes in population, climate, and resource availability.

The relationship between life and the planetary environment is deeply recursive. For example, photosynthetic organisms on Earth have played a crucial role in regulating the composition of the atmosphere by converting carbon dioxide into oxygen, creating the conditions necessary for the evolution of complex aerobic life. This recursive feedback loop between biological processes and atmospheric composition has shaped the planet's climate and ecosystems over billions of years. Similarly, human activities today, such as fossil fuel consumption and deforestation, are creating new feedback loops that are altering the climate and biodiversity, illustrating the ongoing recursive relationship between life and the environment.

The universe's capacity to generate complexity arises from the recursive interactions between matter, energy, and fundamental forces. From the formation of stars and galaxies to the development of planetary systems and the emergence of life, recursive feedback loops drive the self-organisation of increasingly intricate structures. Life itself is a recursive phenomenon, evolving through continuous feedback between genetic variation and environmental selection, and interacting with planetary systems to create dynamic ecosystems.

Recursive Realism provides a framework for understanding how complexity emerges through self-organisation and recursive processes, revealing that the universe is not a static collection of objects but a dynamic, evolving system that continuously generates new forms of order and complexity. By viewing the universe through the lens of recursion, we gain insight into the interconnectedness of cosmic and biological processes, where complexity emerges from simple interactions over time.

In this chapter, we have explored how recursion operates at both the quantum and cosmic scales, shaping the structure and evolution of the universe. From the behaviour of quantum systems to the formation of galaxies and the emergence of life, recursive feedback loops are fundamental to the dynamic processes that drive the universe. By understanding the universe as a recursive system, we gain a deeper appreciation for the complexity and order that emerge from these interactions, revealing a cosmos that is self-organising, evolving, and continuously generating new forms of reality.

References

Bell, J. S. (1964). On the Einstein Podolsky Rosen Paradox. *Physics Physique Физика*, 1(3), 195-200.

Bekenstein, J. D. (1973). Black Holes and Entropy. *Physical Review D*, 7(8), 2333-2346.

Bohm, D. (1952). A Suggested Interpretation of the Quantum Theory in Terms of "Hidden" Variables I. *Physical Review*, 85(2), 166-179.

Dirac, P. A. M. (1930). *The Principles of Quantum Mechanics*. Oxford University Press.

Einstein, A., Podolsky, B., & Rosen, N. (1935). Can Quantum-Mechanical Description of Physical Reality Be Considered Complete? *Physical Review*, 47(10), 777-780.

Feynman, R. P. (1982). Simulating Physics with Computers. *International Journal of Theoretical Physics*, 21, 467-488.

Guth, A. H. (1981). Inflationary Universe: A Possible Solution to the Horizon and Flatness Problems. *Physical Review D*, 23(2), 347-356.

Hawking, S. W. (1974). Black Hole Explosions? *Nature*, 248(5443), 30-31.

Hubble, E. (1929). A Relation Between Distance and Radial Velocity Among Extra-Galactic Nebulae. *Proceedings of the National Academy of Sciences*, 15(3), 168-173.

Linde, A. D. (1982). A New Inflationary Universe Scenario: A Possible Solution of the Horizon, Flatness, Homogeneity, Isotropy, and Primordial Monopole Problems. *Physics Letters B*, 108(6), 389-393.

Penrose, R. (1996). *Shadows of the Mind: A Search for the Missing Science of Consciousness*. Oxford University Press.

Schrödinger, E. (1935). Die gegenwärtige Situation in der Quantenmechanik. *Naturwissenschaften*, 23, 807-812, 823-828, 844-849.

Shor, P. W. (1994). Algorithms for Quantum Computation: Discrete Logarithms and Factoring. *Proceedings 35th Annual Symposium on Foundations of Computer Science*, 124-134.

Susskind, L. (1995). The World as a Hologram. *Journal of Mathematical Physics*, 36(11), 6377-6396.

Thorne, K. S. (1994). *Black Holes and Time Warps: Einstein's Outrageous Legacy*. W.W. Norton & Company.

Weinberg, S. (1972). *Gravitation and Cosmology: Principles and Applications of the General Theory of Relativity*. Wiley.

Zurek, W. H. (2003). Decoherence, Einselection, and the Quantum Origins of the Classical. *Reviews of Modern Physics*, 75(3), 715-775.

Smolin, L. (1997). *The Life of the Cosmos*. Oxford University Press.

Tegmark, M. (2003). Parallel Universes. *Scientific American*, 288(5), 40-51.

Livio, M. (2004). *The Golden Ratio: The Story of Phi, the World's Most Astonishing Number*. Broadway Books.

Kauffman, S. A. (1993). *The Origins of Order: Self-Organization and Selection in Evolution*. Oxford University Press.

De Duve, C. (1995). *Vital Dust: Life as a Cosmic Imperative*. Basic Books.

Chapter 6: Recursive Realism and the Nature of Time

The Emergence of Time in the Physical Universe

Time has long been a central question in both philosophy and physics. Traditionally, time is viewed as a linear progression, moving forward from past to future. However, the recursive nature of physical processes reveals that time itself may be an emergent phenomenon, arising from the interactions of matter, energy, and the fundamental forces that govern the universe. In this section, we will explore how time emerges through recursive processes in both quantum mechanics and cosmology, and how these interactions shape our understanding of time as a dynamic, evolving construct.

In the quantum world, time behaves very differently from how it does in the classical world. Quantum systems often exist in a superposition of states, where events do not follow a simple cause-and-effect relationship. The concept of quantum entanglement further challenges our classical understanding of time, as entangled particles can affect each other instantaneously, regardless of the distance between them. These phenomena suggest that at the quantum level, time may emerge through recursive feedback loops, where the state of one particle influences the state of another, creating a dynamic interaction that transcends linear time.

The temporal behaviour of quantum systems can be viewed as a recursive process, where past, present, and future states are intertwined. The act of measurement or observation collapses the superposition of quantum states, creating a recursive feedback loop that determines the state of the system. This feedback loop implies that time itself may not be an independent, absolute dimension but rather an emergent property of the recursive interactions between quantum systems.

Another intriguing feature of quantum mechanics is time symmetry, where the laws governing quantum systems remain the same whether time is moving forward or backward. This symmetry challenges our everyday experience of time as flowing in one direction. The recursive nature of quantum processes suggests that time may be an artefact of our observation, where the forward progression of time is simply a product of the recursive feedback loops between observers and quantum systems.

At the cosmological level, time is closely linked to the evolution of the universe. The expansion of the universe, driven by forces like dark energy, creates a dynamic system where time is not fixed but evolves alongside the structure of the cosmos. This section will explore how cosmic recursion shapes the flow of time, from the Big Bang to the present and beyond.

One of the most well-known concepts in physics is the arrow of time, which is often associated with the increase of entropy in the universe. Entropy, a measure of disorder, increases over time in isolated systems, giving time its forward direction. However, this process of entropy increase can be understood as a recursive interaction between matter and energy, where each interaction contributes to the overall increase in disorder. This recursive process generates the forward flow of time, creating the arrow of time that we experience.

Time and Cosmic Expansion: The expansion of the universe also plays a role in shaping our experience of time. As the universe expands, the gravitational forces that govern the structure of matter evolve, influencing the rate at which time flows. For instance, in regions of intense gravity, such as near black holes, time flows more slowly than in regions of weaker gravity, a phenomenon known as gravitational time dilation. This suggests that time is not a universal constant but is shaped by recursive interactions between matter, energy, and space.

Time in Human Perception and Consciousness

While time emerges through physical processes at the quantum and cosmic levels, our experience of time is shaped by human cognition and consciousness. Time is not only an external phenomenon but also an internal construct, shaped by recursive feedback loops in the brain that influence how we perceive the passage of time, remember the past, and anticipate the future. In this section, we will explore how recursive processes in the brain give rise to our experience of time and how these processes shape memory, anticipation, and the sense of continuity.

The human brain processes time through a network of recursive feedback loops that integrate sensory information, memory, and expectation. This recursive interaction allows the brain to create a coherent sense of time, where the past, present, and future are continuously being updated and reinterpreted.

Time perception is governed by recursive neural circuits in the brain, particularly in regions such as the prefrontal cortex and the hippocampus, which are involved in memory and decision-making. These circuits allow the brain to compare current sensory input with stored memories, creating a recursive loop that informs our perception of the passage of time. For example, when we are waiting for something to happen, our sense of time may slow down as the brain recursively checks and rechecks whether the expected event has occurred, creating the subjective experience of time dilation.

The brain's ability to anticipate future events is also a product of recursive thinking. By using past experiences to predict what will happen next, the brain creates a recursive feedback loop where memory informs expectation, and expectation shapes perception. This recursive anticipation allows us to navigate complex environments, where the brain is constantly revising its predictions based on incoming information.

Memory is one of the most important factors in shaping our experience of time, as it provides the continuity that links the past with the present and future. However, memory is not a static record of the past; it is dynamic and recursive, constantly being reshaped and reinterpreted in light of new experiences.

Memory is formed through recursive feedback loops between different regions of the brain, where sensory experiences are encoded, stored, and later retrieved. When we recall a memory, the act of remembering involves a recursive process of reconstruction, where the brain integrates past experiences with present context. This means that memory is not a perfect reflection of the past but a recursive reinterpretation, where each act of remembering subtly changes the memory itself.

Our sense of self is closely tied to our ability to construct a narrative identity, a coherent story that links our past, present, and future experiences. This narrative is shaped by recursive processes of memory and reflection, where we continuously revise our understanding of the past in light of new experiences. The recursive nature of memory allows us to create a sense of continuity over time, even as our individual memories change and evolve.

Our perception of time can also be dramatically altered in different states of consciousness, such as during dreams, meditation, or the influence of psychoactive substances. These altered states reveal that time is not an objective, fixed experience but is shaped by recursive interactions between neural processes and conscious awareness.

In dreams, time can seem to stretch or compress, with events that would take hours in waking life unfolding in mere minutes or even seconds. This time dilation occurs because the brain's recursive feedback loops process information differently in dreams, creating a subjective experience of time that is disconnected from external reality. In this state, the brain's recursive circuits are more focused on internal stimuli, allowing for the rapid compression of events and the fluid merging of past, present, and future.

During meditative states, individuals often report a sense of timelessness, where the usual progression of time seems to dissolve. This experience can be understood as a disruption of the brain's recursive time-processing circuits, where the usual feedback loops that compare sensory input with memory and expectation are quieted. As a result, the experience of time as a linear progression fades, and a more fluid, expansive sense of presence emerges.

The Emergence of Time in the Physical Universe

Time is one of the most fundamental yet mysterious aspects of reality. In classical physics, time is often treated as an absolute dimension, flowing in one direction from the past to the future. However, modern physics, particularly in the realms of quantum mechanics and cosmology, suggests that time may not be as straightforward as it appears. Instead, time may emerge through recursive processes, where the interactions of matter, energy, and the fundamental forces give rise to the experience of temporal progression.

Time and Quantum Recursion

In the quantum world, time operates very differently from our everyday experience. At this scale, particles exist in a state of superposition, meaning they occupy multiple states at once, and their interactions can defy classical notions of time.

The Role of Superposition in Time: When a quantum system is not observed, it exists in a superposition of states, where all possible outcomes coexist. The act of measurement collapses this superposition into a single outcome, creating a feedback loop between the observer and the quantum system. This recursive interaction means that the state of a quantum system is not fixed until it is observed, suggesting that time itself may emerge through this process of measurement and collapse.

Entangled particles behave in ways that challenge classical understandings of time. When two particles are entangled, measuring the state of one particle instantly determines the state of the other, regardless of the distance between them. This suggests that time is not a linear progression but is shaped by recursive feedback loops between entangled particles. The nonlocal nature of entanglement hints at a deeper, timeless level of reality, where interactions occur outside the constraints of time as we understand it.

Through the Recursive Realism, time can be seen as an emergent property that arises from recursive feedback loops between quantum systems and the observers interacting with them. At the cosmological level, time is shaped by the dynamic interactions between matter, energy, and the expansion of the universe. By understanding time as a recursive process, we move beyond the classical view of time as a linear flow and begin to appreciate the complex, dynamic nature of temporal reality.

Time in Human Perception and Consciousness

While time exists as a fundamental aspect of the physical universe, the way we perceive and experience time is shaped by complex neural processes that create a subjective sense of temporal flow. Time is not merely a sequence of external events but is constructed through recursive feedback loops within the brain, where memory, anticipation, and awareness interact to form our understanding of past, present, and future. Recursive Realism suggests that our perception of time emerges from the brain's ability to integrate information over different time scales, creating a coherent narrative of experience.

The Brain's Recursive Perception of Time

The human brain processes time through a network of neural circuits that constantly interact to integrate sensory information, predict future events, and revisit past experiences. This recursive processing enables the brain to create a fluid sense of time, where moments flow seamlessly from one to the next, allowing us to navigate the world and make sense of our experiences.

Time perception relies on recursive feedback loops between different regions of the brain, such as the cerebral cortex, the hippocampus, and the cerebellum. These brain regions work together to integrate sensory information over time, creating a continuous stream of consciousness. For example, the hippocampus plays a key role in forming and retrieving memories, allowing us to connect past experiences with the present moment. The cerebellum, which is traditionally associated with motor control, also contributes to the timing of events, helping the brain to synchronize its internal clock with external stimuli. These recursive interactions enable the brain to create a temporal framework, where events are organized in relation to one another, allowing us to experience the passage of time.

One of the challenges the brain faces is to maintain a sense of coherence across different time scales. For instance, when we watch a movie, our brain must integrate the rapid succession of frames into a continuous visual experience. This integration requires recursive processing, where the brain constantly compares incoming sensory data with recently stored information to create a smooth transition between moments. This ability to bind sensory inputs across time is crucial for maintaining a stable perception of the world, where changes in the environment are perceived as continuous rather than disjointed.

The brain's ability to predict future events is another key aspect of its recursive processing of time. By using past experiences to anticipate what will happen next, the brain creates a temporal horizon that extends beyond the immediate present. This predictive ability allows us to prepare for upcoming events, such as catching a ball or planning a conversation. The brain's recursive loops enable it to adjust these predictions in real time, creating a dynamic interplay between expectation and reality. For example, when we walk down a busy street, our brain constantly predicts the movement of people and objects, allowing us to navigate smoothly. If a prediction is incorrect, such as someone suddenly stepping in our path, the brain rapidly adjusts its expectations, demonstrating the flexibility of its recursive temporal processing.

Memory and the Recursive Nature of Time

Memory is central to our experience of time, as it allows us to retain past experiences and integrate them into our ongoing narrative. However, memory is not a static record of past events; it is dynamic and recursive, constantly being reshaped through reflection and reinterpretation. This recursive nature of memory allows us to maintain a sense of continuity over time, even as our memories change and evolve.

When we recall a memory, we do not simply retrieve a snapshot of the past; instead, the brain engages in a recursive process of reconstruction, where it integrates fragments of past experiences with current knowledge and emotions. This means that each act of remembering involves updating the memory, often altering details to fit our present understanding. For example, a person reflecting on a childhood experience might reinterpret the event based on new insights into their family dynamics, changing how they remember the experience. This recursive nature of memory allows us to adapt our sense of self and past to our current perspective.

Our sense of time is closely tied to our ability to construct a narrative identity, a story that links our past, present, and future into a coherent whole. This narrative is shaped by recursive processes of self-reflection and memory reconstruction, where we continuously revisit past experiences to make sense of our present circumstances and future goals. For example, when people experience significant life events, such as graduation, marriage, or loss, they often revisit past memories to understand how those events have shaped their identity. This recursive process of re-evaluation allows individuals to integrate new experiences into their narrative, creating a dynamic story of who they are and how they have evolved over time.

Memory also plays a crucial role in maintaining a sense of temporal continuity, where our experiences are linked together into a continuous stream. This continuity is essential for our ability to navigate the world, as it allows us to understand the cause-and-effect relationships between events. For example, remembering where we placed our keys yesterday helps us find them today. This continuous thread of memory is created through recursive interactions between the hippocampus and other brain regions, which allow us to track events across time and maintain a sense of coherence in our daily lives.

The Experience of Time in Altered States

Our perception of time can change dramatically in altered states of consciousness, such as during dreams, meditation, or the influence of psychoactive substances. These altered states reveal that time is not an objective experience but is shaped by recursive interactions between neural processes and conscious awareness.

In dreams, the brain's usual mechanisms for processing time are altered, leading to experiences where time seems to stretch or compress. Events that would take hours in waking life can unfold in a few minutes in a dream. This compression of time occurs because the brain's recursive feedback loops focus more on internal narratives and less on external sensory input. As a result, the brain is free to create experiences that are not bound by the usual constraints of time. For example, a person might dream of an entire journey across a continent in the space of a few minutes. This ability to manipulate time in dreams highlights the flexibility of the brain's recursive time-processing mechanisms.

Meditative states often involve a sense of timelessness, where the usual flow of time seems to slow down or disappear altogether. This experience can be understood as a result of changes in the brain's recursive processing of time. During meditation, the brain's focus shifts from external stimuli to internal awareness, disrupting the usual feedback loops that track the passage of time. As a result, meditators often experience a sense of expansive presence, where time feels fluid and unbounded. This state of timelessness is a reflection of how the brain's recursive processes can be altered to create different experiences of temporal reality.

Psychoactive substances, such as hallucinogens or psychedelics, can also dramatically alter time perception. Under the influence of these substances, time can seem to slow down, speed up, or become disjointed. This occurs because psychoactive substances affect the brain's neurotransmitter systems, altering the feedback loops that integrate sensory input and memory. For example, a person on LSD might perceive a minute as lasting for an hour, experiencing each moment with intense clarity. This altered time perception reveals the role of recursive neural processes in creating a coherent sense of time and how changes in these processes can create novel experiences of temporal reality.

Our experience of time is not a passive reflection of external events but is actively constructed through recursive feedback loops in the brain. Through processes of memory reconstruction, anticipation, and temporal integration, the brain creates a fluid, dynamic sense of time that allows us to navigate the world, plan for the future, and reflect on the past. Altered states of consciousness reveal the flexibility of these recursive time-processing mechanisms, showing that our perception of time can be stretched, compressed, or even suspended under different conditions.

Recursive Realism offers a framework for understanding how time is both an external reality and an internal construct, shaped by the dynamic interplay between physical processes and conscious awareness. By viewing time as an emergent phenomenon of recursive interactions, we gain a deeper appreciation for the complexity and fluidity of temporal experience.

In this chapter, we have explored how time emerges through recursive processes at both the cosmic and cognitive levels. Time is not a simple linear flow but a dynamic construct shaped by feedback loops between quantum systems, cosmic forces, and the brain's perception. Recursive Realism reveals that time is a multi-layered phenomenon, where physical interactions and subjective experience intertwine to create the rich tapestry of temporal reality.

By understanding time as a recursive process, we can move beyond simplistic notions of past, present, and future, appreciating instead the intricate, evolving nature of time as it unfolds through the universe and within the human mind.

References

Barbour, J. (1999). *The End of Time: The Next Revolution in Physics*. Oxford University Press.

Bohm, D. (1980). *Wholeness and the Implicate Order*. Routledge.

Einstein, A. (1905). On the Electrodynamics of Moving Bodies. *Annalen der Physik*, 322(10), 891-921.

Ellis, G. F. R., & Hawking, S. W. (1973). *The Large Scale Structure of Space-Time*. Cambridge University Press.

Feynman, R. P., Leighton, R. B., & Sands, M. (1963). *The Feynman Lectures on Physics: Volume 1*. Addison-Wesley.

Gisin, N. (2020). *Quantum Chance and Non-Locality: Probability and Nonlocality in the Interpretations of Quantum Mechanics*. Springer.

Hawking, S. W., & Penrose, R. (1996). *The Nature of Space and Time*. Princeton University Press.

Hossenfelder, S. (2018). *Lost in Math: How Beauty Leads Physics Astray*. Basic Books.

Kline, M. (1972). *Mathematical Thought from Ancient to Modern Times*. Oxford University Press.

Maudlin, T. (2012). *Philosophy of Physics: Space and Time*. Princeton University Press.

Penrose, R. (2004). *The Road to Reality: A Complete Guide to the Laws of the Universe*. Jonathan Cape.

Pöppel, E. (1997). A Hierarchical Model of Temporal Perception. *Trends in Cognitive Sciences*, 1(2), 56-61.

Prigogine, I. (1996). *The End of Certainty: Time, Chaos, and the New Laws of Nature*. Free Press.

Rovelli, C. (2004). *Quantum Gravity*. Cambridge University Press.

Rovelli, C. (2018). *The Order of Time*. Riverhead Books.

Smolin, L. (2013). *Time Reborn: From the Crisis in Physics to the Future of the Universe*. Houghton Mifflin Harcourt.

Strogatz, S. H. (2018). *Nonlinear Dynamics and Chaos: With Applications to Physics, Biology, Chemistry, and Engineering*. Westview Press.

Tegmark, M. (2014). *Our Mathematical Universe: My Quest for the Ultimate Nature of Reality*. Alfred A. Knopf.

Wigner, E. P. (1960). The Unreasonable Effectiveness of Mathematics in the Natural Sciences. *Communications on Pure and Applied Mathematics*, 13(1), 1-14.

Chapter 7: Recursive Realism and the Architecture of Reality

Recursion and the Foundations of Physical Laws

The universe is governed by a set of fundamental laws that describe the behaviour of matter, energy, and forces. These laws, from quantum mechanics to general relativity, reveal a universe that is structured yet dynamic, where complex interactions emerge from simple principles. Recursive Realism suggests that these fundamental laws themselves may be recursive in nature, shaped by feedback loops that give rise to the patterns and regularities observed in the physical world. This section explores how recursion is embedded in the laws of physics and how it contributes to the self-organising nature of the universe.

The Recursion of Symmetry

Symmetry is a fundamental concept in physics that describes how certain properties of a system remain unchanged under transformation. From the symmetry of space and time in Einstein's theory of relativity to the gauge symmetries of quantum field theory, symmetry plays a central role in defining the invariants that govern physical systems. However, symmetry is not static; it often involves recursive interactions that allow for self-similarity across different scales. For example, fractal patterns found in nature, such as the branching of trees or the structure of snowflakes, exhibit symmetry through recursive processes, where each part mirrors the whole. This recursive symmetry extends to the fundamental forces, where the interplay of symmetry and symmetry-breaking processes creates the diversity of particles and interactions observed in the universe.

Feedback in the Laws of Thermodynamics

The laws of thermodynamics, which govern the flow of energy and the behaviour of systems, also reveal a recursive nature. For example, the second law of thermodynamics, which states that entropy tends to increase over time, can be understood as a recursive interaction between energy exchange and probability distributions. In closed systems, entropy increases as the system moves toward a state of greater disorder, but in open systems, recursive feedback loops can create local pockets of order, such as in living organisms or stars. This dynamic between order and disorder is governed by recursive processes, where energy flows and feedback mechanisms drive the emergence of complexity within a framework that ultimately trends toward higher entropy.

Quantum Field Theory and Recursion

Quantum field theory (QFT), which describes the interactions of particles through fields, is inherently recursive. In QFT, particles are seen as excitations of underlying fields, and the interactions between these particles create feedback loops that influence the properties of the fields themselves. For example, the self-interactions of particles in a quantum field can lead to renormalisation, a recursive process where the parameters of the theory are adjusted to account for the effects of particle interactions at different energy scales. This recursive nature of QFT allows for the emergence of fundamental constants and the behaviour of elementary particles in ways that are consistent with observed physical phenomena, illustrating how recursion is embedded in the most fundamental descriptions of reality.

Cognitive Models and the Recursion of Knowledge

Just as the physical universe is structured through recursive interactions, the human mind also operates through recursive processes that shape perception, understanding, and knowledge. The cognitive architecture of the mind is not linear but is organised through feedback loops that enable us to integrate new information, reflect on our thoughts, and reorganise our mental models. Recursive Realism offers a perspective on how the mind constructs its understanding of reality through self-referential processes that mirror the recursive nature of the cosmos.

Recursive Learning and Concept Formation

Learning is a recursive process that involves continuously revising mental models in response to new experiences. As individuals encounter new information, they engage in a process of integration and re-evaluation, where prior knowledge is updated to accommodate new insights. This recursive process allows for the formation of abstract concepts, where repeated exposure to patterns and regularities leads to a deeper understanding of underlying principles. For example, a child learning about gravity might first understand it as objects falling downward, but through recursive learning and experimentation, they come to understand gravity as a universal force that governs planetary orbits and the structure of the universe.

Self-Reflection and Metacognition

Self-reflection is a uniquely recursive aspect of human cognition, where the mind becomes aware of itself and its own processes. Through metacognition, individuals can reflect on their thoughts, emotions, and beliefs, creating a recursive feedback loop that allows for self-improvement and cognitive growth. This recursive self-awareness is fundamental to problem-solving and creativity, as it allows individuals to reframe challenges, question assumptions, and explore new perspectives. Recursive Realism suggests that this self-reflective capacity mirrors the self-organising properties of the universe, where systems continuously adjust and evolve in response to internal and external feedback.

Knowledge as a Recursive Network

Human knowledge is not a static collection of facts but a dynamic network of interconnected ideas that evolve through recursive interaction. Each new piece of knowledge is integrated into an existing framework, where it is compared, contrasted, and synthesised with prior understanding. This recursive interaction creates hierarchical structures of knowledge, where general principles connect with specific details to create a coherent worldview. For example, the understanding of biological evolution is built upon recursive interactions between genetic studies, fossil records, and theoretical models, creating a complex, multi-layered understanding of life's history. Recursive Realism sees knowledge as a living structure, constantly evolving through feedback loops that integrate new discoveries and challenge old paradigms.

Recursive Realism and the Redefinition of Perception

Recursive Realism offers a new way of thinking about perception and reality, where the observer and the observed are not separate but are engaged in a continuous dialogue through recursive interactions. This perspective challenges the traditional view of objectivity, suggesting that perception is inherently co-constructed through the interplay of mind and world. By understanding reality as a recursive process, we can develop a more nuanced approach to knowledge, one that embraces complexity and interconnectedness rather than relying on reductionism and linear logic.

The Co-Creation of Reality

In Recursive Realism, reality is not simply discovered by observers; it is co-created through the recursive interaction between the mind's expectations and the properties of the external world. This perspective aligns with constructivist theories of perception, where individuals actively construct their understanding of the world based on sensory input and cognitive processes. For example, when scientists develop a theory to explain a natural phenomenon, they are not merely uncovering pre-existing truths but are engaged in a recursive process of hypothesis formation, experimentation, and revision, where the theory evolves in response to new evidence and interpretations. Recursive Realism suggests that this co-creation extends to all forms of knowledge, where reality is understood as an emergent product of recursive dialogue between the observer and the observed.

Complexity and Holism in Perception

Recursive Realism challenges the reductionist approach to understanding reality, which seeks to break complex systems down into their simplest components. Instead, it emphasises a holistic approach, where the interactions between parts are seen as fundamental to the behaviour of the whole. This holistic perspective is particularly important in fields like ecology, systems biology, and cognitive science, where the recursive relationships between components create emergent properties that cannot be understood by studying the components in isolation. For example, the behaviour of a neural network in the brain cannot be fully explained by analysing individual neurons; it is the pattern of interactions between neurons that gives rise to thoughts, emotions, and consciousness.

Recursive Realism also invites us to rethink the concept of objectivity. Traditional notions of objectivity assume that reality exists independently of the observer, and that truth can be known by eliminating subjective influences. However, Recursive Realism suggests that subjectivity and objectivity are intertwined, as the act of perception is always shaped by recursive feedback between the mind's expectations and the world's responses. This does not mean that truth is purely subjective, but rather that our understanding of truth is shaped by dynamic processes that involve both the observer's perspective and the structure of reality. By embracing this relational view of knowledge, we can develop a more nuanced understanding of how truth emerges from the interaction between thought and world.

Recursion lies at the heart of the laws of physics, shaping the behaviour of particles, fields, and cosmic structures. From the recursive nature of symmetry in quantum mechanics to the self-organising feedback loops of thermodynamics, recursion is a fundamental principle that governs the structure of reality. Understanding these recursive processes allows us to see the universe not as a static collection of fixed laws, but as a dynamic system that evolves through continuous interaction.

Cognitive Models and the Recursion of Knowledge

The way humans understand the world is shaped by cognitive processes that are inherently recursive. Knowledge is not simply accumulated but is continually revised through feedback loops between perception, memory, and reasoning. These recursive interactions allow individuals to integrate new information into their existing frameworks, adapt to novel situations, and reflect on their understanding of reality. Recursive Realism offers a perspective on how knowledge is constructed through self-referential processes, suggesting that the mind's capacity for recursion mirrors the self-organising nature of the cosmos.

Recursive Learning and Concept Formation

Learning is a recursive process where new information is continuously integrated and compared to existing knowledge. This process allows humans to move beyond surface-level understanding and develop abstract concepts that capture the underlying principles of phenomena. Recursive Realism suggests that this ability to generalise and abstract arises from the mind's recursive capacity to revisit and refine its mental models.

Learning involves a cycle of feedback, where new experiences challenge existing beliefs, prompting the mind to revise its understanding. For instance, when learning about natural phenomena like gravity, students may begin with simple observations, such as objects falling. As they encounter more complex ideas, like orbital mechanics or relativity, their understanding of gravity evolves through recursive interactions between experience, theory, and reflection. This iterative process allows for the refinement of concepts, enabling learners to build increasingly sophisticated mental models.

The ability to recognise patterns is central to learning, as it allows individuals to identify regularities and relationships within their experiences. This process is inherently recursive, as the mind revisits previous encounters to find similarities and differences across different contexts. For example, the concept of a triangle is formed through recursive abstraction, where the mind recognises that certain shapes share the property of having three sides, regardless of their orientation or size. This ability to abstract and generalise from specific instances is a key aspect of human cognition, enabling individuals to understand the world in terms of principles rather than isolated facts.

The development of scientific theories and philosophical frameworks is also driven by recursive processes, where hypotheses are continually tested, refined, and adjusted in light of new evidence. For example, the shift from Newtonian mechanics to Einstein's theory of relativity involved a recursive re-evaluation of the principles governing space and time, where observations of gravitational effects and light prompted a revision of established theories. This recursive process of theory refinement allows for the evolution of knowledge, as scientists and philosophers engage in an ongoing dialogue between conceptual models and empirical data.

Self-Reflection and Metacognition

Self-reflection is a uniquely human capability that allows individuals to examine their thoughts, emotions, and beliefs, creating a recursive loop between conscious awareness and cognitive processes. This self-referential capacity is crucial for problem-solving, emotional regulation, and creative thinking, as it enables individuals to step outside their immediate thoughts and consider their underlying assumptions.

Metacognition refers to the mind's ability to think about its own thinking. This recursive process allows individuals to monitor and regulate their cognitive strategies, improving their ability to learn and adapt. For instance, when studying a difficult subject, a student may reflect on the effectiveness of different study techniques, adjusting their approach based on what they find most effective. This recursive loop between action and self-evaluation enables learners to optimise their strategies, making them more efficient at acquiring and retaining information.

Self-reflection is also essential for solving complex problems, as it allows individuals to re-evaluate their assumptions and explore alternative approaches. When faced with a challenging problem, effective problem-solvers engage in a recursive process where they test hypotheses, evaluate outcomes, and adjust their strategies based on feedback. This iterative approach prevents rigidity in thinking, allowing for greater flexibility and creativity. For example, an engineer designing a new product might need to repeatedly prototype, test, and redesign their solution, with each iteration bringing them closer to the optimal outcome.

The Role of Reflection in Emotional Intelligence: Recursive reflection also plays a crucial role in emotional intelligence, as it allows individuals to process and understand their emotions. Through reflection, people can identify the causes of their feelings and adjust their responses to achieve greater emotional balance. For instance, after a stressful interaction, a person might reflect on their reaction and consider how they could respond differently in the future. This recursive process helps to build resilience and self-regulation, making it easier to navigate interpersonal challenges and maintain healthy relationships.

Knowledge as a Recursive Network

Knowledge is not a static repository of facts but a dynamic network of interconnected ideas that evolves through recursive interaction. Each new insight or discovery is integrated into a larger framework, where it is compared, synthesised, and re-evaluated in relation to what is already known. Recursive Realism suggests that this networked structure of knowledge mirrors the self-organising nature of the cosmos, where complexity emerges through interactions between simpler elements.

Knowledge is organised into hierarchical structures, where general principles provide a foundation for more specific details. This hierarchical organisation allows individuals to navigate between different levels of understanding, moving from broad concepts to fine-grained details and back again. For example, a biologist might study ecological systems at different levels of complexity, from individual organisms to populations and ecosystems. This recursive ability to zoom in and zoom out enables a deeper understanding of how individual components interact to form a coherent whole.

Analogy is a key cognitive tool that allows for recursive connections between different domains of knowledge. By recognising similarities between seemingly unrelated concepts, individuals can apply insights from one area to solve problems in another. For instance, the analogy between biological evolution and cultural evolution has led to new ways of understanding how ideas and technologies spread through societies. This ability to transfer knowledge across domains reflects the recursive nature of thought, where connections between ideas create a networked understanding of reality.

The scientific method itself is a recursive process, where hypotheses are tested, results are compared, and theories are refined over time. This ongoing cycle of experiment and revision allows science to build a cumulative body of knowledge that evolves in response to new discoveries. For example, the development of climate science has involved a recursive integration of atmospheric data, computer models, and theoretical frameworks, leading to a more accurate understanding of global warming and its impacts. Recursive Realism suggests that this iterative approach to knowledge is essential for capturing the dynamic complexity of the natural world, where understanding evolves through continuous feedback between theory and observation.

Human knowledge is not static but dynamic, evolving through recursive processes that shape how we perceive, understand, and interact with the world. From learning and self-reflection to the networked structure of scientific inquiry, recursion enables the mind to adapt to new information, revise its mental models, and integrate disparate ideas into a coherent framework. Recursive Realism offers a perspective that sees knowledge as a living system, one that mirrors the self-organising principles of the cosmos and allows for a deeper, more interconnected understanding of reality.

Recursive Realism and the Redefinition of Perception

Perception is not merely the passive reception of external stimuli; it is an active process shaped by recursive feedback between the mind's expectations and the world's responses. Recursive Realism suggests that perception and reality are co-constructed through dynamic interactions, where the observer plays a crucial role in shaping their understanding of the world. This perspective challenges traditional notions of objectivity, offering a relational view where the boundary between subject and object becomes more fluid.

In Recursive Realism, reality is seen as an emergent property of the recursive interactions between observers and the phenomena they observe. This perspective aligns with constructivist theories, where knowledge is not simply discovered but is actively constructed through perception, interpretation, and reflection. For example, when astronomers observe distant galaxies, their understanding is shaped not only by the light that reaches their telescopes but also by the theories and models they use to interpret that light. This interplay between observation and theory creates a recursive loop that refines our understanding of the cosmos, illustrating how knowledge emerges from the interaction between mind and world.

Traditional reductionism seeks to understand systems by breaking them down into their simplest components. Recursive Realism, however, emphasises a holistic approach, where the interactions between parts are seen as fundamental to the behaviour of the whole. This perspective is crucial in fields like ecology and systems biology, where emergent properties cannot be fully understood by analysing individual components. For example, the behaviour of an ecosystem depends on the recursive interactions between plants, animals, microorganisms, and climatic factors. By focusing on the feedback loops between these elements, Recursive Realism offers a way to understand the dynamic balance that characterises complex systems.

Recursive Realism challenges the traditional concept of objectivity, suggesting that subjectivity and objectivity are inherently intertwined. In this view, the act of observation is always influenced by the observer's prior knowledge, cultural background, and cognitive framework. This does not mean that truth is purely subjective but rather that our understanding of truth emerges through recursive interactions between subjective experience and objective reality. For example, scientific paradigms evolve as new observations challenge existing theories, leading to paradigm shifts that reshape our understanding of fundamental concepts like space, time, and matter.

Recursive Realism invites us to rethink the relationship between mind and world, where perception and reality are seen as co-created through recursive interactions. By moving beyond simplistic notions of objectivity and embracing a relational approach to knowledge, Recursive Realism allows for a deeper appreciation of the complexity and interconnectedness of reality. This perspective offers a more nuanced understanding of how truth and understanding emerge from the dynamic dialogue between observers and the phenomena they study.

Throughout this chapter, we have explored how recursion shapes both the physical structure of the universe and the cognitive processes that define human understanding. By recognising the recursive nature of physical laws, knowledge, and perception, Recursive Realism offers a framework for understanding the dynamic interplay between mind and cosmos. This perspective moves beyond reductionist approaches, embracing a holistic view where complexity emerges through feedback loops and self-organising principles. Recursive Realism provides a new lens through which to view reality, one that acknowledges the active role of recursion in shaping both the universe and our understanding of it.

References

Barrow, J.D. (1998). *Impossibility: The Limits of Science and the Science of Limits*. Oxford University Press.

Barrow, J.D. and Tipler, F.J. (1986). *The Anthropic Cosmological Principle*. Oxford University Press.

Bohm, D. (1980). *Wholeness and the Implicate Order*. Routledge.

Cohen, J. and Stewart, I. (1994). *The Collapse of Chaos: Discovering Simplicity in a Complex World*. Penguin Books.

Davies, P.C.W. (1983). *God and the New Physics*. Penguin Books.

Einstein, A. (1916). The Foundation of the General Theory of Relativity. *Annalen der Physik*, 354(7), pp. 769-822.

Feynman, R.P., Leighton, R.B. and Sands, M. (1963). *The Feynman Lectures on Physics: Volume 1*. Addison-Wesley.

Kauffman, S.A. (1995). *At Home in the Universe: The Search for the Laws of Self-Organization and Complexity*. Oxford University Press.

Kuhn, T.S. (1962). *The Structure of Scientific Revolutions*. University of Chicago Press.

Penrose, R. (1989). *The Emperor's New Mind: Concerning Computers, Minds, and the Laws of Physics*. Oxford University Press.

Prigogine, I. (1996). *The End of Certainty: Time, Chaos, and the New Laws of Nature*. Free Press.

Rovelli, C. (2018). *The Order of Time*. Riverhead Books.

Smolin, L. (2013). *Time Reborn: From the Crisis in Physics to the Future of the Universe*. Houghton Mifflin Harcourt.

Strogatz, S.H. (2018). *Nonlinear Dynamics and Chaos: With Applications to Physics, Biology, Chemistry, and Engineering*. Westview Press.

Tegmark, M. (2014). *Our Mathematical Universe: My Quest for the Ultimate Nature of Reality*. Alfred A. Knopf.

Wheeler, J.A. (1983). Law without Law. In: J.A. Wheeler and W.H. Zurek, eds., *Quantum Theory and Measurement*. Princeton University Press, pp. 182-213.

Wigner, E.P. (1960). The Unreasonable Effectiveness of Mathematics in the Natural Sciences. *Communications on Pure and Applied Mathematics*, 13(1), pp. 1-14.

Chapter 8: Recursive Realism and the Boundaries of Consciousness

Recursion as the Foundation of Thought and Behaviour

One of the most profound features of Recursive Realism is its capacity to explain how thought and behaviour emerge through self-referential processes, which lie at the heart of human identity formation and the interaction between consciousness and the environment. To comprehend human thought as a recursive phenomenon is to understand how individuals continually revise their self-concept through layers of reflection, memory, experience, and social feedback. This recursive model allows for adaptability, growth, and the dynamic adjustment of the self in response to internal and external stimuli.

Recursion can be observed in both simple and complex feedback loops within the mind, which manifest in the way we revise our perceptions, actions, and interpretations of the world around us. These feedback loops are not limited to immediate experiences but are continuously influenced by past learning and expectations about the future. In this way, the recursive nature of thought enables individuals to respond not just to current situations, but to long-term goals and abstract ideas about reality.

The Self-Reflective Nature of Thought

At its core, self-reflection is a recursive activity. When we reflect on an event, we are not only thinking about the occurrence itself but also how we interpret and feel about it. This leads to a revisiting of memories and emotions, creating a loop where each instance of reflection modifies the previous one. Over time, these loops deepen, leading to the formation of complex mental constructs such as values, beliefs, and self-identity.

Recursive Realism posits that the process of self-reflection is essential for identity formation. Our thoughts and behaviours are continually shaped and reshaped through recursive interactions between experience, expectation, and self-concept. This idea challenges the notion of a fixed identity, instead proposing that identity is in a constant state of flux, evolving through the feedback loops of internal self-reflection and external social interaction.

In this sense, thought itself becomes an emergent phenomenon, one that arises not from linear causality but from recursive cycles where each reflection on a thought, emotion, or experience influences the next. This recursive process allows humans to develop a nuanced understanding of themselves and the world, leading to adaptation and personal growth.

Feedback Loops and Behavioural Adaptation

Human behaviour, much like thought, is deeply rooted in recursive processes. Every action we take generates feedback from the environment, whether it be in the form of social reactions, physical outcomes, or internal emotional responses. This feedback, in turn, becomes part of the recursive loop that informs future behaviour. Recursive Realism suggests that this constant interaction between action and feedback is what allows individuals to adapt their behaviours to changing circumstances.

For instance, when we make a decision, the outcome of that decision is evaluated through the lens of both immediate consequences and long-term goals. If the outcome is positive, it reinforces the behaviour, creating a loop that solidifies the action as part of our repertoire of behaviours. If the outcome is negative, it prompts a recalibration, leading us to adjust our actions in future similar scenarios. This adaptability is crucial for survival and personal development, as it allows individuals to modify their behaviour in response to the ever-changing environment.

In essence, Recursive Realism highlights that behavioural change is not a simple cause-and-effect process but rather a complex recursion of thought, reflection, and feedback. This allows for fine-tuning of actions, where individuals learn to navigate the subtleties of their environment and social contexts, making recursive adaptation a key driver of personal success and social integration.

Identity as a Recursive Construct

Building on the recursive nature of thought and behaviour, Recursive Realism posits that identity itself is a recursive construct. The sense of self is not a static entity but a dynamic system that evolves through continual self-reference and social feedback. Each interaction with the world, whether through relationships, work, or personal reflection, leads to incremental changes in how we perceive ourselves, shaping our overall identity.

This model of identity suggests that who we are is continually redefined through recursive loops of self-perception and external validation. For instance, our experiences in social groups provide feedback about our social roles and personal characteristics, which we then internalize and use to revise our self-concept. At the same time, our internal self-reflections, our thoughts about our thoughts, contribute to the ongoing redefinition of our sense of self.

In this way, identity is an emergent property of recursive interactions between the self and the world. This dynamic nature of identity means that individuals can constantly evolve, adapt, and grow based on new experiences and feedback from their environment. Recursive Realism provides a framework for understanding how these recursive processes enable identity formation to be both stable and fluid, capable of adjusting to new information while maintaining a core sense of continuity.

The Self as a Recursive Construct

The self is often perceived as a core essence, a stable identity that persists throughout a person's life. However, the recursive nature of cognition and memory suggests that the self is not fixed but is continuously reconstructed through reflection, memory, and social feedback. Recursive Realism offers a framework for understanding the self as an ongoing process, where identity is shaped by recursive interactions between the inner world of thoughts and the outer world of social roles and relationships. This perspective allows us to see the self as dynamic, adaptive, and capable of transformation in response to new experiences.

The Construction of Identity Through Recursive Feedback

Identity is built through a continuous loop between internal self-concept and external feedback. From early childhood, individuals form a sense of self by observing how others respond to them, reflecting on these interactions, and adjusting their self-concept accordingly. This recursive process allows individuals to develop social roles, personal traits, and a sense of continuity over time, even as they adapt to new environments and social expectations.

Social feedback plays a crucial role in shaping the self-concept, as it provides validation or challenge to an individual's sense of who they are. This feedback loop between internal self-perception and external response allows individuals to refine their self-identity. For example, a child who receives positive reinforcement for their academic achievements may internalise a self-concept of being smart or capable, which becomes a central part of their identity. As they continue to receive feedback throughout life, this sense of self-worth evolves, adapting to new roles and challenges. This dynamic nature of identity highlights how the self is not static but is continuously shaped by recursive interactions with the social environment.

The process of narrative identity involves creating a story that links past experiences, present understanding, and future aspirations. This narrative is shaped through recursive self-reflection, where individuals revisit key life events and re-evaluate their significance. This allows for a coherent sense of self that evolves over time. For instance, an adult reflecting on their childhood struggles may reinterpret those experiences as sources of resilience or strength, integrating them into a narrative that shapes their current self-concept. This recursive construction of identity allows individuals to maintain a sense of continuity while being open to self-transformation.

Recursive processes enable individuals to adapt their identity in response to changing circumstances. As people face new life stages, such as graduation, career shifts, or parenthood, they engage in a process of self-evaluation, where they reassess their goals, values, and sense of purpose. This recursive loop between reflection and action allows the self to remain flexible and responsive to new challenges, ensuring that identity evolves along with the person's external environment. For example, a professional who transitions into a leadership role may reframe their sense of self to include qualities like mentorship and responsibility, even as they draw on past experiences that shaped their approach to leadership.

Memory and the Recursive Self

Memory is central to the construction of the self, as it provides the continuity that links past experiences to present identity. However, memory is not a static archive; it is a dynamic, recursive process where each act of remembering involves reconstruction and reinterpretation. This recursive nature of memory allows for personal growth, where the self is continuously updated in response to new insights and reflective processes.

Each time a memory is recalled, the brain reconstructs it, integrating fragments of the original experience with current thoughts, emotions, and new perspectives. This means that memories are not perfect snapshots of the past but are malleable, capable of evolving over time. For example, a person reflecting on a childhood friendship might reinterpret the significance of certain interactions based on what they have learned about human relationships as an adult. This recursive nature of memory allows the self to adapt, as each new reflection reshapes the understanding of past events, making memory a tool for personal transformation rather than a static record.

Autobiographical memory, the ability to recall personal experiences, is central to maintaining a sense of identity over time. This form of memory is inherently recursive, as individuals use it to create a coherent narrative that explains who they are. For example, when a person describes key moments in their life, such as graduation, marriage, or a career change, they are engaging in a recursive process where memories are revisited and reorganised to form a cohesive story. This process helps individuals understand how they have changed, what they have learned, and how past experiences shape their current self-concept. It also allows for the integration of new experiences into the narrative, ensuring that the self remains adaptable and evolving.

The emotional tone of memories also plays a role in shaping the self. Positive or negative emotions associated with memories can influence how individuals interpret their past and understand their personal development. This emotional dimension is shaped through recursive feedback loops, where revisiting a memory can change the emotional response to it. For example, a person reflecting on a challenging period of their life might initially feel regret or sadness. However, through recursive reflection, they might come to see those experiences as sources of strength or wisdom, altering the emotional significance of the memory. This recursive process allows for emotional healing and growth, as the self integrates past experiences in ways that support resilience and well-being.

The Role of Social Interaction in the Recursive Self

The self is not constructed in isolation but is shaped through continuous social interaction. Recursive Realism emphasises the importance of dialogue between the self and others, where conversations, relationships, and social roles provide the context in which the self evolves. This social dimension of recursion allows individuals to adapt their self-concept to different environments, balancing internal values with external expectations.

The responses and feedback we receive from others play a crucial role in shaping our self-concept. These interactions create a recursive loop where the self is constantly revised in response to social feedback. For example, a teacher might receive feedback from students and colleagues that reinforces their self-perception as a knowledgeable and supportive educator. This feedback, in turn, influences how they approach their role and interact with others, creating a cycle of validation and adaptation. The recursive nature of these interactions ensures that the self is responsive to social dynamics, allowing individuals to maintain a coherent sense of identity even as their roles and relationships change.

Culture provides a broader context in which the recursive self is shaped. Cultural norms and values act as external feedback loops that influence how individuals perceive themselves and construct their identity. For example, in cultures that emphasise individualism, people may develop a sense of self that prioritises personal achievements and independence. In contrast, in cultures that emphasise community and collective values, individuals may define themselves more in terms of their relationships and social roles. The recursive interplay between cultural expectations and personal reflection allows the self to adapt to the social environment, ensuring that identity remains relevant and responsive to cultural contexts.

Throughout life, individuals take on different roles, such as student, parent, leader, or mentor, each of which requires a re-evaluation of self-concept. This adaptation involves a recursive dialogue between internal values and the demands of each role, where individuals find ways to integrate their core identity with the expectations of new roles. For example, a person who becomes a parent may need to integrate their sense of responsibility and care into their self-concept, while also balancing their personal interests and career aspirations. This recursive process allows for flexibility, enabling individuals to adapt to new challenges while maintaining a sense of continuity in their identity.

The self is not a fixed entity but a dynamic construct that evolves through recursive feedback loops between memory, social interaction, and self-reflection. By understanding the self as a recursive process, we gain insight into how identity is continuously updated, adapted, and transformed in response to new experiences. Recursive Realism offers a perspective on the fluidity of identity, where the self is capable of maintaining continuity while remaining open to growth and change.

The Unconscious Mind and Recursive Processes

While much of human thought and behaviour is shaped by conscious awareness, the unconscious mind plays a crucial role in influencing perception, cognition, and emotion. The unconscious mind operates through recursive processes that process information below the level of conscious awareness, shaping our automatic responses, intuition, and implicit biases. Recursive Realism suggests that these unseen loops within the unconscious mind are central to understanding how the mind integrates, organises, and interprets experience, creating a foundation for the conscious self.

The Dynamics of Unconscious Processing

The unconscious mind is not a passive repository of forgotten memories; it is an active system that continually processes information through recursive feedback loops. These loops allow the mind to perform routine tasks, respond to familiar situations, and make quick decisions without the need for conscious deliberation. This ability to process information automatically is essential for efficiency and survival, allowing the conscious mind to focus on novel or complex problems.

Habits are a prime example of how the unconscious mind uses recursion to streamline behaviour. When learning a new skill, such as riding a bicycle or playing an instrument, initial efforts require conscious attention and deliberate practice. However, with repetition, the behaviour becomes automatic, as the brain's basal ganglia create neural loops that allow for effortless execution. This transition from conscious practice to unconscious habit is a recursive process, where feedback from each attempt refines the neural pathways, making the action smoother and more efficient. Once a habit is established, it operates unconsciously, allowing individuals to perform complex actions with minimal cognitive effort.

Intuition, the ability to make quick judgments based on limited information, is another product of the recursive processing of the unconscious mind. Intuitive decisions often draw on patterns and associations that have been reinforced through past experiences, creating a database of implicit knowledge that guides behaviour. For example, a doctor diagnosing a patient may rely on intuition to recognise symptoms that suggest a particular condition, even if they cannot immediately articulate the reasoning behind their judgment. This intuitive ability is a result of the brain's pattern recognition systems, which operate through recursive loops that integrate sensory data with previous experiences, allowing for rapid and effective decision-making.

The unconscious mind also influences perception and behaviour through priming, a phenomenon where exposure to certain stimuli influences responses to subsequent stimuli. Priming occurs through recursive loops that create associations between concepts, even when these associations are not consciously recognised. For example, a person who sees the word "bread" may be more likely to recognise the word "butter" more quickly due to the association between these concepts. This process is automatic and often outside of awareness, yet it shapes how individuals interpret and respond to the world around them. Recursive Realism suggests that these unconscious processes reveal the hidden architecture of the mind, where automatic associations are formed through repeated exposure and contextual cues.

Dreams as Recursive Narratives

Dreaming offers a unique window into the recursive processes of the unconscious mind. During sleep, the brain engages in complex narrative construction, where memories, emotions, and symbolic representations are reprocessed in a non-linear and often metaphorical manner. Dreams can be understood as recursive loops, where the mind revisits and reinterprets daily experiences, exploring alternative scenarios and emotional resolutions.

One of the primary functions of dreams is to process emotional experiences, especially those that are unresolved or intense. For example, a person who has experienced a stressful event may dream of situations that symbolically represent the emotional conflict, allowing the mind to explore different responses and interpretations. This recursive process helps to integrate the emotional experience into the broader narrative of the self, allowing for a sense of closure or understanding. By revisiting emotions in a symbolic form, the unconscious mind can process feelings that might be too complex or painful to confront directly in waking life.

Lucid dreaming occurs when an individual becomes aware that they are dreaming, allowing them to exert control over the dream narrative. This state involves a recursion of consciousness into the dream state, where the dreamer is able to reflect on the nature of their experience while still immersed in the dream. Lucid dreaming highlights the recursive capacity of the mind to create nested layers of awareness, where self-reflection can occur even within altered states of consciousness. This ability to navigate between different levels of awareness demonstrates the flexibility of the mind's recursive processes, allowing individuals to engage with their unconscious in a more active and intentional way.

Dreams often involve symbols and archetypes that carry multiple layers of meaning. These symbols emerge through recursive associations between memories, cultural motifs, and personal experiences. For example, dreaming of water might carry associations with emotion, subconscious depth, and life changes, depending on the context and cultural background of the dreamer. The mind's ability to create multi-layered narratives in dreams reflects the recursive nature of thought, where images and ideas are continuously recontextualised and reinterpreted. This process allows for a rich inner world where different aspects of experience are integrated into a cohesive whole.

Recursive Influence of Unconscious Biases

The unconscious mind also shapes perception, judgment, and decision-making through cognitive biases that operate below the level of awareness. These biases are often the result of evolutionary adaptation and cultural conditioning, creating implicit associations that influence how individuals interpret the world. Recursive Realism suggests that these biases are not simply flaws in reasoning but are products of recursive feedback loops that guide adaptive behaviour.

Implicit biases are shaped through repeated exposure to cultural norms and social stereotypes, creating automatic associations that influence thought and action. For example, a person who has grown up in a society with gender stereotypes may unconsciously associate certain traits with men and women, even if they consciously reject these stereotypes. These biases are reinforced through recursive loops of social interaction and media exposure, making them deeply ingrained. Understanding these biases as recursive patterns allows for strategies to challenge and retrain the mind, as individuals can engage in reflective practices that interrupt these loops and create new associations.

Confirmation bias is a cognitive tendency to seek out information that supports one's existing beliefs while ignoring evidence that contradicts them. This bias operates through a recursive feedback loop, where each selective perception reinforces the original belief, making it more entrenched. For example, a person with a strong political belief may only consume media that aligns with their views, creating a feedback loop that strengthens their worldview while dismissing opposing perspectives. Recognising confirmation bias as a recursive process highlights the importance of deliberate exposure to diverse viewpoints as a way to break these loops and expand one's understanding.

The unconscious mind often relies on heuristics, mental shortcuts that simplify decision-making by using past experiences as a guide. These heuristics are the result of recursive pattern recognition, where familiar situations trigger automatic responses. For example, when faced with a snap decision, a person might rely on a gut feeling based on similar past experiences, even if they cannot articulate the reasoning behind it. While heuristics are efficient, they can also lead to systematic errors if the patterns they rely on are misleading or incomplete. Understanding heuristics as a product of recursive processes allows for a more nuanced view of their role in cognition, recognising both their utility and their limitations.

The unconscious mind plays a vital role in shaping human behaviour and cognition, operating through recursive loops that influence automatic responses, intuitive judgments, and emotional processing. By understanding the hidden feedback mechanisms of the unconscious, Recursive Realism offers insights into how the mind integrates implicit knowledge and patterns into the fabric of thought. This perspective allows us to appreciate the depth and complexity of mental processes that lie beyond the reach of conscious awareness, revealing a rich interplay between conscious and unconscious dimensions of the mind.

In this chapter, we have explored how recursion shapes the nature of consciousness, the construction of identity, and the dynamics of the unconscious mind. Recursive Realism offers a framework for understanding the mind as a dynamic, self-organising system, where feedback loops between thought, memory, and emotion shape our experience of reality. By recognising the recursive nature of self-awareness, identity, and intuition, we gain new insights into the interconnectedness of the inner and outer worlds, revealing a continuum between conscious and unconscious processes.

References

Baars, B.J. (1997). *In the Theater of Consciousness: The Workspace of the Mind*. Oxford University Press.

Bandura, A. (1986). *Social Foundations of Thought and Action: A Social Cognitive Theory*. Prentice-Hall.

Bartlett, F.C. (1932). *Remembering: A Study in Experimental and Social Psychology*. Cambridge University Press.

Bruner, J. (1990). *Acts of Meaning*. Harvard University Press.

Conway, M.A. and Pleydell-Pearce, C.W. (2000). The Construction of Autobiographical Memories in the Self-Memory System. *Psychological Review*, 107(2), pp. 261–288.

Dennett, D.C. (1991). *Consciousness Explained*. Little, Brown & Co.

Fischer, K.W. and Bidell, T.R. (2006). Dynamic Development of Action and Thought. In: W. Damon and R.M. Lerner (Eds.), *Handbook of Child Psychology: Theoretical Models of Human Development*. John Wiley & Sons, pp. 313–399.

Gazzaniga, M.S. (2011). *Who's in Charge? Free Will and the Science of the Brain*. HarperCollins.

Hohwy, J. (2013). *The Predictive Mind*. Oxford University Press.

James, W. (1890). *The Principles of Psychology*. Henry Holt and Company.

Kahneman, D. (2011). *Thinking, Fast and Slow*. Farrar, Straus and Giroux.

LeDoux, J.E. (2002). *Synaptic Self: How Our Brains Become Who We Are*. Viking.

Mead, G.H. (1934). *Mind, Self, and Society: From the Standpoint of a Social Behaviorist*. University of Chicago Press.

Metcalfe, J. and Shimamura, A.P., eds. (1994). *Metacognition: Knowing About Knowing*. MIT Press.

Pinker, S. (1997). *How the Mind Works*. Norton & Company.

Schacter, D.L. (1996). *Searching for Memory: The Brain, the Mind, and the Past*. Basic Books.

Sperry, R.W. (1981). Some Effects of Disconnecting the Cerebral Hemispheres. *Science*, 217(4566), pp. 1223–1226.

Tulving, E. (1983). *Elements of Episodic Memory*. Oxford University Press.

Vygotsky, L.S. (1978). *Mind in Society: The Development of Higher Psychological Processes*. Harvard University Press.

Zimmerman, B.J. and Schunk, D.H., eds. (2001). *Self-Regulated Learning and Academic Achievement: Theoretical Perspectives*. Lawrence Erlbaum Associates.

Chapter 9: Recursive Realism and the Philosophy of Knowledge

Recursion and the Nature of Knowledge

The pursuit of knowledge has always been central to human endeavour, driving the development of science, philosophy, and culture. Traditionally, knowledge has been understood as a process of discovery, where truths about the world are uncovered through observation and reasoning. However, Recursive Realism suggests that knowledge is not merely discovered but is actively constructed through recursive interactions between mind and world. This perspective challenges static notions of knowledge, revealing it to be a dynamic process shaped by continuous feedback loops between theory, experience, and reflection.

The Constructive Nature of Knowledge

Recursive Realism posits that knowledge is not a fixed entity but a constructive process where understanding evolves through recursive loops of hypothesis formation, testing, and revision. This view aligns with constructivist approaches in philosophy and science, where the act of knowing is seen as an interaction between the observer and the observed.

Knowledge can be seen as a self-organising system, where ideas and concepts evolve through recursive feedback loops between theory and observation. Just as ecological systems adapt to changes in their environment, knowledge systems evolve as new evidence challenges existing frameworks, prompting revisions and refinements. For example, the transition from Newtonian mechanics to Einstein's theory of relativity was not a simple replacement but a recursive refinement of our understanding of space and time, where the limitations of Newtonian models led to the development of a more comprehensive framework. This process illustrates how knowledge evolves through continuous adaptation, reflecting the dynamic nature of reality itself.

Models play a crucial role in the construction of knowledge, as they provide conceptual frameworks that guide our understanding of complex phenomena. However, models are inherently simplifications of reality, and they must be revised and adjusted as new evidence emerges. This process of model revision is a recursive loop, where each iteration brings us closer to a more accurate understanding of the phenomenon. For example, in climate science, models of global warming are continually refined as new data about atmospheric changes and ocean currents is integrated into the model, leading to a deeper understanding of the dynamics of Earth's climate system. Recursive Realism suggests that this iterative process of model-building is essential for capturing the complexity of the natural world, as it allows knowledge to evolve in response to new insights.

Knowledge is not linear but is organised as a network of interconnected ideas. Each concept or theory is linked to others through logical relationships, analogies, and metaphors, creating a web of understanding. This networked structure allows for the transfer of insights across different domains of knowledge, where analogies between biology and computer science, for instance, have led to the development of artificial intelligence. The recursive nature of this network means that new discoveries can reshape entire fields of knowledge, creating ripples that spread through the web, influencing how other concepts are understood. This interconnectedness allows for a holistic approach to knowledge, where complexity and interdependence are central to understanding reality.

Epistemological Questions and Recursive Realism

Recursive Realism raises important epistemological questions about the limits of knowledge and the nature of truth. It challenges the idea that knowledge can be absolute or final, suggesting instead that understanding is always provisional, shaped by recursive processes of inquiry and re-evaluation.

One of the classic challenges in epistemology is the problem of infinite regress, where every justification for a belief requires another justification, leading to an endless chain. Recursive Realism offers a way to reframe this problem, suggesting that knowledge systems operate through self-referential loops rather than linear chains. Instead of seeking a foundational starting point for knowledge, Recursive Realism suggests that understanding emerges from the coherence of the system itself. For example, in scientific inquiry, the validity of a theory is not determined by a single fundamental truth but by its ability to cohere with empirical evidence and other established theories. This perspective allows for a more flexible approach to knowledge, where truth is seen as a property of consistent networks of understanding rather than an unchangeable absolute.

Recursive Realism also embraces the role of uncertainty and ambiguity in knowledge. It recognises that many phenomena, especially those involving complex systems, cannot be perfectly predicted or understood. For example, chaos theory demonstrates that even simple systems can exhibit unpredictable behaviour due to sensitivity to initial conditions. Rather than seeking absolute certainty, Recursive Realism suggests that uncertainty is an inherent aspect of dynamic systems, where feedback loops create unpredictable outcomes. This perspective encourages an adaptive approach to knowledge, where probabilistic models and contingency plans are used to navigate uncertainty rather than relying on fixed certainties.

From the perspective of Recursive Realism, truth is understood as an emergent property that arises from the coherence of a knowledge system. This means that truth is not merely about correspondence with an external reality but involves the internal consistency and integration of concepts. For example, a scientific theory is considered true not only because it matches observations but because it provides a coherent explanation that integrates diverse phenomena into a unified framework. This recursive understanding of truth allows for paradigm shifts, where new frameworks can replace or expand existing ones by providing a deeper coherence with the empirical world. This approach recognises that understanding evolves, and that truth is a dynamic achievement rather than a static endpoint.

Recursive Realism and the Limits of Understanding

Recursive Realism suggests that there are inherent limits to what can be known, not because of any fundamental barrier but because of the complexity and recursive nature of reality itself. These limits are not failures of human inquiry but reflections of the multi-layered structure of the universe, where understanding is always a work in progress.

Just as the horizon of the universe expands as we look further into space, the horizon of knowledge is constantly shifting as we explore new questions and phenomena. This expanding horizon reflects the recursive nature of inquiry, where each new discovery opens up further questions and areas of investigation. For example, the discovery of quantum mechanics in the early 20th century opened up a new domain of subatomic phenomena that continues to challenge our understanding of reality. Recursive Realism suggests that this expanding horizon is a natural aspect of knowledge systems, where the depth and breadth of understanding grow through iterative exploration.

Recursive Realism also highlights the paradox of self-knowledge, where the mind's attempt to understand itself is inherently recursive. As we reflect on our own thoughts and conscious experiences, we create a loop where the observer becomes the observed. This creates a paradoxical situation, where understanding the nature of consciousness is both possible and limited by the very mechanisms we use to think. For example, efforts to map the neural basis of consciousness reveal much about how brain processes correlate with experience, but the subjective nature of qualia, the raw feeling of experience, remains elusive. Recursive Realism suggests that this paradox is not a flaw but a fundamental characteristic of self-referential systems, where the limits of knowledge are a reflection of the depth of inquiry itself.

While Recursive Realism acknowledges the limits of understanding, it also recognises that mystery plays a crucial role in driving human curiosity and discovery. The unknown is not a barrier but a stimulus for exploration, where the gaps in our understanding become opportunities for creative thought and innovation. For example, the mysteries of dark matter and dark energy challenge physicists to develop new theories that expand our understanding of the cosmic structure. Recursive Realism suggests that embracing the unknown allows for a dynamic relationship with knowledge, where mystery and insight are part of an ongoing dialogue between mind and universe.

Recursive Realism offers a dynamic perspective on knowledge, where understanding is seen as a constructive process that evolves through recursive feedback loops. By recognising the self-organising nature of knowledge systems, we can appreciate the complexity of how truth, coherence, and uncertainty interact to shape our understanding of reality. This perspective challenges static notions of knowledge and embraces a holistic approach that values the interconnectedness of ideas and the emergence of new insights through iterative exploration.

Recursive Realism and the Evolution of Ideas

Throughout history, the development of ideas has followed a recursive pattern, where new concepts emerge through dialogue and refinement of previous theories. Ideas evolve not in isolation but through feedback loops between individuals, communities, and cultural contexts. This recursive evolution allows for the integration of new discoveries, the re-examination of old assumptions, and the transformation of paradigms over time. Recursive Realism suggests that this process is fundamental to the nature of intellectual progress, where each iteration of thought deepens our understanding of reality.

Intellectual Evolution Through Recursion

The evolution of thought is characterised by recursion, where each new generation of thinkers builds upon the insights and limitations of their predecessors. This recursive process allows ideas to become more refined, nuanced, and inclusive over time, as philosophies and scientific theories are tested against empirical evidence and cultural shifts.

Philosophy is inherently recursive, as it involves a continuous dialogue with previous ideas and theories. Thinkers engage with the work of those who came before them, critiquing, refining, and sometimes reinterpreting their concepts. For example, Immanuel Kant's philosophy emerged as a response to the rationalism of Descartes and Leibniz and the empiricism of Hume, creating a synthesis that recognised the role of both sensory experience and innate structures of the mind. This recursive engagement allowed Kant to develop a more comprehensive framework that addressed the strengths and weaknesses of each approach. Similarly, existentialist thinkers like Jean-Paul Sartre and Simone de Beauvoir built upon phenomenology while critiquing its limitations, adding layers of complexity to our understanding of freedom, authenticity, and existence.

In science, theories evolve through a recursive process of hypothesis testing, empirical verification, and theoretical refinement. Thomas Kuhn's concept of paradigm shifts illustrates this process, where periods of normal science are interrupted by crises that lead to radical changes in the theoretical framework. For example, the shift from geocentrism to heliocentrism in astronomy, or from classical mechanics to quantum mechanics in physics, involved a recursion of inquiry, where anomalies in the existing framework prompted a re-evaluation and reformulation of fundamental principles. Recursive Realism suggests that these shifts are not just replacements of one theory by another but are iterations that integrate new insights while preserving the valuable aspects of previous knowledge. This recursive evolution allows science to adapt to new discoveries while maintaining a continuity of intellectual progress.

Ideas do not evolve in isolation from cultural contexts; they are deeply influenced by the social norms, values, and historical events of their time. Artistic movements, political ideologies, and scientific theories often reflect the cultural currents of their era, creating a recursive interaction between thought and culture. For example, the Enlightenment was both a cultural movement and a philosophical shift that emphasised reason, individualism, and scientific inquiry, leading to profound changes in political structures and social thought. This recursive dialogue between cultural change and intellectual evolution allows ideas to remain relevant and adaptive, as they are continually reshaped to address the challenges of their time.

The Role of Dialogue in Recursive Knowledge

Dialogue is central to the recursive nature of intellectual evolution, as it creates a space for ideas to be tested, challenged, and refined. This dialogue occurs not only between individual thinkers but also between disciplines, traditions, and perspectives. Recursive Realism suggests that the cross-pollination of ideas through dialogue is essential for the growth and maturation of knowledge.

Some of the most profound advancements in knowledge have occurred through interdisciplinary dialogue, where ideas from different fields are brought together to create new syntheses. For example, the development of cognitive science involved the integration of psychology, neuroscience, philosophy, and computer science, creating a recursive interaction between empirical research and theoretical models of the mind. This interdisciplinary recursion has led to new ways of understanding consciousness, learning, and artificial intelligence, expanding the horizons of what is possible in understanding the mind. Recursive Realism suggests that such boundary-crossing dialogues are crucial for intellectual innovation, as they allow for the creation of conceptual frameworks that are richer and more flexible.

Cross-cultural dialogue has also played a significant role in the evolution of ideas, as it allows for the exchange of worldviews and philosophical traditions. For example, the encounter between Greek philosophy and Islamic thought during the Islamic Golden Age led to a synthesis of Aristotelian logic and Islamic metaphysics, producing advancements in astronomy, medicine, and mathematics. Later, the translation and study of these works in medieval Europe contributed to the Renaissance and the Scientific Revolution. This recursive exchange of ideas allowed for the preservation and transformation of knowledge, demonstrating how dialogue between cultural traditions can create new intellectual paradigms.

Philosophical debates are often seen as oppositional, but they can also be understood as recursive dialogues, where opposing perspectives push each other to greater clarity and depth. For example, the debates between empiricism and rationalism in early modern philosophy helped to clarify the strengths and limitations of each approach, leading to more sophisticated theories of knowledge and perception. Dialectical methods, such as those used by Hegel, involve a recursive process where thesis and antithesis interact to create a synthesis that transcends the limitations of both. Recursive Realism sees these debates as essential to the growth of knowledge, as they allow for the exploration of complex questions from multiple angles, producing a richer understanding of reality.

Recursive Realism and the Shaping of Worldviews

Worldviews are the cognitive frameworks through which individuals and cultures interpret the world. These frameworks are not fixed but are shaped by recursive interactions between beliefs, experiences, and social influences. Recursive Realism offers a perspective on how worldviews evolve through reflection, challenge, and adaptation, allowing them to remain responsive to changing contexts.

A worldview is a complex network of values, assumptions, and interpretations that shape how people understand existence and purpose. These worldviews evolve through recursive processes of reflection and reinterpretation, where new experiences or paradigm shifts prompt a re-evaluation of core beliefs. For example, the transition from a mythological worldview to a scientific worldview during the Enlightenment involved a shift in how nature, knowledge, and human agency were understood. This shift was not instantaneous but involved a recursive process of debate, experimentation, and philosophical inquiry that reshaped the intellectual landscape of Europe. Recursive Realism suggests that worldviews are not rigid structures but adaptive systems that change as they encounter new challenges and insights.

Self-reflection is crucial for the evolution of worldviews, as it allows individuals to question their assumptions and reframe their understanding of reality. For example, a person who has a rigid belief about morality may experience a crisis when confronted with cultural differences or moral dilemmas that challenge their views. Through a process of recursive self-reflection, they may come to reconstruct their worldview to incorporate a more nuanced understanding of ethics and cultural diversity. This recursive ability to reflect on and revise one's worldview is essential for personal growth and intellectual maturity, allowing individuals to adapt to the complexity of the world.

Cultures develop paradigms, shared ways of understanding reality, that guide collective beliefs and practices. These paradigms evolve through recursive dialogue between tradition and innovation, where new ideas are tested against cultural values. For example, the shift from a mechanistic view of the universe in the 19th century to a more dynamic and interconnected view in the 20th century was influenced by advances in quantum physics, systems theory, and ecology. These shifts were not merely scientific discoveries but involved a cultural transformation in how interconnectedness and complexity were understood. Recursive Realism suggests that cultural paradigms are flexible and adaptive, capable of evolving in response to new ideas and perspectives.

The evolution of ideas is a recursive process, where thought and understanding are continuously shaped by dialogue, reflection, and cultural exchange. Recursive Realism offers a framework for understanding how intellectual progress is not a linear path but a dynamic interplay between tradition and innovation, where each new iteration of thought builds upon the foundations of the past. By embracing the recursive nature of knowledge, we can appreciate the depth and richness of human inquiry, where ideas evolve through a continual process of engagement with reality and with each other.

In this chapter, we have explored how Recursive Realism reshapes our understanding of knowledge as a dynamic process, where truth, understanding, and worldviews evolve through recursive interactions. By recognising the self-organising nature of intellectual evolution, Recursive Realism challenges reductionist approaches and embraces a more holistic view of how knowledge is constructed and refined. This perspective offers new insights into the interconnectedness of science, philosophy, and culture, allowing us to appreciate the complexity of human understanding in an ever-changing world.

References

Aristotle (1998). *The Metaphysics*. Translated by H. Lawson-Tancred. Penguin Classics.

Bachelard, G. (1984). *The New Scientific Spirit*. Beacon Press.

Bartlett, F.C. (1932). *Remembering: A Study in Experimental and Social Psychology*. Cambridge University Press.

Bhaskar, R. (1998). *The Possibility of Naturalism: A Philosophical Critique of the Contemporary Human Sciences*. Routledge.

Bohm, D. (1980). *Wholeness and the Implicate Order*. Routledge.

Chalmers, A.F. (1999). *What is This Thing Called Science?*. Open University Press.

Einstein, A. (1920). *Relativity: The Special and General Theory*. Methuen.

Feyerabend, P. (1993). *Against Method*. Verso.

Gadamer, H.-G. (2004). *Truth and Method*. Continuum.

Hacking, I. (1983). *Representing and Intervening: Introductory Topics in the Philosophy of Natural Science*. Cambridge University Press.

Heidegger, M. (1962). *Being and Time*. Translated by J. Macquarrie and E. Robinson. Harper & Row.

Hume, D. (2000). *An Enquiry Concerning Human Understanding*. Edited by T.L. Beauchamp. Oxford University Press.

Kant, I. (1998). *Critique of Pure Reason*. Translated by P. Guyer and A.W. Wood. Cambridge University Press.

Kuhn, T.S. (1970). *The Structure of Scientific Revolutions*. University of Chicago Press.

Lakatos, I. and Musgrave, A., eds. (1970). *Criticism and the Growth of Knowledge*. Cambridge University Press.

Popper, K. (2002). *The Logic of Scientific Discovery*. Routledge.

Russell, B. (1995). *A History of Western Philosophy*. Routledge.

Toulmin, S. (2001). *Return to Reason*. Harvard University Press.

Vygotsky, L.S. (1978). *Mind in Society: The Development of Higher Psychological Processes*. Harvard University Press.

Whitehead, A.N. (1929). *Process and Reality: An Essay in Cosmology*. Macmillan.

Chapter 10: Recursive Realism and the Limits of Perception

The Recursive Nature of Sensory Perception

Perception is often thought of as a direct window into the world, where sensory organs gather information and transmit it to the brain for interpretation. However, Recursive Realism suggests that perception is a dynamic process involving continuous feedback loops between sensory input and cognitive interpretation. Rather than passively receiving information, the mind actively constructs a coherent experience through recursive interactions between expectation and sensation. This section explores how recursion is embedded in the act of perceiving, allowing the mind to integrate fragmentary sensory data into a continuous and meaningful experience.

The Feedback Loops of Sensory Processing

Perception involves complex feedback loops where the brain continuously compares incoming sensory data with prior knowledge and expectations. These loops allow the mind to filter, adjust, and synthesise sensory information, creating a dynamic interplay between what is seen, heard, or felt and how it is interpreted.

Sensory perception relies on a balance between bottom-up processes, which involve the reception of raw sensory data, and top-down processes, where the brain uses prior knowledge to interpret this data. This interaction is inherently recursive, as top-down expectations influence how sensory information is processed, while bottom-up signals can update and refine those expectations. For example, when reading blurred text, the brain uses contextual clues to fill in missing information, allowing for a coherent understanding of the words. If the brain encounters a discrepancy between what is expected and what is actually seen, it adjusts its interpretation accordingly. Recursive Realism suggests that this interplay between expectation and sensation is central to how we construct our experience of the world.

Predictive coding is a theory of perception that highlights the brain's tendency to create predictions about sensory input and then compare these predictions with actual sensory data. The brain operates as a predictive engine, where predictions about the environment are continuously tested and refined through recursive feedback loops. This allows the brain to anticipate changes in the environment and minimise sensory surprises. For example, when walking through a familiar room in the dark, the brain predicts the location of furniture and adjusts its movements accordingly. If a chair has been moved, the discrepancy between prediction and sensation prompts the brain to update its model of the environment. Recursive Realism sees this prediction-error minimisation as a fundamental aspect of how perception operates, where understanding emerges through a continuous feedback loop between expectation and reality.

The brain's ability to integrate information from different sensory modalities, such as vision, hearing, and touch, is another example of recursive processing. Each sensory system provides partial data about the environment, which the brain synthesises into a unified experience. This integration requires constant feedback between sensory channels, allowing for the adjustment of perception based on context. For instance, when watching a ventriloquist act, the brain integrates visual and auditory information to perceive the sound as coming from the puppet's mouth, even though the sound originates from a different location. This ability to integrate multisensory data reflects the brain's recursive capacity to cross-reference and harmonise different streams of information, creating a cohesive perception of reality.

Perceptual Illusions and the Limits of Sensory Processing

The limitations of perception are revealed through illusions, which demonstrate how the brain's interpretations of sensory data can diverge from physical reality. These anomalies highlight the recursive nature of interpretation, where expectations and context shape how sensory information is perceived.

Visual illusions, such as the Müller-Lyer illusion, where two lines of equal length appear different due to the orientation of arrowheads, illustrate how the brain's assumptions about perspective and depth influence perception. In this case, the brain uses top-down processing to interpret the lines in a way that aligns with its understanding of three-dimensional space, even though this interpretation is misleading. The illusion persists because the brain's recursive loops continuously apply these assumptions, creating a consistent interpretation despite the conflicting sensory data. This demonstrates how perceptual systems rely on cognitive shortcuts to create coherent experiences, even when these shortcuts lead to distortions.

Auditory illusions, such as the Shepard tone, where a series of tones creates the illusion of a continuously rising pitch, reveal how the brain's temporal integration of sounds can create perceptual effects that do not correspond to the physical stimulus. The brain's ability to synthesize and compare auditory inputs over time allows it to construct temporal patterns, but this recursive process can also lead to misinterpretations when the patterns do not align with reality. These illusions highlight the limits of perception, where the brain's attempt to create a coherent narrative of sound can result in anomalous experiences.

Context plays a crucial role in how sensory data is interpreted, as the brain continuously adjusts its perceptual framework based on environmental cues. This adjustment is a recursive process, where past experiences and current context influence how new sensory inputs are processed. For example, the colour of an object may appear different depending on the lighting conditions or the surrounding colours, a phenomenon known as colour constancy. The brain's recursive ability to compensate for these variations allows for a more stable perception of the world, but it also reveals the subjective nature of sensory experience, where contextual factors shape our interpretation of objective stimuli.

The Brain's Construction of Temporal Experience

Our perception of time is also shaped by recursive processes, as the brain integrates sensory events into a temporal framework. This ability to construct time is essential for coordinating actions, anticipating events, and maintaining a continuous sense of self. Recursive Realism suggests that the experience of time emerges through feedback loops between past memories, present awareness, and future anticipation.

The brain's ability to bind events together in time allows for the perception of continuous motion and causality. For example, when watching a movie, the brain integrates the rapid succession of frames into a seamless visual experience. This requires temporal binding, where discrete moments are synchronised into a continuous flow. This process is inherently recursive, as the brain must constantly compare the current frame with previous frames to create a coherent narrative of what is happening. Recursive Realism suggests that this integration is essential for constructing a stable perception of reality, where changes in the environment are understood as progressions rather than isolated events.

Time perception is not uniform; it is influenced by attention, emotions, and context. For instance, time seems to slow down during moments of intense focus or danger, while it appears to speed up during routine tasks. These variations in time perception reflect the recursive interactions between cognitive processing and internal states, where the brain's interpretation of events adjusts based on the significance of each moment. For example, when experiencing novelty, the brain's attention to details creates a rich temporal experience, making time feel longer. This adaptability highlights the plasticity of temporal experience, where perception of time is shaped by the recursive loops that link attention, emotion, and memory.

The ability to anticipate future events is another aspect of the brain's temporal recursion. By creating mental models of what is likely to happen, the brain can prepare responses and actions in advance. This is seen in everyday activities, such as crossing the street, where the brain predicts the movement of oncoming cars and adjusts behaviour accordingly. These predictive models are refined through experience, allowing for more accurate predictions over time. Recursive Realism suggests that this ability to simulate future possibilities is a key aspect of human cognition, as it enables planning, problem-solving, and the pursuit of goals. This recursive engagement with past experiences and future possibilities creates a temporal depth in human consciousness, where the present moment is understood within a continuum of past and future.

Perception is not a passive reception of stimuli but a dynamic process shaped by recursive interactions between expectation and sensory input. By understanding perception as a recursive dialogue between mind and environment, Recursive Realism provides a framework for appreciating the richness and complexity of how we experience the world. This perspective reveals the flexibility and adaptability of the perceptual system, where context, memory, and prediction create a coherent reality from fragmented sensory data.

The Limits of Perception and the Boundaries of Reality

Human perception is shaped by biological limitations that define what we can and cannot sense. While our sensory systems provide a rich array of experiences, they also filter out vast amounts of information that lie beyond our cognitive capacity. Recognising these limits is crucial for understanding the boundaries of what we can know about reality. Recursive Realism suggests that these perceptual limits are not failings but necessary constraints that enable the mind to construct a coherent world from overwhelming sensory data. By exploring how technological tools extend these limits and what this means for our philosophical understanding, we gain insight into how perception shapes our worldview.

Sensory Limits and the Filtering of Reality

Our senses provide access to a narrow slice of the electromagnetic spectrum, sound frequencies, and chemical signals that make up the world. This selective filtering of reality is both a strength and a limitation, as it allows us to focus on relevant information while leaving out extraneous data.

Each sensory system is tuned to a specific range of stimuli, allowing us to perceive certain aspects of reality while remaining blind to others. For example, the human eye can detect wavelengths from approximately 400 to 700 nanometres, perceiving these as visible light. Beyond this range, ultraviolet and infrared wavelengths are invisible to us. Similarly, our ears can detect sound frequencies between 20 Hz and 20,000 Hz, leaving us deaf to ultrasounds and infrasounds that other animals can hear. These limitations mean that our sensory experience is inherently partial, shaped by the biological constraints of our perceptual apparatus. Recursive Realism suggests that these constraints are not a flaw but a necessary adaptation, allowing us to focus on the information that is most relevant for survival and interaction within our environment.

Even within the range of what we can perceive, our brain actively filters and selects sensory information, prioritising some aspects while ignoring others. This filtering occurs through processes like attention, where the mind focuses on salient stimuli while background details fade into peripheral awareness. For instance, when engaged in a conversation at a noisy party, we can focus on the voice of the person we are speaking with while tuning out the surrounding chatter, a phenomenon known as the cocktail party effect. This selective attention is a recursive process, where the mind continually adjusts its focus based on context and expectation, creating a stable field of awareness even amidst a chaotic environment. Recursive Realism highlights that this filtering is a form of cognitive efficiency, enabling us to navigate complexity without becoming overwhelmed by sensory overload.

The limits of our sensory systems create blind spots in our perception, areas of reality that we cannot access directly. For example, the blind spot in our retina, where the optic nerve exits the eye, results in a small gap in our visual field, which the brain fills in with surrounding information. Similarly, phenomena like magnetic fields, ultrasonic communication, and subatomic particles are beyond the reach of our naked senses, requiring specialised instruments for detection. Recursive Realism suggests that these cognitive gaps are not just missing data but defining features of how we construct reality. By being selectively attuned to certain aspects of the world, our minds can build a simplified model that enables action and understanding, even if it does not capture the full complexity of the universe.

The Role of Technology in Expanding Perceptual Limits

While our biological senses are limited, technology extends our perceptual reach, allowing us to explore dimensions of reality that would otherwise remain hidden. This technological extension creates a recursive loop between human cognition and instrumental observation, where tools become extensions of our senses, reshaping how we understand the universe.

Seeing the Unseen: The invention of the microscope and the telescope revolutionised our understanding of scale, revealing microcosms and macrocosms that are invisible to the naked eye. Microscopes allow us to see cells, bacteria, and molecular structures, providing insight into the building blocks of life. Meanwhile, telescopes extend our vision to the vast reaches of the universe, showing us galaxies, nebulae, and cosmic phenomena. These instruments create a recursive loop between observation and theory, where new discoveries lead to new questions and revisions of existing models. For example, the observation of microbial life under a microscope challenged pre-existing notions about the origin of diseases, leading to the development of germ theory. Recursive Realism suggests that these technological extensions of perception allow for a deeper engagement with reality, revealing layers that were previously beyond our grasp.

Time-lapse photography, high-speed cameras, and astronomical observations allow us to perceive temporal scales that are either too slow or too fast for human perception. For instance, time-lapse images of flowering plants or weather patterns reveal the dynamic changes that occur over days or months, while high-speed cameras can capture the split-second movements of insects or explosions. Astronomical time scales, such as the lifespan of stars or the expansion of galaxies, are revealed through long-term observations made possible by telescopes and space probes. These technologies create a recursive dialogue between human experience and the cosmic scale, allowing us to see our place in the universe within a broader temporal context. Recursive Realism suggests that these temporal expansions reshape our concepts of time, creating new ways of understanding change and transformation.

Virtual reality (VR) technologies provide a new way to manipulate and explore perceptual experience, allowing users to enter simulated worlds that are both immersive and interactive. These virtual environments create a recursive loop between cognitive processes and digital input, where the brain is tricked into experiencing spaces that do not physically exist. For example, VR simulations can recreate ancient cities, outer space, or microscopic environments, offering a first-person experience of places that are otherwise inaccessible. Recursive Realism suggests that VR represents a new frontier in the co-creation of reality, where perception and imagination converge in a recursive interplay that expands the boundaries of experience.

The Philosophical Implications of Perceptual Limits

The recognition of our perceptual limitations raises philosophical questions about the nature of knowledge, reality, and the relationship between appearance and essence. Recursive Realism offers a framework for navigating these epistemological challenges, suggesting that the gaps in perception are not obstacles to understanding but fundamental features of how we engage with reality.

A Recursive Challenge: Philosophers have long grappled with the distinction between appearance and reality, questioning whether our perceptions reflect an objective world or are filtered through the limitations of our senses. Plato's allegory of the cave symbolises this tension, where the shadows on the cave wall represent the limits of human perception, while the outside world signifies deeper truths beyond our immediate reach. Recursive Realism suggests that this dichotomy is not absolute; rather, our understanding of reality emerges through recursive engagement with the world, where appearances serve as provisional models that can be refined through experience and reflection. For instance, our initial perception of the sun's movement across the sky led to the geocentric model, which was later revised through astronomical observation to reveal a heliocentric cosmos. This recursive refinement demonstrates that while perception may be limited, it can serve as a stepping stone toward deeper understanding.

The idea of objective reality is challenged by the recognition that our sensory systems and cognitive frameworks shape how we interpret the world. Recursive Realism does not deny the existence of objective truths but suggests that these truths are always mediated through the processes of interpretation and contextualisation. For example, the laws of physics are consistent and universal, yet our understanding of these laws is shaped by the tools and conceptual models we use to observe them. The shift from classical mechanics to quantum mechanics illustrates how different perspectives on reality can coexist, each revealing aspects of the universe that were previously hidden. Recursive Realism suggests that objectivity is a moving target, where each iteration of knowledge brings us closer to a comprehensive view of reality, even if the ultimate essence remains beyond our grasp.

The limits of perception invite us to accept that certain aspects of existence may remain mysterious or inaccessible to human understanding. Rather than seeing this as a defeat, Recursive Realism embraces mystery as a motivating force for exploration. The unknown becomes a source of wonder, driving curiosity and speculation. For instance, the nature of dark matter, consciousness, or the origins of the universe remains elusive, but these mysteries provide fertile ground for theoretical exploration and philosophical reflection. Recursive Realism suggests that by embracing the limits of our perception, we can cultivate a humble approach to knowledge that values inquiry over certainty, and openness over dogmatism.

The limits of human perception shape our understanding of reality in profound ways, creating boundaries that define what we can observe, comprehend, and imagine. Recursive Realism offers a perspective that recognises these constraints as part of the dynamic process through which we construct our understanding of the world. By embracing the recursive interplay between sensory input, cognitive interpretation, and technological extension, we gain a deeper appreciation of the richness and complexity of perception, where the known and the unknown exist in a continuous dialogue.

In this chapter, we have explored how perception operates through recursive feedback loops, shaping our understanding of reality and defining the boundaries of what we can know. Recursive Realism offers a framework for understanding perception as a constructive process, where limitations are not obstacles but essential features of how we engage with the complexity of the world. By recognising the dynamic nature of sensory experience, we can appreciate the role of tools, reflection, and imagination in expanding our perceptual horizons, allowing us to navigate the mysteries that lie beyond our immediate cognitive reach.

References

Baars, B.J. (1988). *A Cognitive Theory of Consciousness*. Cambridge University Press.

Barlow, H.B. (1972). 'Single Units and Sensation: A Neuron Doctrine for Perceptual Psychology,' *Perception*, 1(4), pp. 371–394.

Clark, A. (2013). *Surfing Uncertainty: Prediction, Action, and the Embodied Mind*. Oxford University Press.

Dennett, D.C. (1991). *Consciousness Explained*. Little, Brown & Co.

Feynman, R.P., Leighton, R.B., and Sands, M. (2011). *The Feynman Lectures on Physics, Vol. 1: Mainly Mechanics, Radiation, and Heat*. Basic Books.

Friston, K. (2010). 'The Free-Energy Principle: A Unified Brain Theory?' *Nature Reviews Neuroscience*, 11(2), pp. 127–138.

Gregory, R.L. (1997). *Eye and Brain: The Psychology of Seeing*. 5th edn. Oxford University Press.

Helmholtz, H. von (1867). *Handbuch der Physiologischen Optik*. Leipzig: Voss. [Translated as *Helmholtz's Treatise on Physiological Optics*, 1924, Optical Society of America.]

Hohwy, J. (2013). *The Predictive Mind*. Oxford University Press.

Kandel, E.R., Schwartz, J.H., and Jessell, T.M. (2000). *Principles of Neural Science*. 4th edn. McGraw-Hill.

Lee, T.S. and Mumford, D. (2003). 'Hierarchical Bayesian Inference in the Visual Cortex,' *Journal of the Optical Society of America A*, 20(7), pp. 1434–1448.

Llinás, R. and Ribary, U. (1993). 'Coherent 40-Hz Oscillation Characterizes Dream State in Humans,' *Proceedings of the National Academy of Sciences*, 90(5), pp. 2078–2081.

Merleau-Ponty, M. (1945). *Phénoménologie de la Perception*. Paris: Gallimard. [Translated as *Phenomenology of Perception*, 2012, Routledge.]

Pessoa, L., Thompson, E., and Noë, A. (1998). 'Finding out About Filling-In: A Guide to Perceptual Completion for Visual Science and the Philosophy of Perception,' *Behavioral and Brain Sciences*, 21(6), pp. 723–802.

Pylyshyn, Z.W. (1999). 'Is Vision Continuous with Cognition? The Case for Cognitive Impenetrability of Visual Perception,' *Behavioral and Brain Sciences*, 22(3), pp. 341–365.

Ramachandran, V.S. and Hirstein, W. (1998). 'The Perception of Phantom Limbs: The D.O. Hebb Lecture,' *Brain*, 121(9), pp. 1603–1630.

Schrödinger, E. (1944). *What is Life? The Physical Aspect of the Living Cell*. Cambridge University Press.

Von Neumann, J. (1958). *The Computer and the Brain*. Yale University Press.

Zeki, S. (1993). *A Vision of the Brain*. Blackwell Science.

These references provide a foundation for the exploration of sensory perception, predictive coding, and the recursive nature of cognition as discussed in this chapter.

Chapter 11: Recursive Realism and the Nature of Conscious Experience

The Architecture of Thought and Recursion

The mind is a self-organising system that generates a continuous stream of thoughts, ideas, and associations. These thoughts do not emerge in isolation but are part of a dynamic web of self-referential loops, where each idea leads to another, and memories interact with perception to create a coherent mental landscape. Recursive Realism offers a perspective on how cognitive processes are structured through recursive feedback, where reflection and reinterpretation shape the flow of inner experience.

Recursion and the Flow of Thought

Thought is often experienced as a stream of consciousness, where ideas flow from one to the next, forming narratives and chains of association. This flow is not linear but recursive, involving continuous loops of reflection and reinterpretation.

One of the most immediate examples of recursion in thought is self-talk, where individuals engage in an internal dialogue with themselves. This dialogue involves a recursive loop between internal voices, where the mind questions, replies, reflects, and debates within itself. For example, when making a difficult decision, a person might internally weigh the pros and cons, anticipating possible outcomes and revisiting past experiences. This recursive conversation allows for deeper processing of information and a more nuanced understanding of the situation. Recursive Realism suggests that this self-referential process is fundamental to problem-solving and introspection, enabling the mind to refine its thoughts through continuous feedback.

Associative thinking involves the mind's ability to connect ideas, memories, and concepts through similarities or analogies. This process is inherently recursive, as each new association prompts the mind to revisit past memories and reinterpret them in light of new experiences. For instance, a smell that evokes a memory of childhood triggers a cascade of related memories, creating a network of interconnected associations. This recursive ability to navigate memory allows for a rich mental landscape, where ideas are not fixed but are continuously recontextualised based on current awareness. Recursive Realism suggests that associative loops are essential for creativity and imagination, as they enable the mind to blend different concepts into novel insights.

Reflection involves a recursive process where the mind revisits previous thoughts, examining them from different angles and perspectives. This recursive capacity allows for metacognition, thinking about one's own thinking, which is crucial for self-awareness and intellectual growth. For example, after reading a complex philosophical text, a person might reflect on the arguments, considering how they align or conflict with their existing beliefs. This recursive reflection allows for a deeper understanding of both the text and one's own perspective, creating a dialogue between external ideas and internal frameworks. Recursive Realism suggests that reflective thought is not merely a mental exercise but a transformative process that shapes how individuals understand themselves and the world.

Emotion as a Recursive Experience

Emotions are often perceived as immediate reactions to external events, but they also involve a recursive process of interpretation, modulation, and integration. Recursive Realism offers a perspective on how emotions are dynamic processes that evolve through feedback loops between bodily sensations, cognitive appraisal, and social context.

Emotional experiences involve a cycle of appraisal, where the mind interprets a stimulus, assigns it emotional significance, and then revisits this interpretation in light of new information. For example, the initial shock of hearing bad news might evolve into a more measured response as the mind processes the implications and context of the situation. This recursive process allows for the modulation of emotional responses, creating a more balanced and adaptive reaction. Recursive Realism suggests that emotions are not static but are shaped through ongoing feedback between the body's physiological state and the mind's cognitive interpretations.

The ability to regulate emotions is closely tied to the mind's recursive capacity to reframe experiences. Reframing involves shifting perspective to see a situation in a new light, changing its emotional impact. For example, after a disappointment, a person might reflect on the positive aspects of the situation, finding lessons or new opportunities that transform their initial reaction into acceptance or optimism. This recursive process of reappraisal allows for greater emotional flexibility, helping individuals to cope with stress and adapt to change. Recursive Realism suggests that emotional regulation is not just a matter of suppressing feelings but involves revisiting and reinterpreting them in a way that integrates them into the broader narrative of the self.

Empathy, the ability to understand and share the feelings of others, also relies on recursive processes. Empathy involves perspective-taking, where individuals imagine what another person is experiencing, creating a mental simulation of their thoughts and emotions. This simulation is inherently recursive, as it requires the mind to shift between self-awareness and other-awareness, continuously adjusting its understanding based on cues from the other person. For example, when comforting a friend, an empathetic person might reflect on their own similar experiences while adjusting their support based on the friend's responses. Recursive Realism suggests that empathy is a feedback loop between self and other, where understanding evolves through mutual interaction.

Self-Awareness and the Recursive Nature of the Self

Self-awareness is one of the most profound aspects of conscious experience, allowing individuals to perceive themselves as distinct entities with thoughts, emotions, and intentions. Recursive Realism offers a perspective on how self-awareness is a recursive phenomenon, where the mind turns inward, creating a loop between observation and reflection.

Self-awareness involves a recursive loop where the mind observes its own thoughts and experiences, creating a sense of self as observer. This meta-cognitive ability allows individuals to evaluate their own actions, intentions, and desires, creating a deeper understanding of their inner world. For example, a person who recognises their anger during a conflict can step back and reflect on its source, considering how past experiences or unmet needs might be influencing their reaction. This recursive capacity for self-reflection is central to personal growth, as it enables individuals to learn from their experiences and adapt their behaviour in the future. Recursive Realism suggests that self-awareness is not a fixed state but a dynamic process of self-examination, where the mind continually revisits and refines its understanding of itself.

The self is often experienced as a narrative, where individuals construct a story that links their past, present, and future into a coherent identity. This narrative construction is inherently recursive, as each new experience prompts a re-evaluation of the self-story, leading to adjustments and revisions. For example, a person who has always seen themselves as introverted might reconsider this self-concept after thriving in a public-speaking role, integrating this new experience into their identity. This recursive storytelling allows the self to remain coherent while being open to change, creating a sense of continuity even amidst transformation. Recursive Realism suggests that identity is not a static essence but a self-organising narrative that evolves through reflection and re-interpretation.

The boundary between the self and the world is not fixed but is shaped through recursive interactions with the environment. As individuals engage with their surroundings, they create internal representations that allow them to navigate and interpret the world. This recursive boundary-making allows for a sense of agency, where individuals perceive themselves as active agents capable of influencing their environment. For instance, the ability to plan and execute goal-directed actions is based on a recursive understanding of intention and outcome, where the mind anticipates potential results and adjusts its actions accordingly. Recursive Realism suggests that the self is co-created through these dynamic exchanges with the world, where each interaction shapes the boundary between inner experience and external reality.

The nature of thought, emotion, and self-awareness is fundamentally recursive, where each aspect of conscious experience is shaped by feedback loops between reflection, interpretation, and self-referential observation. Recursive Realism offers a perspective that sees inner experience as a dynamic process, where the mind's complexity is a result of its ability to continuously revisit and refine its own processes. This understanding provides a framework for exploring the depth of human consciousness, where cognitive patterns and emotional responses are woven into a cohesive narrative that evolves through self-reflection.

Recursive Realism and the Depths of Consciousness

While much of our awareness is shaped by conscious thought and intentional focus, the unconscious mind plays a crucial role in shaping behaviour, perception, and creativity. Recursive Realism suggests that the interaction between conscious and unconscious processes is a recursive relationship, where hidden mental processes continuously influence the stream of consciousness. By understanding this dynamic interplay, we can gain insight into how subconscious patterns shape the texture of our inner world.

The Subconscious Mind as a Recursive System

The subconscious mind is often described as the reservoir of automatic thoughts, unconscious drives, and implicit memories. However, Recursive Realism suggests that the subconscious is not merely a passive storage space but an active system that processes information through recursive loops.

The subconscious mind is adept at recognising patterns that the conscious mind might overlook. These patterns often manifest as intuitive hunches or gut feelings, where the mind detects subtle regularities in the environment without explicit awareness. For example, a person may have a feeling that something is wrong in a familiar environment without being able to identify what has changed. This ability arises from the brain's recursive capacity to scan the environment for anomalies, comparing current sensory input with stored memories to detect discrepancies. Recursive Realism suggests that these intuitive insights are products of deep cognitive loops, where the subconscious mind processes sensory data and past experiences to generate a pre-reflective understanding of the world.

Dreaming offers a unique window into the subconscious mind, where thoughts, emotions, and memories are reassembled into vivid narratives. Dreams can be understood as recursive loops, where the mind revisits fragments of daily experiences, reinterpreting them through symbolic imagery and emotional themes. For example, a person dealing with stress at work might dream of being chased or trapped, symbolically representing their emotional state. These dreams allow the mind to process unresolved emotions and explore alternative scenarios, creating a psychological space for subconscious conflicts to be played out. Recursive Realism suggests that dreams are not merely random images but are structured narratives that reflect the recursive processing of subconscious themes, providing a way for the mind to integrate experiences into the broader story of the self.

The subconscious mind exerts influence over conscious decisions through priming, where exposure to certain stimuli influences how we perceive and interpret later information. For instance, seeing a smiling face can make a person more likely to interpret a neutral expression as friendly. These subtle shifts are the result of unconscious feedback loops, where prior experiences shape the interpretation of new stimuli without conscious awareness. Recursive Realism suggests that this priming effect demonstrates how the subconscious mind continually interacts with the conscious mind, creating a fluid dialogue between automatic responses and deliberate thought.

Altered States of Consciousness and Recursive Patterns

Altered states of consciousness, such as meditation, hypnosis, and psychedelic experiences, reveal the plasticity of cognition, where the boundaries of consciousness can be expanded or transformed. These states often involve a heightened awareness of internal processes, offering a deeper insight into the recursive nature of thought and perception.

Meditation practices, such as mindfulness or transcendental meditation, involve a recursive focus on the present moment, where attention is directed toward breathing, bodily sensations, or thoughts as they arise. This self-referential loop creates a space where mental patterns can be observed without judgment, allowing for a detachment from automatic reactions. For example, by observing the flow of thoughts without becoming attached to them, a meditator can become aware of the transient nature of anger, anxiety, or desire. Recursive Realism suggests that meditation is a process of recursive refinement, where the mind's habitual loops are brought into conscious awareness and restructured, leading to greater emotional regulation and self-awareness.

Hypnosis involves a state of focused attention and increased suggestibility, where the mind becomes more receptive to suggestions that bypass conscious filters. This state can be understood as a recursive alteration of mental loops, where hypnotic suggestions create new patterns that influence perception and behaviour. For example, a person under hypnosis might be able to recall forgotten memories or experience sensations that are inconsistent with reality, such as feeling warm when told they are in a hot room. This recursive capacity to alter perception through suggestion demonstrates the flexibility of conscious awareness, where mental loops can be reprogrammed to create new experiences. Recursive Realism suggests that hypnosis reveals the depth of the mind's plasticity, where subconscious influences can be brought to the surface and manipulated through focused attention.

Psychedelic substances, such as LSD, psilocybin, and DMT, can create profound alterations in perception, thought, and emotion. These experiences often involve a breakdown of the normal boundaries between self and world, where sensory inputs and thoughts become interconnected in novel and unexpected ways. For example, a person might experience synaesthesia, where sounds are perceived as colours, or a sense of oneness with their surroundings. Recursive Realism suggests that these experiences reflect a loosening of the mind's habitual loops, allowing for new patterns of perception to emerge. This temporary expansion of conscious awareness can lead to insights into the interconnectedness of thought and reality, offering a glimpse into the depth of the mind's potential.

The Interaction Between Conscious and Unconscious Mind

The relationship between the conscious and unconscious mind is one of reciprocal influence, where conscious intentions shape unconscious patterns and unconscious drives influence conscious thought. Recursive Realism offers a framework for understanding this dynamic interaction, where the mind's layers are seen as interdependent loops that continuously shape each other.

Dreams provide a space where conscious concerns and unconscious themes come into dialogue. While REM sleep is associated with vivid dreaming, the content of dreams often reflects the preoccupations of waking life, transformed into symbolic imagery. For example, a person who is anxious about an upcoming presentation might dream of being unprepared or exposed in front of a crowd. This dream narrative allows the unconscious mind to process fears and anxieties in a safe space, creating a recursive loop between daytime experiences and night-time reflections. Recursive Realism suggests that dreams serve as a feedback mechanism that integrates conscious and unconscious concerns, allowing the mind to find resolutions and insights through symbolic play.

Sigmund Freud and Carl Jung both recognised the importance of the unconscious mind, but they differed in their interpretations of its nature. Freud saw the unconscious as a repository of repressed desires and conflicts, while Jung viewed it as a source of archetypal imagery and collective symbolism. Recursive Realism synthesises these views by suggesting that the unconscious operates through layered loops, where personal memories and cultural archetypes interact in a dynamic system. For example, a person might experience a recurring dream of being lost in a forest, which reflects both a personal fear and a universal symbol of uncertainty. This recursive interaction between individual and collective themes allows the unconscious mind to create rich narratives that inform self-understanding.

Creativity often involves a dialogue between conscious effort and unconscious inspiration, where ideas seem to emerge from nowhere after a period of incubation. This phenomenon is seen in artistic endeavours, scientific breakthroughs, and problem-solving, where the mind shifts between focused concentration and subconscious rumination. Recursive Realism suggests that the creative process involves a loop between intentional exploration and unconscious pattern recognition, where insights arise from the interaction between these layers. For example, a mathematician might struggle with a problem for days, only to find the solution arriving in a flash of insight during a moment of relaxation. This interplay between conscious analysis and unconscious synthesis highlights the depth of the mind's recursive structure, where innovation emerges from the integration of different cognitive modes.

The depths of consciousness reveal a complex interplay between subconscious processes, altered states, and the dialogue between conscious and unconscious mind. Recursive Realism offers a perspective that sees inner life as a dynamic network of feedback loops, where dreams, intuition, and creativity arise from the reciprocal influence of mental layers. By understanding these recursive interactions, we gain insight into the mystery and richness of the human mind, where the boundaries between awareness and unconscious patterns are fluid and interconnected.

In this chapter, we have explored how Recursive Realism provides a framework for understanding the complex architecture of consciousness, where thought, emotion, and unconscious processes are shaped by recursive feedback loops. This perspective reveals the dynamic nature of inner experience, where the self is continually constructed and reconstructed through reflection, imagination, and emotional integration. By recognising the recursive nature of the mind, we can appreciate the depth of subjective experience, where the boundaries of awareness are fluid and adaptive, allowing for growth, creativity, and self-transformation.

References

Baars, B.J. (1988). *A Cognitive Theory of Consciousness*. Cambridge University Press.

Bargh, J.A. and Chartrand, T.L. (1999). 'The Unbearable Automaticity of Being,' *American Psychologist*, 54(7), pp. 462–479.

Chalmers, D.J. (1996). *The Conscious Mind: In Search of a Fundamental Theory*. Oxford University Press.

Dennett, D.C. (1991). *Consciousness Explained*. Little, Brown & Co.

Dretske, F. (2000). *Perception, Knowledge and Belief: Selected Essays*. Cambridge University Press.

Edelman, G.M. (1989). *The Remembered Present: A Biological Theory of Consciousness*. Basic Books.

Freud, S. (1915). 'The Unconscious,' in Strachey, J. (ed.) *The Standard Edition of the Complete Psychological Works of Sigmund Freud, Volume XIV (1914-1916): On the History of the Psycho-Analytic Movement, Papers on Metapsychology and Other Works*. London: Hogarth Press, pp. 159–215.

Jung, C.G. (1969). *The Archetypes and the Collective Unconscious*. Princeton University Press.

Kandel, E.R., Schwartz, J.H., and Jessell, T.M. (2000). *Principles of Neural Science*. 4th edn. McGraw-Hill.

Llinás, R. and Paré, D. (1991). 'Of Dreaming and Wakefulness,' *Neuroscience*, 44(3), pp. 521–535.

Maturana, H.R. and Varela, F.J. (1980). *Autopoiesis and Cognition: The Realization of the Living*. D. Reidel Publishing Company.

Metzinger, T. (2003). *Being No One: The Self-Model Theory of Subjectivity*. MIT Press.

Penrose, R. (1994). *Shadows of the Mind: A Search for the Missing Science of Consciousness*. Oxford University Press.

Revonsuo, A. (2000). 'The Reinterpretation of Dreams: An Evolutionary Hypothesis of the Function of Dreaming,' *Behavioral and Brain Sciences*, 23(6), pp. 877–901.

Tononi, G. (2008). 'Consciousness as Integrated Information: A Provisional Manifesto,' *Biological Bulletin*, 215(3), pp. 216–242.

Varela, F.J., Thompson, E., and Rosch, E. (1991). *The Embodied Mind: Cognitive Science and Human Experience*. MIT Press.

Wegner, D.M. (2002). *The Illusion of Conscious Will*. MIT Press.

Wittgenstein, L. (1953). *Philosophical Investigations*. Blackwell.

Zeki, S. (1993). *A Vision of the Brain*. Blackwell Science.

Chapter 12: Recursive Realism and the Architecture of Time

Time as a Recursive Construct

Time is one of the most perplexing aspects of human experience, shaping everything from memory and anticipation to our sense of identity and continuity. While physics offers objective models of time, human experience of time is deeply subjective, influenced by perception, emotion, and reflection. Recursive Realism suggests that time is not simply a linear sequence of events but is constructed through recursive interactions between memory, anticipation, and present awareness. This perspective allows us to explore how temporal experience is woven into the fabric of consciousness, where the mind's recursive loops create a sense of continuity amid the flux of change.

Memory and the Construction of the Past

Memory is central to our experience of time, as it allows us to reconstruct the past and anchor our sense of self within a narrative of continuity. However, memory is not a static recording of past events but a dynamic, recursive process, where each act of remembering involves reconstruction and reinterpretation.

When we recall a memory, we do not simply retrieve a fixed image from the past; instead, the act of remembering involves reinterpreting the memory in light of current experiences and emotional states. For example, a childhood memory of a family trip might be recalled with nostalgia or regret, depending on the individual's current relationship with their family. Each time a memory is revisited, the emotional tone and narrative context can shift, creating a recursive loop where the present reshapes the past. Recursive Realism suggests that this fluidity of memory is essential for personal growth, as it allows individuals to integrate past experiences into their evolving sense of identity.

Autobiographical memory, the ability to recall personal experiences, is a key aspect of how we construct a narrative of the self over time. This form of memory is inherently recursive, as individuals use it to create a coherent story that explains who they are and how they have changed. For example, a person might reflect on formative moments in their life, such as graduation, career achievements, or relationships, and connect these events to their current goals and aspirations. This recursive process allows for a sense of continuity in identity, even as individuals adapt to new challenges and transformations. Recursive Realism suggests that autobiographical memory is a self-organising narrative, where each new reflection updates the story of the self, allowing for a dynamic understanding of the past.

Forgetting is often seen as a loss of information, but it also plays a crucial role in the architecture of memory. By selectively forgetting, the mind can prioritise certain experiences over others, creating a curated narrative of the past. This selective memory is a recursive process, where forgotten details are replaced by generalised themes that shape the story of our lives. For example, a person might forget the mundane details of daily life while retaining the emotional highlights of significant events. This allows for a simplified narrative that focuses on key turning points, creating a sense of coherence and meaning. Recursive Realism suggests that forgetting is not a failure but a necessary adaptation, allowing the mind to simplify the complexity of experience into a manageable story of the past.

Anticipation and the Projection of the Future

While memory allows us to reconstruct the past, anticipation allows us to project into the future, creating a mental model of what is to come. This ability to simulate future events is a recursive capacity that draws on past experiences and current knowledge to predict and plan.

Anticipation involves imagining future scenarios, evaluating possible outcomes, and adjusting actions based on predicted consequences. This process is inherently recursive, as each new prediction is refined through feedback from experience. For example, when planning a vacation, a person might imagine different itineraries, adjusting their plans based on weather forecasts, financial considerations, and personal preferences. This recursive loop allows for flexibility and adaptation, enabling individuals to navigate uncertainty with a sense of preparedness. Recursive Realism suggests that anticipation is not merely a cognitive exercise but a way of constructing the future, where mental simulations shape how we approach challenges and seize opportunities.

Emotional states like hope and anxiety are deeply tied to our sense of the future, reflecting how expectations shape emotional responses. Hope involves a positive anticipation of future possibilities, creating a forward-looking optimism that motivates goal-directed behaviour. In contrast, anxiety involves a preoccupation with uncertainty and potential threats, leading to a sense of apprehension about what might happen. Both of these states are recursive, as the mind continually re-evaluates its expectations in light of new information. For example, a person who is anxious about a job interview might imagine different scenarios and prepare for potential questions, adjusting their strategy with each new insight. Recursive Realism suggests that hope and anxiety are not just emotional reactions but recursive processes that shape how we engage with the uncertainty of the future.

The ability to plan is closely tied to the mind's recursive capacity to project into the future while drawing on past experiences. Planning involves creating a mental map of actions and outcomes, where each step is adjusted based on feedback from previous actions. For example, a scientist designing an experiment might anticipate potential challenges, create contingency plans, and revise their approach as new data emerges. This iterative process allows for strategic thinking, where the future is seen as a space of possibilities that can be shaped through deliberate action. Recursive Realism suggests that planning is not simply a linear process but a recursive dialogue between intentions and outcomes, where each new insight shapes the trajectory of future efforts.

Present Awareness as a Recursive Synthesis

While memory connects us to the past and anticipation projects us into the future, present awareness is the moment-to-moment experience that ties these temporal dimensions together. Recursive Realism suggests that present awareness is not a static point in time but a dynamic synthesis of past memories, current sensations, and future expectations.

Present awareness involves synthesising sensory input with memory and anticipation, creating a coherent experience of now. This synthesis is inherently recursive, as the mind continuously updates its awareness based on new sensory data and internal states. For example, when listening to music, the mind integrates the melody, rhythm, and emotional resonance into a continuous flow that is both immediate and shaped by past associations with similar sounds. This ability to create a unified present allows for engagement with the world in a way that is fluid and adaptive. Recursive Realism suggests that present awareness is a constructive process, where the mind's recursive loops weave together the threads of time into a seamless experience.

Mindfulness practices focus on enhancing present awareness by directing attention to breathing, bodily sensations, or thoughts as they arise. This practice involves a recursive loop where the mind observes its own processes, allowing for a deeper immersion in the present moment. For example, a person practising mindful breathing may become aware of the sensations of air entering and leaving their lungs, creating a sense of connection to their body and the moment. This recursive focus allows for a deeper sense of inner stillness, where the mind's habitual loops are quieted and new insights can arise. Recursive Realism suggests that mindfulness allows for a recalibration of the temporal mind, creating a space where the past and future recede, and the present is experienced in its fullness.

Flow states occur when individuals are fully absorbed in a challenging task, where awareness of time seems to disappear and action becomes effortless. These states are characterised by a perfect alignment between attention, skill, and environment, creating a sense of timeless immersion. For example, an artist lost in painting or a surfer riding a wave might experience a blurring of past, present, and future, where all focus is on the moment of creation. Recursive Realism suggests that flow states are a form of temporal synthesis, where the mind's recursive loops are fully harmonised, allowing for a direct engagement with the present that transcends ordinary time-consciousness.

The experience of time is shaped by recursive interactions between memory, anticipation, and present awareness, where the past is continually reinterpreted, the future is simulated, and the present is synthesised into a coherent whole. Recursive Realism offers a perspective that sees time not as a linear progression but as a dynamic process, where the mind creates a sense of continuity through feedback loops that integrate multiple temporal dimensions.

The Philosophy of Time and Recursive Realism

Time has long been a subject of philosophical inquiry, raising profound questions about its nature, its relation to change, and its role in shaping the human experience. While physics describes time as a dimension that can be measured and quantified, philosophers and mystics have often viewed time as a more mysterious and elusive phenomenon, intimately connected to the flow of consciousness. Recursive Realism offers a perspective that bridges these approaches, suggesting that time is understood through recursive engagement with reality, where the mind continuously shapes its experience of past, present, and future through self-referential loops.

The Nature of Time: Between Physics and Phenomenology

The scientific and philosophical approaches to time offer contrasting perspectives that highlight different aspects of temporal reality. Recursive Realism suggests that time is best understood as a recursive interplay between objective processes and subjective experience, where theories of time and lived time are interwoven.

A Dimension of Change: In physics, time is often treated as a dimension similar to space, where events occur along a timeline that can be measured with precision. Relativity theory, developed by Albert Einstein, revealed that time is not an absolute flow but is relative to the observer's frame of reference, influenced by speed and gravitational fields. For example, time dilation means that clocks run slower in strong gravitational fields or for objects moving close to the speed of light. This understanding of time as a physical variable allows for the mathematical modelling of cosmic phenomena, such as the expansion of the universe or the evolution of stars. Recursive Realism recognises the value of this scientific perspective but suggests that it is incomplete without considering how time is experienced by conscious beings.

Lived Experience and Duration: Phenomenology, a philosophical movement led by thinkers like Edmund Husserl and Maurice Merleau-Ponty, focuses on how time is experienced in consciousness as a flow or duration. Unlike the objective time of physics, phenomenological time is deeply subjective, experienced as a continuum of moments that are linked through memory and anticipation. For example, the experience of waiting for an event feels longer than the same amount of time spent in enjoyable activity, reflecting the elasticity of lived time. Recursive Realism suggests that phenomenological time is constructed through recursive loops where the mind continually relates its present awareness to past memories and future possibilities, creating a sense of continuity amidst the flux of change.

Recursive Realism seeks to reconcile the scientific and phenomenological perspectives on time by recognising that objective time and subjective time are different aspects of the same reality. While physics describes the structural framework of time as it relates to matter and energy, conscious experience reveals how time is interpreted and lived through the recursive structure of the mind. For instance, while physics might describe a day as a fixed 24-hour period, human experience of that day can vary widely, depending on attention, mood, and activity. Recursive Realism suggests that both models of time are essential for a full understanding, as they reveal the interplay between the mind's perception of time and the physical processes that govern the universe.

The Flow of Time and the Illusion of Continuity

One of the central questions in the philosophy of time is whether time is a continuous flow or a series of discrete moments. Recursive Realism offers a perspective that suggests time appears continuous due to the mind's recursive ability to connect individual experiences into a seamless whole.

Human experience of time often feels like a river, flowing smoothly from one moment to the next. This sense of continuity is created through temporal integration, where the mind binds discrete sensory events into a cohesive experience. For example, when listening to music, each note is perceived as part of a melodic sequence, rather than as isolated sounds. This ability to connect moments into a fluid experience is a result of the brain's recursive processing, where memory and expectation shape the ongoing interpretation of sensory input. Recursive Realism suggests that this continuous flow of time is not an illusion but a construct of the mind's recursive architecture, which allows us to experience temporal coherence in a changing world.

In contrast to the smooth flow of experienced time, quantum physics suggests that time may be discrete at the smallest scales, broken into quanta or moments of change. For example, quantum events occur in jumps, where particles transition between states without passing through intermediate phases. This quantum discreteness challenges the traditional view of time as a smooth continuum and suggests that at the fundamental level, time may be a series of discrete events. Recursive Realism suggests that the mind's interpretation of time as continuous is a macro-level phenomenon, emerging from the recursive integration of discrete moments into a coherent narrative. Just as movies create the illusion of movement through rapidly changing frames, the brain constructs the experience of time by knitting together individual sensory inputs.

The concept of the arrow of time refers to the one-way flow of time from the past to the future, often associated with the increase of entropy as described in the second law of thermodynamics. As systems move from order to disorder, time seems to progress, creating a sense of irreversibility. However, at the microscopic level, the laws of physics are often time-symmetric, meaning that they do not distinguish between forward and backward directions. Recursive Realism suggests that the arrow of time is a macro-level construct, arising from the mind's interaction with entropy and irreversibility in the physical world. For example, our memories only store past events, creating a mental asymmetry that aligns with the physical arrow of increasing entropy. This perspective allows for a deeper understanding of how the experience of time's directionality is rooted in both physical processes and the mind's recursive structure.

Time, Existence, and the Nature of Reality

The relationship between time and existence has been a central question in metaphysics, raising issues about the nature of being, causality, and the structure of reality. Recursive Realism offers a way to reframe these questions by exploring how time is constructed through reciprocal interactions between mind and world.

Existence is often tied to the continuity of time, where being is understood as a process that unfolds over time. Recursive Realism suggests that existence is not a static state but a dynamic process where identity and reality are shaped through recursive loops that integrate past, present, and future. For example, a tree exists not just as a physical entity in space but as a temporal being that grows, ages, and changes over time. This temporal dimension allows us to understand existence as a process of becoming, where continuity is maintained through recursive interactions with the environment. Recursive Realism suggests that time is the medium through which existence manifests, creating a sense of persistence and identity in a changing world.

Causality is often seen as a linear chain, where causes lead to effects in a sequence. However, Recursive Realism suggests that causal relationships are more complex, involving feedback loops where effects can influence their causes through recursive interactions. For example, in ecological systems, the behaviour of a species can alter the environment, which in turn shapes the future behaviour of the species. This circular causality challenges the traditional view of time as a one-way progression and suggests that reality is shaped by interconnected feedback loops. Recursive Realism proposes that causality is better understood as a web of interactions, where past, present, and future influence each other in dynamic patterns.

Some philosophical traditions, such as mysticism and Eastern philosophies, suggest that beyond the flow of time, there is a timeless reality, a ground of being that transcends temporal change. Recursive Realism acknowledges this perspective, suggesting that time-bound experience is a manifestation of deeper principles that are outside of time. For example, in mystical experiences, individuals often report a sense of oneness with the universe, where the boundaries of self dissolve, and time becomes irrelevant. This timeless dimension can be understood as a state of pure recursion, where the mind's feedback loops reach a point of equilibrium, transcending the distinctions between past, present, and future. Recursive Realism suggests that this transcendent perspective is not in conflict with the temporal experience but represents a deeper layer of reality that underlies the flow of time.

Time is a multi-faceted phenomenon, shaped by both objective measurements and subjective experiences. Recursive Realism offers a framework for understanding time as a recursive construct, where the mind's engagement with memory, anticipation, and present awareness creates a rich temporal experience. By recognising the interplay between physical time and phenomenological time, Recursive Realism allows us to appreciate the depth and complexity of temporal existence, where reality is woven through layers of perception, reflection, and self-awareness.

In this chapter, we have explored how Recursive Realism reshapes our understanding of time as a dynamic construct, where the mind's recursive processes create a sense of continuity and direction. By bridging the scientific and philosophical perspectives on time, Recursive Realism offers a way to understand temporal experience as a dialogue between objective processes and subjective interpretations. This perspective allows us to see time not just as a dimension of the universe but as a central aspect of consciousness, where the past, present, and future are continually redefined through reciprocal engagement with the world.

References

Barbour, J. (1999). *The End of Time: The Next Revolution in Physics*. Oxford University Press.

Bergson, H. (1910). *Time and Free Will: An Essay on the Immediate Data of Consciousness*. Macmillan.

Block, R.A. (1990). *Cognitive Models of Psychological Time*. Psychology Press.

Davies, P. (1995). *About Time: Einstein's Unfinished Revolution*. Simon & Schuster.

Einstein, A. (1916). *Relativity: The Special and General Theory*. Methuen & Co.

Eliade, M. (1954). *The Myth of the Eternal Return: Cosmos and History*. Princeton University Press.

Gleick, J. (1999). *Faster: The Acceleration of Just About Everything*. Pantheon Books.

Heidegger, M. (1962). *Being and Time*. Harper & Row.

Husserl, E. (1964). *The Phenomenology of Internal Time-Consciousness*. Indiana University Press.

Kant, I. (1781/1929). *Critique of Pure Reason*. Translated by N. Kemp Smith. Macmillan.

McTaggart, J.M.E. (1908). 'The Unreality of Time,' *Mind*, 17(68), pp. 457–474.

Merleau-Ponty, M. (1945/2013). *Phenomenology of Perception*. Translated by D.A. Landes. Routledge.

Prigogine, I. (1980). *From Being to Becoming: Time and Complexity in the Physical Sciences*. W.H. Freeman.

Rovelli, C. (2018). *The Order of Time*. Riverhead Books.

Sacks, O. (1995). *An Anthropologist on Mars: Seven Paradoxical Tales*. Alfred A. Knopf.

Smolin, L. (2013). *Time Reborn: From the Crisis in Physics to the Future of the Universe*. Houghton Mifflin Harcourt.

Stiegler, B. (1998). *Technics and Time, 1: The Fault of Epimetheus*. Stanford University Press.

Varela, F.J., Thompson, E., and Rosch, E. (1991). *The Embodied Mind: Cognitive Science and Human Experience*. MIT Press.

Wittgenstein, L. (1921). *Tractatus Logico-Philosophicus*. Routledge.

Zohar, D. (1990). *The Quantum Self: Human Nature and Consciousness Defined by the New Physics*. William Morrow.

Chapter 13: Recursive Realism and the Emergence of Complexity

The Role of Recursion in Natural Systems

Complexity is a hallmark of natural systems, where patterns and structures emerge from the interactions of simple components. From the formation of galaxies to the evolution of life and the dynamics of ecosystems, recursion plays a crucial role in shaping the self-organisation of nature. Recursive Realism offers a perspective that sees complexity as the result of iterative feedback loops, where local interactions give rise to global patterns that, in turn, influence the local dynamics.

Self-Organisation in Physical Systems

The concept of self-organisation describes how order and structure can emerge spontaneously from the interactions of individual elements without the need for a centralised control. This process is inherently recursive, as the system's structure evolves through repeated interactions between its constituents.

Pattern formation is seen in many physical phenomena, such as the hexagonal cells of a beehive, the spiral shapes of galaxies, and the branching patterns of river networks. These patterns arise through recursive interactions between elements, where local processes, like the gravitational pull between stars or the flow dynamics of water, generate large-scale structures. For example, snowflakes form intricate geometric patterns as water molecules crystallise around microscopic imperfections, with each layer of growth following the template set by previous layers. Recursive Realism suggests that such natural patterns are not random but result from recursive rules that guide how simple components combine to create complex forms.

Nonlinear systems are those in which small changes in initial conditions can lead to large-scale effects, a phenomenon often described as the butterfly effect. These systems exhibit sensitive dependence on initial conditions, where feedback loops amplify tiny variations over time. For example, the formation of clouds and weather patterns depends on microscopic changes in temperature and humidity, leading to unpredictable variations in climate systems. Recursive Realism suggests that chaos theory reveals the underlying order within seemingly random phenomena, where complex behaviour emerges from the recursive interplay of interconnected elements. This perspective allows for a deeper understanding of how complexity arises from simple rules, where order and disorder coexist in a delicate balance.

Fractals are geometric shapes that exhibit self-similarity at different scales, where patterns repeat in a recursive manner. Examples of fractal structures include coastlines, mountain ranges, and vegetation patterns. Mandelbrot's fractal sets, such as the Mandelbrot set, are mathematical representations of how simple equations can generate infinitely complex shapes through recursive iteration. Recursive Realism suggests that fractals reveal a fundamental principle of nature, where recursive patterns shape the organisation of matter and energy at multiple scales, from the microscopic to the cosmic. This perspective allows us to see recursion as a unifying principle in natural complexity, where local interactions give rise to global forms that maintain structural coherence.

Recursion in Biological Systems

Life itself is a product of recursive processes, where genetic information, cellular interactions, and evolutionary dynamics create the rich diversity of biological forms. Recursive Realism suggests that the emergence of biological complexity can be understood through the interplay of recursive feedback loops that shape organisms, ecosystems, and evolution.

DNA is a molecule that encodes genetic instructions through a recursive structure of nucleotide pairs arranged in a double helix. This recursive pattern allows for the replication of genetic information, where each strand serves as a template for its complementary pair. The processes of transcription and translation, where DNA is converted into RNA and then into proteins, involve recursive loops that ensure accuracy and fidelity in gene expression. For example, the feedback mechanisms that regulate gene expression allow for adaptive responses to environmental changes, ensuring that proteins are produced in the right amounts when needed. Recursive Realism suggests that DNA is a symmetry line maker, where the recursive nature of genetic encoding allows for the maintenance and evolution of complex traits across generations.

The development of a multicellular organism from a single fertilised egg involves recursive interactions between genes, cells, and tissues. This process is guided by morphogenetic fields, where chemical gradients and mechanical forces shape the patterning of organs and body structures. For example, the formation of limbs in vertebrates involves recursive signalling pathways, where genes interact with growth factors to create symmetrical structures. Recursive Realism suggests that developmental processes are self-organising systems, where local interactions between cells and genes produce emergent patterns of form and function. This perspective reveals how recursive symmetry is fundamental to biological organisation, where the mind of DNA creates functional structures through iterative adaptation to the environment.

Ecosystems are complex webs of interaction where species, climate, and geology influence each other in dynamic ways. These networks are shaped by reciprocal feedback loops, where predators and prey, plants and pollinators, and nutrient cycles create a self-sustaining system. For example, the presence of a keystone species, such as wolves in Yellowstone National Park, can reshape the entire ecosystem, influencing vegetation growth, river flow, and biodiversity through cascading effects. Recursive Realism suggests that ecosystems are self-regulating systems, where local interactions give rise to global stability through recursive feedback loops. This perspective allows for a deeper understanding of how nature organises itself through patterns that emerge from interdependent relationships.

Recursion in Evolutionary Dynamics

Evolution is a process of adaptation and variation, where species change over generations through natural selection, mutation, and genetic drift. Recursive Realism offers a perspective on how evolution can be seen as a recursive process, where genetic changes interact with environmental pressures to create complex adaptive systems.

Natural selection involves a feedback loop between organisms and their environment, where genetic variations that enhance survival are amplified through reproduction. This process is inherently recursive, as each generation creates a new set of variations that interact with the changing environment, leading to evolutionary change over time. For example, the evolution of moths during the Industrial Revolution in England showed how dark-coloured variants became more common as soot darkened the landscape, making them less visible to predators. Recursive Realism suggests that adaptive evolution is a process of continuous refinement, where genes and environment interact in feedback loops that produce novel traits and species diversity.

While mutations are often seen as random events, their integration into the genome depends on recursive processes that determine whether a mutation will persist or disappear. Recursive Realism suggests that the success of certain mutations is influenced by directional pressures from the environment, where a series of aligned mutations can lead to the emergence of complex traits. For example, the evolution of the eye required multiple genetic changes that, over time, produced increasingly sophisticated structures capable of light detection and image formation. While each step in this evolutionary process may have seemed random, the reciprocal feedback between mutations and environmental conditions created a directional path toward functional adaptation. Recursive Realism suggests that nature's intelligence is not a conscious force but a self-organising system, where recursion guides the emergence of adaptations that align with environmental realities.

The formation of new species, or speciation, often involves a recursion of genetic divergence and environmental adaptation, where populations become isolated and develop distinct traits. This process can be seen in island ecosystems, where geographic isolation leads to the emergence of unique species adapted to local conditions. For example, Darwin's finches in the Galápagos Islands evolved a variety of beak shapes suited to different food sources, reflecting how local adaptations lead to divergent evolutionary paths. Recursive Realism suggests that speciation is not a sudden event but a gradual process where genetic variations interact with environmental pressures, creating new forms of life through a recursive interplay of differentiation and integration.

Complexity in nature emerges from the recursive interactions between simple components, where local processes give rise to global patterns that, in turn, shape local dynamics. Recursive Realism offers a framework for understanding how natural systems organise themselves through feedback loops that connect matter, energy, and information into coherent structures. This perspective allows us to see recursion as a unifying principle in biology, ecology, and evolution, where patterns of self-organisation reveal the depth and interconnectedness of natural complexity.

Recursion in Social Systems and Human Innovation

Human societies are among the most complex systems on Earth, characterised by rapid cultural changes, technological advancements, and intricate networks of relationships. Unlike biological evolution, where changes occur over millions of years, cultural evolution unfolds on a much shorter timescale, driven by reciprocal interactions between individuals, groups, and institutions. Recursive Realism suggests that the evolution of social complexity is shaped by recursive processes, where ideas, behaviours, and technological innovations feed back into each other, creating self-organising structures that adapt to changing environments.

Cultural Evolution and the Recursion of Ideas

Culture is a dynamic system that evolves through the exchange of ideas, beliefs, and practices among individuals and groups. This process is inherently recursive, as new ideas are adopted, modified, and transmitted across generations, leading to the emergence of shared norms and traditions.

Memes are units of cultural information, such as phrases, symbols, or rituals, that spread from person to person through imitation and communication. The concept of memetics, popularised by Richard Dawkins, suggests that memes evolve through a process analogous to natural selection, where some ideas become more popular or influential than others. Recursive Realism sees memes as part of a recursive feedback loop, where cultural information is continuously reinvented through collective discourse. For example, a popular meme might change meaning over time as it is adapted to different contexts and subcultures, reflecting how ideas evolve through dialogue and reinterpretation. This perspective allows for a deeper understanding of how cultural trends and social movements emerge from the bottom-up interactions of individuals, creating complex networks of shared meaning.

Language is one of the most powerful tools for cultural transmission, allowing humans to encode, share, and store knowledge across generations. The structure of language is inherently recursive, with rules of grammar and syntax that allow for the infinite generation of meaningful sentences. For example, the ability to embed clauses within sentences allows for complex expression and nuanced communication, making it possible to convey abstract ideas and subtle distinctions. Recursive Realism suggests that language is a key factor in the evolution of cultural complexity, as it enables collective thought and the refinement of concepts through dialogue. This recursive capacity for linguistic exchange creates a shared cognitive space, where knowledge and values are continuously shaped through conversation and storytelling.

Traditions represent the collective memory of cultural practices that have been passed down over time. These traditions are adapted to new social contexts, creating a recursive relationship between past practices and present needs. For example, rituals associated with harvest festivals may have originated as agrarian practices but have evolved into symbolic celebrations that maintain cultural identity even in urbanised societies. Recursive Realism suggests that tradition is not static but is continually reinterpreted, where historical practices are reintegrated into modern life through a dialogue between memory and innovation. This perspective allows for a nuanced understanding of how cultural resilience is maintained through the adaptive reuse of symbols and practices, creating a sense of continuity within cultural evolution.

Technological Innovation as a Recursive Process

Technology is a driving force in the transformation of societies, shaping everything from economic structures to social relations. The development of new tools and technological systems involves a recursive process, where inventions build on existing knowledge, leading to exponential growth in human capabilities.

Technological progress is rarely a linear process; instead, it involves a cycle of iteration, where ideas are tested, refined, and reimagined through feedback from users and communities. For example, the development of the internet involved decades of collaborative research, where computer scientists and engineers built on each other's work to create increasingly sophisticated networks. Recursive Realism suggests that innovation is a self-reinforcing process, where each new advancement opens up possibilities for further improvements. This recursive nature of technological evolution allows for the cumulative expansion of human knowledge and the creation of tools that reshape our relationship with the world.

The rise of digital technologies has created complex networks where information is shared, amplified, and manipulated through social media, search engines, and data algorithms. These networks operate through recursive loops, where user interactions generate data that is then used to refine algorithms, creating a feedback cycle that shapes user behaviour. For example, recommendation systems on platforms like YouTube or Spotify learn from user preferences and engagement, creating personalised content that guides future choices. Recursive Realism suggests that digital networks are self-organising systems, where complex patterns of communication and influence emerge from the recursive interactions between individuals and algorithms. This perspective highlights the power of recursion in shaping the dynamics of modern society, where technology and culture are interwoven in an ongoing cycle of adaptation.

Artificial Intelligence (AI) represents a new frontier in recursive systems, where machines learn through feedback loops that allow them to improve over time. Machine learning algorithms, such as neural networks, operate through recursive layers that process information and adjust their parameters based on errors and successes. For example, deep learning models used in image recognition learn to distinguish between objects by iteratively refining their filters through exposure to large datasets. Recursive Realism suggests that AI systems mirror the cognitive recursion seen in biological systems, where adaptation and learning arise from continuous feedback. This perspective allows us to explore how artificial intelligence can augment human cognition, creating new possibilities for problem-solving and creative expression through the recursive co-evolution of human and machine intelligence.

Societal Complexity and Recursive Governance

As societies grow larger and more interconnected, the challenges of coordination and governance become increasingly complex. Recursive Realism offers a perspective on how social systems can self-regulate through recursive processes, where rules, institutions, and social norms adapt to changing conditions through feedback loops.

Governance systems, including laws, policies, and social norms, operate through feedback loops, where the effects of policies are monitored and adjusted based on outcomes. For example, a government might introduce a policy to address pollution, only to adjust it based on public response and environmental impact. This recursive process allows for flexibility and adaptation, where rules evolve to better align with social needs. Recursive Realism suggests that governance is a self-organising system, where policies and institutions are refined through continuous interaction with citizens and environments. This perspective allows for a deeper understanding of how democratic processes can create resilient societies, where complex challenges are met with adaptive solutions.

Cultural resilience refers to the ability of communities to maintain their identity and function in the face of crises and disruptions. This resilience is often the result of recursive adaptations, where traditions and values are reinterpreted to fit new circumstances. For example, the rise of remote work during the COVID-19 pandemic forced many companies and institutions to adapt their practices and rituals, leading to innovations in virtual communication and digital collaboration. Recursive Realism suggests that cultural adaptation is not about abandoning tradition but about recontextualising it through a dialogue with the present. This perspective allows for a more nuanced understanding of how societies can maintain their core values while embracing change.

Social movements, such as those for civil rights, environmental protection, or gender equality, often follow a recursive trajectory, where progress is achieved through cycles of action, response, and reflection. For example, a protest movement might gain momentum through initial successes, encounter resistance, and then adapt its strategies based on public discourse and political conditions. Recursive Realism suggests that social change is a feedback process, where movements evolve through reciprocal interactions with institutions, media, and public opinion. This perspective highlights the power of recursion in driving change, where collective efforts are amplified through the iterative refinement of strategies and ideas.

Social systems are shaped by recursive interactions between individuals, communities, and institutions, where cultural norms, technological advancements, and governance structures evolve through self-organising processes. Recursive Realism provides a framework for understanding the complexity of human society, where ideas, technologies, and social movements co-evolve through feedback loops that adapt to changing environments. By recognising the recursive nature of cultural evolution and technological progress, we can appreciate the dynamic forces that shape societal complexity and human innovation.

In this chapter, we have explored how Recursive Realism offers a perspective on the emergence of complexity in natural, biological, and social systems, revealing how recursion shapes the self-organisation of matter, life, and culture. This perspective allows us to see complexity as the result of iterative processes, where local interactions give rise to global patterns that, in turn, influence individual behaviours. By understanding the recursive principles that govern natural evolution, technological innovation, and cultural change, we gain insight into the unifying dynamics that shape the emergent complexity of the world.

References

Baldwin, J.M., 1902. *Development and Evolution.* New York: Macmillan.
Dawkins, R., 1976. *The Selfish Gene.* Oxford: Oxford University Press.
Deacon, T.W., 2012. *Incomplete Nature: How Mind Emerged from Matter.* New York: W.W. Norton.
Gould, S.J., 2002. *The Structure of Evolutionary Theory.* Cambridge, MA: Harvard University Press.
Kauffman, S., 1995. *At Home in the Universe: The Search for Laws of Self-Organization and Complexity.* Oxford: Oxford University Press.
Lewontin, R.C., 1983. *The Organism as the Subject and Object of Evolution.* Scientific American, 239(3), pp.156–167.
Mandelbrot, B.B., 1982. *The Fractal Geometry of Nature.* San Francisco: W.H. Freeman.
Morin, E., 2008. *On Complexity.* Cresskill, NJ: Hampton Press.
Prigogine, I. and Stengers, I., 1984. *Order Out of Chaos: Man's New Dialogue with Nature.* New York: Bantam Books.
Turchin, P., 2003. *Historical Dynamics: Why States Rise and Fall.* Princeton, NJ: Princeton University Press.

Wilson, E.O., 1998. *Consilience: The Unity of Knowledge.* New York: Vintage Books.
Wolfram, S., 2002. *A New Kind of Science.* Champaign, IL: Wolfram Media.

Chapter 14: Recursive Realism and the Patterns of Consciousness

The Recursion of Thought and Inner Narratives

Consciousness is often experienced as a stream of thought, where ideas, images, and memories flow in a continuous sequence. However, this stream is not random but follows patterns that emerge from recursive feedback loops within the brain. Recursive Realism suggests that thought processes are structured through self-referential dynamics, where internal dialogue and imaginative exploration shape the landscape of inner experience.

The Structure of Inner Narratives

One of the defining features of conscious thought is the ability to construct narratives, coherent stories that link past experiences, present awareness, and future intentions. This narrative structure is inherently recursive, as the mind continually revisits and refines its story of self through reflection and reinterpretation.

Self-talk, the inner conversation that occurs within the mind, is a recursive process where the mind acts as both speaker and listener. This internal dialogue allows for critical thinking, decision-making, and self-reflection, as individuals pose questions, weigh options, and evaluate their own thoughts. For example, when faced with a difficult choice, a person might internally debate the pros and cons, considering how each outcome aligns with their values and goals. Recursive Realism suggests that self-talk is a form of cognitive recursion, where the mind continually adjusts its interpretations based on internal feedback, creating a space for self-exploration and personal growth.

Narrative identity is the story that individuals construct about who they are and how they have come to be. This self-narrative evolves over time, incorporating new experiences and shifting perspectives. Recursive Realism suggests that narrative identity is a recursive structure, where each new reflection on past events leads to a re-interpretation of the self-story. For example, a person who has overcome adversity might reframe their challenges as part of a story of resilience, seeing their struggles as stepping stones toward personal growth. This recursive process allows for psychological flexibility, enabling individuals to reorganise their self-concept in response to changing circumstances. Recursive Realism emphasises that identity is not a fixed essence but a self-organising narrative that evolves through continuous engagement with memory, emotion, and future goals.

Memory plays a central role in the construction of inner narratives, as it provides the raw material from which stories of the self are woven. However, memory is not a static record but a dynamic process of recall and reinterpretation. Each time a memory is revisited, it is modified by the context of the present and the emotional state of the individual, creating a recursive loop where the past is continually reshaped. For example, a fond memory of a childhood friendship might take on new meaning after a reunion, where the emotions of the present transform the recollection. Recursive Realism suggests that memory functions as a narrative thread, binding together experiences into a cohesive story that gives meaning to the passage of time. This perspective allows us to see memory as a creative act, where the mind constructs a narrative world that reflects both reality and imagination.

Imagination as a Recursive Exploration

Imagination is the ability to transcend the immediate reality, creating mental images of alternative worlds, future possibilities, and hypothetical scenarios. This capacity is deeply recursive, as it involves recombining elements of experience in novel ways, allowing for creativity and problem-solving.

Imagination allows the mind to simulate different scenarios, exploring how events might unfold under various conditions. This process is inherently recursive, as the mind adjusts its mental models based on new insights and feedback. For example, when imagining a new business venture, an entrepreneur might envision different strategies, evaluate potential challenges, and adjust their plans based on predicted outcomes. This ability to mentally rehearse future possibilities allows for strategic thinking and adaptation. Recursive Realism suggests that imagination is a form of cognitive recursion, where the mind creates virtual worlds to explore uncertainty and innovation, blending knowledge with speculation to create new insights.

Creativity involves the ability to combine familiar elements in unfamiliar ways, generating novel ideas and original works. This process is recursive, as it involves reflecting on existing concepts and rearranging them into new configurations. For example, a writer might draw on personal experiences, myths, and cultural references to create a unique story that resonates with universal themes. Recursive Realism suggests that creativity is a self-organising process, where the mind's recursive loops allow for the recombination of ideas, creating patterns that are both familiar and innovative. This perspective highlights the power of recursion in art and literature, where imaginative exploration leads to the emergence of new forms and narratives.

Daydreams provide a space for imaginative play, where the mind is free to wander through fantasies, memories, and future hopes. This mental wandering is inherently recursive, as the mind loops back to familiar themes, revisits ideas, and transforms them into new scenarios. For example, a person might daydream about travelling to an exotic location, imagining detailed experiences that blend memories of past trips with imagined adventures. Recursive Realism suggests that daydreaming is a form of cognitive exploration, where the mind creates new possibilities by recombining and reinterpreting the elements of experience. This perspective allows for a deeper appreciation of how imagination shapes our inner world, creating a rich tapestry of thought that is both playful and insightful.

Patterns of Awareness and the Dynamics of Attention

Attention is the process of focusing on specific sensory inputs or thoughts, filtering out irrelevant information to create a coherent experience. The patterns of attention are shaped by recursive feedback loops, where the mind continually adjusts its focus based on context, intention, and emotional state.

Selective attention involves directing focus to a particular aspect of the environment, while suppressing distractions. This process is recursive, as the mind continually monitors the quality of attention and readjusts its focus as needed. For example, when reading a book, a person may become distracted by external sounds, only to bring their attention back to the text after a brief moment. Recursive Realism suggests that attention is a dynamic feedback process, where the mind creates a stable field of awareness through recursive adjustments. This perspective allows us to understand how concentration is maintained, even in distracting environments, by reorienting focus in a self-correcting manner.

Conscious awareness can be experienced as a spotlight, where attention is narrowly focused on a specific task, or as a diffuse field, where the mind is open to a broader range of impressions. These modes of awareness are shaped by recursive interactions between top-down goals and bottom-up sensory inputs. For example, a painter might switch between a focused mode while working on fine details and a diffuse mode while considering the overall composition of their artwork. Recursive Realism suggests that attention is a flexible system, where the mind navigates between focused and expansive states through recursive feedback. This perspective highlights how awareness is not a static condition but a dynamic process that adjusts to the needs of the moment.

Mindfulness practices focus on enhancing awareness of the present moment, creating a space where the mind can observe its own thoughts without becoming entangled in them. This self-referential focus is inherently recursive, as it involves watching the mind as it watches itself, creating a meta-awareness of thought and emotion. For example, a person practising mindful breathing might become aware of the flow of thoughts without judging them, creating a sense of detachment from automatic reactions. Recursive Realism suggests that mindfulness allows for a deepened understanding of the patterns of awareness, where the mind becomes aware of its own recursive processes and learns to navigate them with greater clarity. This perspective offers a way to integrate conscious reflection with the flow of experience, creating a state of harmony between self-observation and present engagement.

Consciousness is shaped by recursive patterns that structure our thoughts, imagination, and awareness. Recursive Realism offers a perspective that sees inner experience as a self-organising process, where the mind creates narratives, explores possibilities, and focuses attention through self-referential loops. By recognising the recursive nature of cognition, we gain insight into how thought patterns evolve, how creativity emerges, and how awareness adapts to the complexity of inner and outer worlds.

Recursive Patterns in Emotional Experience and Self-Understanding

Emotions play a crucial role in shaping our perception of reality, influencing how we interpret events, respond to challenges, and relate to others. While emotions are often perceived as immediate reactions to external stimuli, they also involve a recursive dimension, where feelings and cognitive appraisals shape each other through ongoing feedback loops. Recursive Realism suggests that emotional experience is a self-organising process, where inner states and external conditions interact to create complex patterns of response and adaptation.

The Recursion of Feeling and Thought

Emotions and thoughts are often intertwined, creating a recursive interplay where each influences and modifies the other. This interaction is central to how individuals understand and process their feelings, allowing for a more nuanced and adaptive approach to emotional regulation.

Emotional appraisal is the process by which the mind evaluates a situation and assigns it emotional significance. This process is inherently recursive, as the initial emotional response can be re-evaluated based on new information or a shift in perspective. For example, after receiving critical feedback at work, a person might initially feel defensive or upset, but later, upon reflection, recognise the constructive nature of the comments. This recursive reappraisal allows for the transformation of negative emotions into acceptance or motivation, creating a more balanced emotional response. Recursive Realism suggests that emotional regulation involves a loop between automatic reactions and conscious reflection, where the mind learns to adjust its emotional state based on insights gained through self-reflection.

Self-reflection is the ability to turn inward and examine one's own thoughts and emotions, creating a recursive dialogue between inner experience and cognitive interpretation. This process allows individuals to explore the layers of their feelings, gaining a deeper understanding of the factors that influence their emotional states. For example, a person who feels anxious might reflect on past experiences that have shaped their fears, uncovering patterns that reveal how their responses have developed over time. Recursive Realism suggests that self-reflection is a form of cognitive recursion, where the mind revisits its emotional landscape to identify patterns, uncover insights, and adjust behaviours. This perspective allows for a more holistic understanding of emotions, where depth of feeling is seen as the result of self-referential processes that integrate past, present, and future experiences.

Mindset refers to the attitudes and beliefs that shape how individuals interpret experiences and manage emotions. Changes in mindset can create recursive shifts in how emotions are processed, leading to new patterns of response. For example, a person with a growth mindset, the belief that abilities can be developed through effort, might see failures as opportunities for learning, leading to feelings of determination rather than despair. Recursive Realism suggests that mindset operates through a feedback loop, where beliefs influence emotional responses, which in turn reinforce or challenge those beliefs. This recursive dynamic allows for emotional resilience, where individuals learn to adapt their emotional states in alignment with changing perspectives and experiences.

The Dynamics of Empathy and Emotional Resonance

Empathy is the ability to understand and share the feelings of others, creating a connection that transcends individual experience. This capacity involves a recursive process where the mind simulates the emotions of another person and reflects on how those feelings relate to one's own experience.

Empathy involves a recursive process of perspective-taking, where individuals imagine themselves in another person's situation, creating a mental simulation of their thoughts and feelings. This process allows for emotional resonance, where the mind aligns its emotional state with that of the other, creating a sense of shared experience. For example, when comforting a friend who is grieving, a person might recall their own experiences of loss, creating a bridge between their feelings and those of their friend. Recursive Realism suggests that empathy is a recursive loop between self-awareness and other-awareness, where each reflection deepens the understanding of the other's emotions. This perspective highlights the complexity of empathic connections, where emotional understanding is achieved through interwoven layers of self-reference and imaginative engagement.

Emotions can spread through social interactions, creating emotional contagion, the tendency for people to mirror the feelings of those around them. This process is recursive, as the emotional states of individuals influence the group dynamic, which in turn shapes the emotional experience of each member. For example, in a concert setting, the excitement of the crowd can create a feedback loop where individual enthusiasm amplifies the collective energy, creating an intense shared experience. Recursive Realism suggests that emotional contagion is a social feedback process, where individual emotions become intertwined through shared attention and mirroring behaviours. This perspective allows for a deeper understanding of how group emotions emerge from the interplay between personal feelings and social contexts, creating collective experiences that are greater than the sum of their individual parts.

Empathy not only deepens our understanding of others, but it also provides a mirror through which we can understand ourselves. By connecting with the emotions of others, we are often prompted to examine our own responses, leading to greater self-awareness. For example, feeling compassion for someone struggling with self-doubt might prompt a person to reflect on their own insecurities, uncovering parallel struggles. Recursive Realism suggests that empathy is a recursive journey, where understanding others leads to self-reflection and personal growth. This perspective highlights the reciprocal relationship between social understanding and self-knowledge, where the inner and outer worlds are interconnected through emotional resonance.

Recursive Realism and the Integration of Inner and Outer Worlds

The inner world of thoughts and feelings is continuously shaped by the outer world of experiences and interactions. Recursive Realism suggests that the boundary between inner and outer is fluid, defined by recursive exchanges where perceptions and experiences are integrated into a coherent sense of reality.

Cognitive dissonance occurs when there is a conflict between an individual's beliefs, behaviours, or feelings, creating a sense of internal discomfort. This discomfort prompts a recursive process of re-evaluation, where the mind seeks to resolve the inconsistency and restore equilibrium. For example, a person who values honesty but tells a white lie may experience dissonance, leading them to justify their actions or adjust their self-concept. Recursive Realism suggests that cognitive dissonance is a form of inner recursion, where the mind continually adjusts its beliefs to create a more consistent narrative. This perspective allows for a deeper understanding of how psychological conflicts are resolved through self-correcting feedback loops, creating a sense of internal coherence.

Intuition is often described as a gut feeling or immediate insight that arises without conscious reasoning. This implicit knowledge is shaped by subconscious processes that integrate past experiences and current contexts into a holistic judgement. Recursive Realism suggests that intuition is a product of recursive pattern recognition, where the mind detects subtle regularities in the environment and draws on previous encounters to make quick decisions. For example, a skilled chess player might intuitively recognise a winning strategy without being able to explain the exact steps. This perspective highlights how recursion allows for a synthesis of complex information, creating intuitive insights that are deeply grounded in experience.

The connection between mind and body is another domain where recursion plays a vital role. Emotions are not only cognitive states but are also embodied experiences, where physical sensations, such as tension, warmth, or relaxation, influence how we interpret our feelings. For example, stress might manifest as tightness in the chest, creating a feedback loop where the physical sensation amplifies the feeling of anxiety. Recursive Realism suggests that the mind-body connection is a bidirectional feedback system, where mental states and physical conditions continually influence each other. This perspective allows for a holistic view of emotional experience, where thought, feeling, and physical sensation are seen as interconnected layers of a self-organising system.

Emotions and self-awareness are shaped by recursive feedback loops, where feelings, thoughts, and physical sensations are interwoven into a dynamic tapestry of inner life. Recursive Realism provides a framework for understanding the complexity of emotional experience, where self-reflection, empathy, and cognitive processes interact to create rich patterns of response and adaptation. By recognising the recursive nature of emotion and self-understanding, we gain insight into how the mind navigates the depths of feeling and integrates the inner and outer worlds into a cohesive whole.

In this chapter, we have explored how Recursive Realism reshapes our understanding of consciousness, revealing how thought, imagination, and emotional experience are structured through recursive dynamics. By examining the interplay between self-reflection, intuitive insights, and the embodied mind, Recursive Realism provides a holistic perspective on the inner architecture of consciousness, where complexity emerges from the interconnected patterns that define our mental and emotional landscapes.

References

Barrett, L.F., 2017. *How Emotions Are Made: The Secret Life of the Brain*. Boston: Houghton Mifflin Harcourt.
Dennett, D.C., 1991. *Consciousness Explained*. Boston: Little, Brown and Company.
Damasio, A., 1999. *The Feeling of What Happens: Body and Emotion in the Making of Consciousness*. London: Heinemann.
Freud, S., 1923. *The Ego and the Id*. Translated by J. Riviere. London: Hogarth Press.
James, W., 1890. *The Principles of Psychology*. New York: Henry Holt and Company.
Kahneman, D., 2011. *Thinking, Fast and Slow*. London: Penguin Books.
Lakoff, G. and Johnson, M., 1980. *Metaphors We Live By*. Chicago: University of Chicago Press.
LeDoux, J., 1996. *The Emotional Brain: The Mysterious Underpinnings of Emotional Life*. New York: Simon & Schuster.
Maturana, H.R. and Varela, F.J., 1980. *Autopoiesis and Cognition: The Realization of the Living*. Dordrecht: Reidel Publishing Company.
Metzinger, T., 2009. *The Ego Tunnel: The Science of the Mind and the Myth of the Self*. New York: Basic Books.
Ramachandran, V.S., 2011. *The Tell-Tale Brain: A Neuroscientist's Quest for What Makes Us Human*. New York: W.W. Norton & Company.
Singer, T. and Klimecki, O.M., 2014. Empathy and Compassion. *Current Biology*, 24(18), pp. R875–R878.

Thagard, P., 2019. *Mind-Society: From Brains to Social Sciences and Professions*. Oxford: Oxford University Press.

Chapter 15: Recursive Realism and the Nature of Reality

Recursion and the Structure of the Universe

The universe is often described as a vast and intricate system, governed by laws and principles that create a sense of order amid cosmic scale. However, these laws are not simply external constructs; they are interpreted and understood through the recursive processes of the human mind, which seeks to create models and patterns to explain the nature of existence. Recursive Realism suggests that reality itself is shaped by recursive interactions between matter, energy, space, and time, creating a cosmic tapestry that is both ordered and dynamic.

The Laws of Nature as Recursive Patterns

The laws of physics describe how matter and energy behave under various conditions, providing a framework for understanding the mechanics of the universe. Recursive Realism suggests that these laws are manifestations of deeper recursive principles, where repetitive patterns and feedback loops shape the dynamics of natural phenomena.

Symmetry plays a central role in the laws of physics, where conservation laws, such as the conservation of energy, momentum, and charge, are related to the invariance of physical laws under certain transformations. For example, Noether's theorem demonstrates that every symmetry corresponds to a conservation law, revealing how physical principles are rooted in mathematical patterns that repeat across different scales. Recursive Realism suggests that symmetry is a recursive pattern, where the laws of nature maintain consistency through self-similar structures. This perspective allows us to see how conservation is not just a mechanical process but a reflection of the underlying recursion that shapes reality's continuity.

Fractal geometry reveals how self-similar patterns can describe the complexity of natural forms, from clouds and coastlines to galaxies and cosmic structures. For example, the distribution of galaxies in the universe follows a fractal pattern, where clusters and voids repeat at different scales. Recursive Realism suggests that the structure of space itself may be recursive, with patterns of expansion and clustering that mirror the dynamics of growth seen in biological systems. This perspective allows for a deeper understanding of how reality is shaped by recursive geometry, where simple rules generate complex forms that are self-similar across scales.

Quantum mechanics reveals a world where particles exist in superposition, and their states become determinate only through observation. This phenomenon, often described through the wave-function collapse, suggests that reality is influenced by the act of measurement, creating a recursive relationship between the observer and the observed. For example, the double-slit experiment shows that particles exhibit wave-like behaviour until they are measured, at which point they collapse into a definite state. Recursive Realism suggests that quantum phenomena reflect the recursive interplay between potential states and actual outcomes, where the mind and reality are interconnected through a loop of observation. This perspective offers a way to understand how consciousness and physical reality are entangled in a mutual process of co-creation, where reality's nature is shaped through recursive interactions.

The Cosmos as a Self-Organising System

The universe is not just a collection of objects but a self-organising system, where stars, planets, galaxies, and life emerge through interactions that are both predictable and adaptive. Recursive Realism suggests that the cosmic order is shaped by feedback loops between forces, matter, and energy, creating a dynamic balance that allows for the emergence of complex structures.

The process of stellar formation, where clouds of gas and dust collapse under gravity to form stars, is a recursive process, where gravitational attraction creates dense cores that ignite nuclear fusion, leading to the formation of stars and planetary systems. For example, spiral galaxies exhibit patterns of rotation and density waves that reflect the gravitational dance of stars and dark matter, creating a self-sustaining structure. Recursive Realism suggests that the patterns seen in galactic formations are a product of cosmic recursion, where local interactions give rise to global structures that, in turn, influence the behaviour of individual stars. This perspective allows us to see the universe as a living system, where order and complexity emerge through recursive dynamics.

The second law of thermodynamics states that entropy, or disorder, tends to increase over time, leading to a progression toward equilibrium. However, this tendency toward disorder coexists with the emergence of order through self-organisation, where local pockets of complexity arise even as the universe as a whole moves toward thermal equilibrium. For example, the formation of life on Earth represents a local reduction in entropy, made possible by the energy input from the Sun. Recursive Realism suggests that the balance between order and entropy is a recursive process, where patterns of organisation and dissolution interact through feedback loops that maintain cosmic balance. This perspective allows for a deeper appreciation of how the universe creates complexity through dynamic equilibrium, where order and chaos are intertwined through recursion.

The concept of cosmic evolution suggests that the universe has changed over billions of years, evolving from a hot, dense state into a vast expanse filled with galaxies, stars, and planets. This process involves a recursive cycle of expansion, cooling, clumping, and reorganisation, where gravity and thermodynamics shape the formation of structures at different scales. For example, the Big Bang represents the initial conditions of space-time, while the subsequent formation of galaxies and clusters reflects the recursive process of matter organising itself into hierarchical structures. Recursive Realism suggests that cosmic evolution is a recursive unfolding, where the patterns of the universe are continually refined through feedback interactions between space-time and energy. This perspective offers a way to understand the universe as a self-developing system, where reality evolves through recursive cycles of expansion and contraction, order and entropy.

The Observer and the Reality Loop

The role of the observer has been a central question in both philosophy and physics, raising issues about how perception shapes our understanding of reality. Recursive Realism suggests that observation itself is a recursive act, where the act of perceiving modifies the nature of what is perceived, creating a loop between consciousness and existence.

Perception is not a passive reception of sensory data but an active process where the mind creates internal models of the world based on incoming information. This process is inherently recursive, as the mind continually compares its expectations with sensory inputs, adjusting its understanding of reality. For example, when observing a sunset, the mind integrates visual information with emotional responses, creating a layered experience that is both sensory and symbolic. Recursive Realism suggests that observation is a dialogue between mind and world, where reality is co-constructed through the feedback loop of perception and interpretation. This perspective allows for a deeper understanding of how reality is experienced as a relationship between observer and observed.

The interpretation of quantum mechanics often raises the question of whether consciousness plays a role in the collapse of the wave function, where potential states become actualised through observation. Some interpretations, such as the Copenhagen interpretation, suggest that measurement itself is what determines the outcome of a quantum event. Recursive Realism suggests that conscious observation is part of a recursive loop between potentiality and actuality, where the act of perceiving influences the manifestation of reality. For example, the observer effect highlights how scientific experiments cannot be separated from the context in which they are conducted, as the act of measurement becomes part of the phenomenon being measured. This perspective suggests that reality is not a fixed state but a dynamic interplay between potential realities and the conscious mind that interacts with them.

The mind-body problem, the question of how consciousness relates to physical processes, has been a longstanding philosophical challenge. Recursive Realism offers a way to approach this problem by suggesting that consciousness and matter are connected through reciprocal interactions, where mental states and physical states influence each other through recursive loops. For example, neural networks in the brain create patterns of thought through electrical activity, while thought processes shape the brain's structure through plasticity. This perspective suggests that mind and matter are not separate substances but different expressions of a recursively structured reality, where consciousness emerges from the interplay of physical complexity and self-referential processes.

The universe is a self-organising system where recursive patterns shape the emergence of order and complexity. Recursive Realism offers a framework for understanding how natural laws, cosmic structures, and quantum phenomena are connected through reciprocal interactions between matter, energy, and consciousness. By recognising the recursive nature of reality, we gain insight into how the universe unfolds through patterns of self-organisation, where observation and existence are interconnected through dynamic feedback loops.

Reality, Perception, and the Philosophical Implications of Recursive Realism

The nature of reality has been a central question in philosophy, touching upon the relationship between perception, existence, and the structure of the universe. Traditional metaphysical models often attempt to separate the objective world from subjective experience, creating a divide between reality as it is and reality as it appears. Recursive Realism challenges this dualistic approach, suggesting that reality is shaped through reciprocal interactions between mind and world, where self-referential loops integrate observation with the fabric of existence.

Reality as a Constructed Process

Reality, in the framework of Recursive Realism, is not a fixed entity but a dynamic process that is co-created through interaction with consciousness. This perspective suggests that reality's nature is best understood as a recursive construct, where the world's structures are shaped by the processes that perceive, interpret, and adapt to them.

The philosophical tradition of constructivism argues that knowledge and understanding are constructed by the mind, rather than simply discovered as objective truths. This view emphasises the active role of perception in shaping reality, where cognitive frameworks and cultural narratives influence how we interpret the world. Recursive Realism expands on this idea, suggesting that reality itself is shaped through recursive processes, where each act of perception modifies our internal model of the world, which in turn influences how we engage with future experiences. For example, a scientific theory is not just a description of objective facts but a model that evolves through feedback from experimentation and observation, continuously refining our understanding of nature. This perspective allows for a deeper appreciation of how reality is shaped by the reciprocal relationship between human cognition and the phenomena it seeks to explain.

Phenomenal reality refers to the world as it appears to our senses, shaped by the brain's processing of sensory inputs. This filtered reality is inherently subjective, as the mind emphasises certain aspects of experience while downplaying others based on attention, expectation, and cultural conditioning. Recursive Realism suggests that phenomenal reality is a recursive construction, where the mind creates a coherent picture of the world through continuous interpretation of sensory data. For example, the perception of time can slow down during moments of danger or speed up during periods of boredom, reflecting how awareness of time is shaped by reciprocal feedback between internal states and external events. This perspective reveals that phenomenal reality is not a lesser version of the objective world but a complex synthesis of mind and environment, where experience is tailored to the individual's needs and context.

Science is often seen as a methodical search for objective truths, aiming to uncover the laws of nature through empirical observation and mathematical modelling. However, the history of science reveals that theories are often revised or abandoned as new evidence comes to light, creating a recursive process of hypothesis testing and theory refinement. For example, the shift from Newtonian mechanics to relativity and quantum mechanics demonstrates how scientific understanding evolves through a dialogue between theory and observation. Recursive Realism suggests that scientific progress is inherently recursive, as each new discovery prompts a re-evaluation of existing models and a redefinition of reality. This perspective allows for a nuanced understanding of how science constructs reality through a dynamic feedback loop between ideas and experiments, where truth is always provisional and open to revision.

Metaphysics, Recursion, and the Nature of Being

The study of metaphysics seeks to explore the fundamental nature of being, existence, and the structure of reality beyond empirical observation. Recursive Realism offers a perspective that suggests being itself is structured recursively, where the mind's awareness of existence is part of a larger pattern of self-reference within the universe.

Existence is often described as a state of being that is independent of observation, yet conscious awareness reveals that being is deeply intertwined with self-reflection. Recursive Realism suggests that being is not simply a static presence but a dynamic process of self-awareness, where the mind reflects on its own existence through recursive loops of thought and perception. For example, the philosophical question "What does it mean to exist?" involves a recursive exploration of self-awareness, where the mind turns inward to question its own nature. This perspective allows for a more fluid understanding of existence, where being is seen as a self-referential phenomenon that is intertwined with the mind's ability to conceptualise its own reality.

Many philosophical traditions, including Hegelian dialectics and Eastern philosophies like Taoism, recognise that reality often involves a unity of opposites, where contradictory forces are interconnected through a dynamic balance. Recursive Realism suggests that this unity is a recursive relationship, where opposing forces, such as order and chaos, stability and change, are continually reconciled through feedback loops. For example, in ecosystems, predator-prey dynamics create a balance where populations are kept in check through reciprocal interactions. This perspective allows for a deeper understanding of how contradictions within reality can be seen as complementary aspects of a self-organising whole, where tension and harmony are resolved through recursion.

Mystical traditions often speak of a reality that is beyond words, where the limits of language and conceptual thought are transcended in a state of oneness or pure being. Recursive Realism acknowledges this transcendent dimension, suggesting that reality contains levels that are beyond the reach of rational analysis, accessible only through direct experience or intuition. For example, Zen koans use paradoxical statements to break through the habitual thought patterns of the mind, revealing a deeper truth that cannot be articulated. Recursive Realism suggests that mystical experiences reflect the mind's encounter with the recursive structure of reality, where self-reference reaches a point of transcendence. This perspective allows us to explore how reality's nature is both accessible through reason and beyond the grasp of concepts, creating a space for wonder and reverence.

Recursive Realism and the Reconciliation of Dualisms

Philosophy has long grappled with dualistic frameworks, such as mind vs. matter, self vs. other, and subjectivity vs. objectivity. Recursive Realism offers a way to transcend these dualities, suggesting that they are interconnected aspects of a larger recursive system, where opposing concepts are integrated through self-referential processes.

The mind-body problem has traditionally been framed as a conflict between mental phenomena and physical processes. Recursive Realism suggests that mind and matter are not opposites but different manifestations of a recursive reality, where cognition emerges from the complex organisation of matter, and matter is shaped by perception. For example, conscious thought arises from neural networks in the brain, while thought patterns influence bodily states through hormonal feedback and nervous system responses. This perspective allows for a holistic view of consciousness, where mental and physical realms are seen as mutually reflective, integrated through recursive processes that connect mind and body.

Objectivity seeks to understand the world as it is, while subjectivity focuses on the world as experienced. Recursive Realism suggests that these perspectives are interdependent, as objective knowledge is always shaped by subjective frameworks, and subjective experience is informed by objective structures. For example, scientific observation is guided by theories and hypotheses, which in turn are refined based on empirical findings. This reciprocal relationship allows for a more integrated view of truth, where objectivity and subjectivity are seen as different aspects of the same process of understanding reality. Recursive Realism offers a way to bridge these perspectives, allowing for a more complete picture of reality that honours the complexity of both.

Dualistic thinking often creates rigid boundaries between concepts, leading to conflicts such as the debates between God vs. gravity or science vs. spirituality. Recursive Realism suggests that such extremes can be reconciled through a recursive approach, where opposing viewpoints are understood as part of a larger whole that evolves through self-reference. For example, the mystical concept of unity can be seen as a recognition of the interconnectedness of all phenomena, where differences dissolve into a deeper harmony. This perspective offers a way to transcend polarities, allowing for a philosophy of integration that embraces complexity without simplifying the mysteries of existence.

Reality is a complex interplay between mind and world, where self-referential loops shape our understanding of existence and the structure of the universe. Recursive Realism offers a way to see reality not as a fixed state but as a dynamic process, where perception, reflection, and experience create a coherent world through recursive feedback. By recognising the interdependence of objectivity and subjectivity, science and metaphysics, Recursive Realism allows us to explore the depth of reality in a way that honours its mystery and embraces its richness.

In this chapter, we have explored how Recursive Realism reshapes our understanding of reality, offering a framework that sees the universe as a self-organising system shaped by recursive processes. By examining the patterns of nature, the role of observation, and the philosophical dimensions of existence, Recursive Realism provides a way to understand reality as an evolving construct, where mind and world are woven together through self-referential interactions. This perspective allows us to appreciate the dynamic nature of reality, where order and mystery coexist in a recursive dance that defines the essence of existence.

References

Barbour, J., 1999. *The End of Time: The Next Revolution in Physics*. Oxford: Oxford University Press.

Bohm, D., 1980. *Wholeness and the Implicate Order*. London: Routledge.

Capra, F., 1997. *The Web of Life: A New Scientific Understanding of Living Systems*. New York: Anchor Books.

Chalmers, D.J., 1996. *The Conscious Mind: In Search of a Fundamental Theory*. Oxford: Oxford University Press.

Einstein, A., 1920. *Relativity: The Special and General Theory*. London: Methuen & Co Ltd.

Feynman, R.P., Leighton, R.B. and Sands, M., 1964. *The Feynman Lectures on Physics: Volume I*. Reading: Addison-Wesley.

Gleick, J., 1987. *Chaos: Making a New Science.* New York: Viking Penguin.

Greene, B., 2004. *The Fabric of the Cosmos: Space, Time, and the Texture of Reality.* New York: Knopf.

Hawking, S.W., 1988. *A Brief History of Time: From the Big Bang to Black Holes.* New York: Bantam Books.

Klein, E., 2004. *The Universe in a Nutshell.* New York: Bantam Books.

Lakoff, G. and Núñez, R.E., 2000. *Where Mathematics Comes From: How the Embodied Mind Brings Mathematics into Being.* New York: Basic Books.

Prigogine, I., 1997. *The End of Certainty: Time, Chaos, and the New Laws of Nature.* New York: Free Press.

Smolin, L., 2013. *Time Reborn: From the Crisis in Physics to the Future of the Universe.* Boston: Houghton Mifflin Harcourt.

Tegmark, M., 2014. *Our Mathematical Universe: My Quest for the Ultimate Nature of Reality.* New York: Knopf.

Zurek, W.H., 2003. Decoherence, Einselection, and the Quantum Origins of the Classical. *Reviews of Modern Physics*, 75(3), pp.715–775.

Chapter 16: Recursive Realism and the Dynamics of Knowledge

The Recursive Nature of Learning and Discovery

The process of learning is inherently recursive, where new information is integrated with existing knowledge, reshaping our understanding and expanding our horizons. Recursive Realism suggests that knowledge grows through feedback loops, where each discovery prompts a re-evaluation of prior beliefs, leading to a deeper understanding of complex phenomena. This section explores how recursion shapes the growth and organisation of knowledge, from individual learning to the collective evolution of science and philosophy.

The Role of Recursion in Cognitive Development

Human cognition develops through a series of feedback loops, where sensory experiences interact with cognitive frameworks to create a coherent understanding of the world. Recursive Realism suggests that cognitive growth is shaped by the mind's ability to revisit, revise, and expand its mental models, allowing for the integration of new insights into a dynamic system of understanding.

Schema theory describes how the mind organises knowledge into mental structures called schemas, which are adapted through assimilation and accommodation. For example, a child who learns that birds can fly may initially assimilate the concept of a penguin as a bird that does not fly, requiring a revision of their schema for birds. Recursive Realism suggests that schemas are recursive structures, where each new piece of information modifies the existing mental framework, creating a loop between expectation and experience. This perspective allows us to see cognitive development as a self-organising process, where understanding evolves through iterative refinement of concepts and categories.

Metacognition, the ability to think about one's own thinking, is a recursive process that allows individuals to monitor and adjust their cognitive strategies. For example, a student who realises that they learn better through visual aids might adjust their study habits to include more diagrams and charts, leading to improved comprehension. Recursive Realism suggests that metacognition is a form of cognitive recursion, where the mind creates a model of its own mental processes and revises it through self-reflection. This recursive self-awareness allows for adaptive learning, where individuals become more effective at understanding and navigating complex subjects. This perspective highlights the importance of self-reflection in personal growth, where recursive loops between thought and awareness lead to greater cognitive flexibility.

The acquisition of skills, whether in sports, music, or problem-solving, involves a recursive process of practice, feedback, and adjustment. For example, a pianist learning a new piece might focus on tricky passages, receiving feedback from their mistakes and refining their technique until the performance becomes fluid. Recursive Realism suggests that skill development is a self-correcting process, where errors are integrated into a feedback loop that guides improvement. This perspective allows for a deeper understanding of how expertise is developed, where practice is not merely repetition but a recursive cycle of adjustment and mastery. It emphasises the role of recursive feedback in transforming initial attempts into proficient actions.

The Evolution of Scientific Knowledge

Science is a process of inquiry that seeks to understand the natural world through observation, experimentation, and theory-building. Recursive Realism suggests that the evolution of scientific knowledge is inherently recursive, as theories are continually refined through empirical testing and conceptual revision, creating a feedback loop between evidence and understanding.

The scientific method involves a cycle of hypothesis formulation, data collection, analysis, and revision, where theories are adjusted based on new findings. For example, the development of plate tectonics involved decades of data collection on continental drift, seafloor spreading, and geological formations, leading to a unifying theory that reshaped our understanding of earth's structure. Recursive Realism suggests that the scientific method is a recursive process of self-correction, where hypotheses are continually tested against reality, leading to increasingly accurate models. This perspective allows for a nuanced appreciation of how scientific knowledge evolves through cycles of questioning and discovery, where each iteration deepens our understanding of the universe.

Paradigm shifts, as described by Thomas Kuhn, occur when a fundamental change in the underlying framework of a scientific field leads to a new way of understanding phenomena. For example, the shift from a geocentric to a heliocentric model of the solar system transformed astronomy and challenged the cultural worldview of the time. Recursive Realism suggests that paradigm shifts represent a recursive leap in conceptual understanding, where existing models are re-evaluated in light of new evidence, leading to a reconstruction of the intellectual landscape. This perspective highlights the role of recursion in scientific revolutions, where knowledge evolves through a dialectic process of challenge and response, creating new frameworks that better capture the complexity of nature.

The growth of interdisciplinary fields, such as biophysics, neuroscience, and environmental science, reflects how knowledge systems interact through recursive feedback loops, where insights from one discipline inform and transform others. For example, the application of chaos theory in ecology has provided new ways to understand population dynamics and ecosystem stability, drawing on mathematics and physics to solve biological puzzles. Recursive Realism suggests that interdisciplinary knowledge is a self-organising network, where ideas cross-pollinate through recursive dialogue between fields. This perspective allows for a richer understanding of how complex problems are addressed through collaboration, where theories are expanded and refined through shared inquiry.

Knowledge Systems and the Recursion of Culture

Cultural knowledge encompasses the beliefs, stories, and symbols that shape a society's understanding of itself and the world. Recursive Realism suggests that cultural systems evolve through recursive interactions between tradition and innovation, where ideas are continually reinterpreted and reintegrated into a cohesive worldview.

Myths and symbols play a central role in shaping cultural identity, providing a shared language for understanding life's mysteries. These narratives evolve over time, as each generation reinterprets traditional stories in light of new experiences. For example, the myth of Prometheus, which speaks of knowledge and defiance, has been adapted across different eras to reflect changing attitudes toward science, authority, and innovation. Recursive Realism suggests that myths are recursive structures, where meaning is continually reconstructed through collective reflection. This perspective allows for a deeper understanding of how culture is a living system, where symbols are not static relics but dynamic patterns that evolve through dialogue between past and present.

Education is the process of transmitting knowledge across generations, allowing for the preservation and innovation of cultural values. This process involves a recursive relationship between teachers and learners, where ideas are shared, questioned, and refined through interaction. For example, Socratic dialogues involve a recursive process of questioning, where students are encouraged to reflect on their assumptions and arrive at deeper truths. Recursive Realism suggests that education is a recursive dialogue, where knowledge is not simply transferred but transformed through reciprocal engagement. This perspective allows for a more adaptive approach to learning, where students become active participants in the creation of knowledge through reflection and experimentation.

The concept of collective intelligence refers to the shared knowledge and problem-solving capacity of groups, which emerges through collaboration and information exchange. This distributed intelligence evolves through recursive interactions among individuals, where ideas are tested, amplified, and refined through social networks. For example, the open-source software movement allows developers from around the world to collaborate on projects, creating a recursive loop of improvement and innovation. Recursive Realism suggests that collective intelligence is a self-organising phenomenon, where cultural evolution is driven by feedback loops between individual contributions and community adaptation. This perspective highlights how societies can harness recursion to navigate complexity, creating a shared understanding that evolves in response to new challenges.

Knowledge grows through recursive processes of reflection, revision, and integration, where new insights reshape our understanding of the world. Recursive Realism offers a framework for understanding the evolution of knowledge as a dynamic dialogue between individual cognition, scientific inquiry, and cultural traditions, where each discovery prompts a re-evaluation of existing beliefs. By recognising the recursive nature of learning, we gain insight into how understanding evolves through self-correcting cycles, where ideas are continually refined and reimagined.

Recursive Realism, Knowledge, and the Limits of Understanding

As knowledge expands, we are continually confronted with the edges of understanding, where concepts become more abstract, and certainty becomes elusive. Recursive Realism suggests that self-reflective thinking enables us to navigate these boundaries, offering a way to engage with the unknown through a process of iterative refinement and self-correction. This approach allows us to appreciate the dynamic relationship between what is known and what remains beyond our grasp, emphasising that intellectual growth is shaped by reciprocal interactions between certainty and ambiguity.

The Boundaries of Human Knowledge

Human knowledge is vast, encompassing everything from the intricacies of subatomic particles to the expanse of the cosmos. However, there are inherent limitations to what we can understand, shaped by the constraints of perception, language, and the complexity of the universe itself. Recursive Realism suggests that these limitations are not barriers but thresholds that invite deeper exploration through self-referential processes.

The Heisenberg Uncertainty Principle in quantum mechanics states that there are limits to how precisely we can know certain pairs of properties of a particle, such as position and momentum. This fundamental limit is not a matter of technological inadequacy but an inherent feature of the quantum world, where measurement alters the system being measured. Recursive Realism suggests that the uncertainty principle reveals the recursive nature of knowledge, where the act of observation is intertwined with the nature of reality itself. This perspective allows us to understand that uncertainty is not merely a gap in knowledge but a window into the interplay between observer and phenomenon, where reality is shaped through reciprocal interactions.

Gödel's incompleteness theorems demonstrate that in any formal mathematical system, there are statements that are true but cannot be proven within the system itself. This insight reveals that logic has limits, where certainty gives way to unprovable truths. Recursive Realism suggests that Gödel's insights highlight the self-referential nature of mathematical systems, where truth is shaped by internal consistency and external interpretation. This perspective allows for a deeper appreciation of the mystery that lies at the heart of mathematics, where recursion creates a space for intuition, imagination, and the unknown. It suggests that the quest for knowledge is an open-ended journey, where new insights always bring new questions.

In complex systems such as ecosystems, weather patterns, and economies, small variations in initial conditions can lead to large-scale changes, making long-term predictions highly uncertain. This phenomenon, often described as sensitive dependence or the butterfly effect, reveals the limits of predictive models. Recursive Realism suggests that complexity is shaped by recursive feedback loops, where outcomes are continually reshaped by interacting variables. This perspective allows us to appreciate that complexity is not a deficiency in understanding but a fundamental feature of reality, where uncertainty and unpredictability are integral to how systems evolve. It challenges the notion that knowledge is about achieving control and instead emphasises the value of adaptive thinking in navigating complexity.

Recursion, Paradox, and the Nature of Inquiry

As knowledge deepens, we often encounter paradoxes, situations where logic seems to contradict itself, revealing the limits of our conceptual frameworks. Recursive Realism suggests that paradox is not a failure of reason but a signpost that points to the recursive structure of reality, where truth is not always linear or absolute.

Self-reference can create paradoxes, such as the liar's paradox, which involves a statement that declares itself to be false. These logical puzzles challenge the boundaries of truth and meaning, revealing how recursion can create loops that defy simple resolution. Recursive Realism suggests that self-reference is a mirror that reflects the complexity of thought, where statements about truth become intertwined with meta-statements about themselves. This perspective allows us to see paradoxes not as contradictions but as expressions of a reality that is multi-layered, where meaning unfolds through recursive interpretation.

Zeno's paradoxes, such as the Achilles and the tortoise, challenge our understanding of motion and continuity, suggesting that space and time can be infinitely divided. These paradoxes reveal the limits of human intuition when applied to infinite processes. Recursive Realism suggests that Zeno's paradoxes illustrate the recursion inherent in concepts of continuity, where infinite sequences create patterns that are both logical and mysterious. This perspective allows for a more fluid understanding of space-time, where continuity is seen as an emergent property of recursive interactions, challenging the linear view of reality.

Paradoxes have also played a significant role in spiritual traditions, where they are used to dissolve conceptual boundaries and awaken deeper insight. For example, the Zen koan "What is the sound of one hand clapping?" invites a recursive questioning that transcends rational analysis, leading to a state of direct awareness. Recursive Realism suggests that paradox serves as a gateway to non-dual understanding, where self-reference reaches a point of equilibrium that dissolves conceptual distinctions. This perspective allows us to explore how philosophy and spirituality can integrate through a recursive approach, where mystery is embraced as an essential aspect of wisdom.

Recursive Realism and the Pursuit of Wisdom

Wisdom involves the ability to navigate uncertainty, embrace ambiguity, and understand the limitations of knowledge. Recursive Realism offers a framework for pursuing wisdom through self-reflective inquiry, where intellectual humility and curiosity guide the exploration of reality's mysteries.

Intellectual humility is the recognition that knowledge is always partial and that certainty can be elusive. This attitude is central to Recursive Realism, which emphasises the importance of questioning assumptions and being open to new perspectives. For example, a scientist studying the origins of life might remain open to different hypotheses, recognising that the complexity of life's emergence may involve unknown factors. Recursive Realism suggests that humility is a form of recursive thinking, where understanding is continually revised in response to new insights. This perspective allows for a more flexible approach to knowledge, where the mind is attuned to the limits of its own understanding.

Curiosity is the drive to explore, question, and discover, even in the face of uncertainty. Recursive Realism suggests that curiosity is a recursive impulse, where the mind continually seeks to extend its understanding by exploring new questions. For example, the exploration of deep space is driven by a curiosity about what lies beyond the edges of current knowledge, prompting scientists to probe the mysteries of dark matter and cosmic expansion. This perspective allows for a deeper appreciation of how intellectual curiosity is not just a desire for answers but a reciprocal dialogue between known and unknown, where questions create pathways into new realms of understanding.

Wisdom involves balancing the pursuit of knowledge with an acceptance of mystery, recognising that some aspects of reality may always remain beyond comprehension. Recursive Realism suggests that mystery is not a failure of understanding but a source of wonder that enriches the intellectual journey. For example, the nature of consciousness and the origins of the universe remain profound mysteries, yet they invite ongoing reflection and speculation. This perspective allows for a holistic view of knowledge, where certainty and mystery are seen as complementary aspects of a recursive reality, creating a space for wonder alongside understanding.

Knowledge is shaped by recursive processes that explore the boundaries between certainty and ambiguity, where each new insight invites deeper questions. Recursive Realism provides a framework for embracing the complexity of reality, where paradox, uncertainty, and mystery are integral to the growth of understanding. By recognising the recursive nature of intellectual inquiry, we gain insight into how wisdom emerges through self-reflection, curiosity, and a willingness to engage with the unknown.

In this chapter, we have explored how Recursive Realism offers a framework for understanding the growth and limits of knowledge, revealing how recursion shapes the evolution of ideas, the structure of learning, and the philosophical challenges of understanding reality. By examining the recursive interplay between certainty and mystery, self-reflection and exploration, Recursive Realism allows us to appreciate the dynamic nature of intellectual growth, where knowledge is not a fixed body but an ever-evolving process of discovery and inquiry.

References

Aristotle, 350 BCE. *Metaphysics*. Translated by W.D. Ross. Oxford: Clarendon Press.

Bateson, G., 1972. *Steps to an Ecology of Mind*. Chicago: University of Chicago Press.

Bruner, J.S., 1996. *The Culture of Education*. Cambridge, MA: Harvard University Press.

Chalmers, D.J., 1996. *The Conscious Mind: In Search of a Fundamental Theory*. Oxford: Oxford University Press.

Dawkins, R., 1976. *The Selfish Gene*. Oxford: Oxford University Press.

Dennett, D.C., 1991. *Consciousness Explained*. Boston: Little, Brown.

Gödel, K., 1931. Über formal unentscheidbare Sätze der Principia Mathematica und verwandter Systeme. *Monatshefte für Mathematik und Physik*, 38(1), pp.173–198.

Kuhn, T.S., 1962. *The Structure of Scientific Revolutions*. Chicago: University of Chicago Press.

Lakatos, I., 1976. *Proofs and Refutations: The Logic of Mathematical Discovery*. Cambridge: Cambridge University Press.

Maturana, H.R. and Varela, F.J., 1980. *Autopoiesis and Cognition: The Realization of the Living*. Dordrecht: Springer.

Piaget, J., 1950. *The Psychology of Intelligence*. London: Routledge.

Popper, K.R., 1959. *The Logic of Scientific Discovery*. London: Routledge.

Prigogine, I. and Stengers, I., 1984. *Order Out of Chaos: Man's New Dialogue with Nature*. London: Heinemann.

Senge, P.M., 1990. *The Fifth Discipline: The Art and Practice of the Learning Organization*. New York: Doubleday.

Vygotsky, L.S., 1978. *Mind in Society: The Development of Higher Psychological Processes*. Cambridge, MA: Harvard University Press.

Wheeler, J.A., 1990. Information, Physics, Quantum: The Search for Links. In: W.H. Zurek, ed., *Complexity, Entropy, and the Physics of Information*. Redwood City, CA: Addison-Wesley, pp.309–336.

Chapter 17: Recursive Realism and the Architecture of the Mind

Neural Recursion and the Structure of Cognition

The human brain is one of the most complex structures in the known universe, with billions of neurons forming networks that process and transmit information. Recursive Realism suggests that the architecture of the brain is shaped by recursive patterns, where neural circuits create feedback loops that allow for learning, memory, and adaptive behaviour. This section explores how recursion is embedded in the biological processes that underlie cognition, revealing how thought and perception emerge from the self-organising dynamics of neural networks.

The Brain as a Self-Organising System

The brain's complexity arises from its ability to self-organise through feedback loops that connect neurons, circuits, and regions into a coherent system. Recursive Realism suggests that self-organisation in the brain is driven by recursive interactions between sensory inputs, internal states, and external behaviours, creating a dynamic equilibrium that allows for flexibility and adaptation.

Neurons communicate through synaptic connections, where the activity of one neuron influences the firing patterns of others, creating networks that process information. This process is inherently recursive, as neural circuits create loops that amplify, inhibit, or modulate signals based on context and experience. For example, cortical circuits in the visual system process incoming light from the retina, while feedback connections adjust perception based on expectations and attention. Recursive Realism suggests that neural networks operate through reciprocal interactions, where feedforward and feedback pathways integrate sensory data with internal models of the world. This perspective allows us to see the brain as a self-organising system, where cognition emerges from the interplay between external stimuli and internal dynamics.

Homeostasis refers to the brain's ability to maintain internal balance by adjusting physiological processes in response to changes in the environment. This process is inherently recursive, as the nervous system monitors internal states, such as temperature, hydration, and blood sugar levels, and initiates corrective actions to restore equilibrium. For example, if the body becomes too hot, the brain triggers sweating to cool down; if glucose levels drop, it stimulates hunger to prompt eating. Recursive Realism suggests that homeostasis is a self-regulating loop, where the brain continually adjusts its outputs to maintain a stable internal environment. This perspective highlights how survival and adaptation are shaped by recursive feedback between the body and the external world, creating a dynamic balance that sustains life.

Neuroplasticity refers to the brain's ability to reorganise its structure and function in response to experience, learning, and injury. This process is deeply recursive, as neuronal connections are strengthened or weakened based on patterns of activity, creating a feedback loop where use and practice shape the brain's architecture. For example, a musician who practices daily develops stronger connections in regions associated with motor control and auditory processing, while a stroke survivor might retrain other areas of the brain to compensate for damaged regions. Recursive Realism suggests that neuroplasticity is a self-organising process, where the brain refines its structure through reciprocal interactions with learning and environmental feedback. This perspective allows for a deeper understanding of how adaptation is not a passive process but an active reconfiguration of neural circuits, enabling flexibility and resilience.

The Recursive Dynamics of Memory and Learning

Memory is the foundation of human experience, allowing us to store and recall information, integrate past experiences, and adapt to new situations. Recursive Realism suggests that memory is a recursive process, where patterns of experience are continually reorganised and reintegrated into cognitive frameworks, shaping our sense of continuity and identity.

Short-term memory holds information for brief periods, allowing the mind to keep details in focus while performing tasks. Working memory, a more active process, involves manipulating and reorganising information to solve problems or make decisions. This process is recursive, as working memory draws on stored knowledge to interpret new data, while new experiences are encoded into existing frameworks. For example, when learning a new language, working memory is used to practice vocabulary, while connections to prior knowledge help to assimilate grammar rules. Recursive Realism suggests that short-term memory functions as a self-referential loop, where thoughts are refreshed and modified through continuous interaction with long-term memory. This perspective allows for a deeper understanding of how immediate experiences are linked to stored knowledge, creating a seamless flow of awareness.

Long-term memory involves the storage of information over extended periods, from factual knowledge to emotional experiences. The process of recall is inherently recursive, as the act of remembering involves reconstructing memory traces in the context of present thoughts and emotions. For example, recalling a childhood memory may involve sensory impressions, emotional overtones, and related memories, which blend together to create a cohesive recollection. Recursive Realism suggests that memory retrieval is not a simple replay of stored data but a reconstructive process that involves updating and reinterpreting past experiences. This perspective allows us to see memory as a dynamic interplay between past and present, where each act of recall creates a new version of the remembered event.

Learning often involves recognising patterns in the environment and adapting behaviour accordingly, from simple reflexes to complex skills. This process is recursive, as neural circuits create loops that reinforce effective responses through repetition and reward. For example, when learning to ride a bicycle, the brain integrates visual information with motor actions, creating a loop that allows for balance and coordination. Recursive Realism suggests that habit formation is a feedback process, where behavioural patterns become automatised through continuous practice and reinforcement. This perspective highlights how learning is not a linear acquisition of knowledge but a self-organising process, where behavioural patterns emerge through reciprocal interactions between cognition, experience, and reward.

Recursion and the Integration of Sensory Experience

Perception involves the integration of sensory data into a coherent model of the world, allowing us to navigate our environment and interact with reality. Recursive Realism suggests that sensory experience is shaped by recursive feedback loops, where the brain continually updates its internal models based on new inputs and expectations.

Perception involves a balance between bottom-up processing, where sensory data flows upward from the senses to the brain, and top-down processing, where expectations and prior knowledge shape the interpretation of sensory information. For example, when viewing a familiar object like a face, top-down signals help to recognise features based on past experience, while bottom-up data provides details about the current expression. Recursive Realism suggests that perception is a recursive interplay between expectation and observation, where internal models are adjusted through feedback from the sensory world. This perspective allows for a deeper understanding of how perception is an active construction of reality, where the mind continually refines its understanding of the world.

The brain integrates information from multiple senses, such as vision, hearing, touch, and proprioception, to create a unified experience of the environment. This process is inherently recursive, as sensory inputs are continually cross-referenced to verify and clarify perceptions. For example, when catching a ball, the brain uses visual information to track the object's movement, while proprioceptive feedback helps to position the hand accurately. Recursive Realism suggests that multisensory integration is a self-organising process, where sensory channels create a cohesive model of reality through feedback loops. This perspective highlights how perception is not a passive reception of stimuli but an active synthesis of information, where sensory experiences are woven together into a seamless whole.

Attention is the ability to focus on specific aspects of the environment, filtering out irrelevant details to create a clear representation of what matters. This process is recursive, as the mind monitors its focus and shifts attention based on goals, expectations, and sensory inputs. For example, while reading a book, a person might become distracted by a sudden noise but then redirect their focus back to the text. Recursive Realism suggests that attention is a feedback loop between cognitive goals and environmental inputs, where the mind continuously adjusts its focus to maintain a stable awareness. This perspective allows for a deeper understanding of how attention shapes the flow of experience, creating a dynamic balance between concentration and openness to new stimuli.

Cognition is shaped by recursive interactions between neurons, circuits, and sensory systems, creating a self-organising network that allows for learning, memory, and perception. Recursive Realism provides a framework for understanding the brain as a dynamic system, where thought and awareness emerge from feedback loops that integrate experience with internal models. By recognising the recursive nature of neural processes, we gain insight into how the mind creates a coherent experience of the world, where complexity arises from the continuous interplay between internal dynamics and external reality.

Recursive Patterns in Emotion, Thought, and Conscious Awareness

Emotions and thoughts are often seen as separate domains, with emotion linked to instinctual responses and thought associated with rational analysis. However, Recursive Realism suggests that emotion and cognition are deeply intertwined through recursive feedback loops, where feelings shape thoughts, and thoughts modify feelings. This section delves into how recursion in the brain creates a dynamic interplay between emotional states, reflective thinking, and the development of self-awareness.

The Recursion of Emotion and Cognitive Appraisal

Emotional experiences are often the result of cognitive appraisals, where the mind evaluates a situation and assigns it emotional significance. Recursive Realism suggests that emotion is not merely a reaction to external events but a recursive process where feelings and thoughts continually influence each other, creating a complex emotional landscape.

The amygdala, a subcortical structure, plays a key role in processing emotions such as fear, anger, and pleasure, while the prefrontal cortex is involved in cognitive control and rational thinking. These regions are connected through neural pathways that create feedback loops, allowing for emotional regulation and contextual interpretation. For example, a person who feels anxious before a public speech might use the prefrontal cortex to reframe the situation as an opportunity for growth, thereby reducing anxiety. Recursive Realism suggests that the interaction between the amygdala and prefrontal cortex is a self-regulating loop, where emotional intensity is modulated through cognitive reflection. This perspective highlights how emotion and thought are not opposing forces but interconnected processes that shape human experience.

Emotional memories are particularly salient, as they are encoded with feelings that create a stronger imprint on the brain's networks. This process is recursive, as recalling an emotionally charged event can reactivate the original feelings, leading to a reinforcement of the memory trace. For example, remembering a joyful moment from childhood might evoke a sense of warmth, while recalling a traumatic event could trigger anxiety. Recursive Realism suggests that emotional memory operates through a loop between recall and emotional response, where each act of remembering deepens the association between thought and feeling. This perspective allows for a deeper understanding of how memories shape the emotional landscape, creating patterns of response that persist over time.

Emotions serve as motivational signals, guiding behaviour by providing feedback on goals and values. This process is inherently recursive, as emotions influence decisions, and decisions in turn shape emotions. For example, the anticipation of reward might motivate persistence in a challenging task, while disappointment could prompt a re-evaluation of goals. Recursive Realism suggests that emotion is a dynamic system, where internal states and external outcomes create feedback loops that guide motivation and adaptation. This perspective highlights how emotions are not just reactions but adaptive processes that integrate cognitive evaluations with physiological responses, allowing for goal-directed behaviour.

Self-Reflection, Metacognition, and the Emergence of Self-Awareness

Self-awareness is the ability to reflect on one's own thoughts and experiences, creating a sense of identity and a continuous narrative of self. Recursive Realism suggests that self-awareness emerges from recursive processes in the brain, where thoughts turn inward to examine themselves, creating a loop that allows for introspection and self-discovery.

The default mode network (DMN) is a network of brain regions that becomes active during resting states, daydreaming, and self-reflection. This network is involved in thinking about the self, remembering the past, and imagining the future. Recursive Realism suggests that the DMN functions as a self-referential loop, where the mind explores its own thoughts, creating a narrative of identity that integrates past experiences with future goals. For example, during quiet moments, a person might reflect on their values and aspirations, creating a cohesive sense of who they are. This perspective highlights how self-awareness is a recursive process, where the mind's focus turns inward to create a sense of continuity across time.

Metacognition involves thinking about one's own thinking, allowing for self-monitoring and strategic adjustment of mental processes. This ability is deeply recursive, as the mind evaluates its own cognitive strategies, adjusts them, and monitors their effectiveness. For example, a student preparing for an exam might assess which study methods are most effective and change their approach accordingly. Recursive Realism suggests that metacognition is a self-regulating process, where thought becomes a subject of thought, creating a feedback loop that enhances learning and problem-solving. This perspective allows for a more sophisticated understanding of how cognition can adapt to new challenges, where self-reflection guides mental flexibility.

Mirror neurons are a type of neuron that fire both when an individual performs an action and when they observe someone else performing the same action. This neural mirroring creates a basis for empathy and social understanding, allowing individuals to simulate the experiences of others. Recursive Realism suggests that the mirror neuron system is a self-referential mechanism, where the mind creates a virtual representation of another's state through internal reflection. For example, observing a friend in distress might trigger similar neural patterns in one's own brain, leading to a shared sense of sadness. This perspective highlights how empathy is not just emotional resonance but a recursive simulation of others' experiences, creating a bridge between self-awareness and social connection.

Consciousness and the Recursive Integration of Experience

Consciousness is the state of being aware of one's own thoughts, feelings, and surroundings, creating a subjective experience that is both rich and multi-dimensional. Recursive Realism suggests that consciousness is shaped by recursive loops within the brain, where sensory inputs, internal reflections, and cognitive processes are integrated into a coherent stream of awareness.

The binding problem refers to the question of how the brain integrates information from different sensory modalities, such as sight, sound, and touch, into a unified perception. Recursive Realism suggests that neural synchrony, the coordinated firing of neurons across different regions, plays a key role in binding sensory experiences into a coherent whole. For example, when perceiving a moving object, the brain must integrate visual data with motion cues and spatial awareness to create a continuous perception of the object's trajectory. This perspective allows for a deeper understanding of how conscious experience is not the sum of isolated sensations but a self-organising process, where neural networks create feedback loops that unify experience.

Attention is central to conscious experience, as it determines which aspects of the environment are highlighted and which are filtered out. This process is recursive, as attention shifts based on internal priorities and external changes, creating a dynamic balance between focused awareness and background perception. For example, when listening to music, a person might focus on the melody while tuning out the ambient sounds of a crowded room. Recursive Realism suggests that attention operates through recursive feedback loops, where consciousness continually re-evaluates what is most relevant in the moment. This perspective highlights how consciousness is not a passive state but an active process of selective focus, where awareness adapts to changing contexts.

Self-consciousness involves a heightened awareness of oneself as an individual, creating a sense of self that persists across time. This state is inherently recursive, as it involves reflecting on one's own thoughts, actions, and existence. For example, self-conscious moments, such as feeling embarrassed or proud, involve an awareness of how others might perceive us, creating a recursive loop between internal states and external perspectives. Recursive Realism suggests that self-consciousness is a self-referential process, where the mind creates a model of itself and adjusts it through ongoing reflection. This perspective allows us to understand how the sense of identity emerges from recursive interactions, where the self is continually redefined through experience and self-reflection.

Emotions, thoughts, and conscious awareness are shaped by recursive feedback loops within the brain, creating a dynamic interplay between feeling, reflection, and self-perception. Recursive Realism offers a framework for understanding how cognition and emotion are interconnected, revealing how the mind creates a coherent self through recursive patterns of self-reflection and neural integration. By recognising the recursive nature of mental processes, we gain insight into the depth of human experience, where consciousness is not a static state but an evolving process of awareness and understanding.

In this chapter, we have explored how Recursive Realism provides a framework for understanding the architecture of the mind, revealing how neural recursion shapes cognition, emotion, and consciousness. By examining the feedback loops that integrate sensory data, emotional responses, and self-awareness, Recursive Realism allows us to see the mind as a self-organising system, where complexity arises from the recursive interplay between brain structures and experience. This perspective offers a deeper understanding of how the mind creates a coherent sense of self and reality, where awareness is a dynamic process that evolves through reflection and adaptation.

References

Baddeley, A.D. and Hitch, G., 1974. Working memory. *Psychology of Learning and Motivation*, 8, pp.47–89.

Barrett, L.F., 2017 *How Emotions Are Made: The Secret Life of the Brain*. Boston: Houghton Mifflin Harcourt.

Buckner, R.L. and Krienen, F.M., 2013. The evolution of distributed association networks in the human brain. *Trends in Cognitive Sciences*, 17(12), pp.648–665.

Chalmers, D.J., 1996. *The Conscious Mind: In Search of a Fundamental Theory*. Oxford: Oxford University Press.

Damasio, A., 1999. *The Feeling of What Happens: Body and Emotion in the Making of Consciousness*. New York: Harcourt Brace.

Dennett, D.C., 1991. *Consciousness Explained*. Boston: Little, Brown.

Edelman, G.M., 1989. *The Remembered Present: A Biological Theory of Consciousness*. New York: Basic Books.

Friston, K., 2010. The free-energy principle: A unified brain theory? *Nature Reviews Neuroscience*, 11(2), pp.127–138.

Gallese, V., Keysers, C., and Rizzolatti, G., 2004. A unifying view of the basis of social cognition. *Trends in Cognitive Sciences*, 8(9), pp.396–403.

Hebb, D.O., 1949. *The Organization of Behavior: A Neuropsychological Theory*. New York: Wiley.

Kandel, E.R., 2006. *In Search of Memory: The Emergence of a New Science of Mind*. New York: W.W. Norton & Company.

LeDoux, J., 1996. *The Emotional Brain: The Mysterious Underpinnings of Emotional Life*. New York: Simon & Schuster.

Logothetis, N.K., 2008. What we can do and what we cannot do with fMRI. *Nature*, 453(7197), pp.869–878.

Miller, E.K. and Cohen, J.D., 2001. An integrative theory of prefrontal cortex function. *Annual Review of Neuroscience*, 24, pp.167–202.

Pinker, S., 1997. *How the Mind Works*. New York: W.W. Norton & Company.

Raichle, M.E., 2015. The brain's default mode network. *Annual Review of Neuroscience*, 38, pp.433–447.

Rizzolatti, G. and Sinigaglia, C., 2008. *Mirrors in the Brain: How Our Minds Share Actions and Emotions*. Oxford: Oxford University Press.

Rolls, E.T., 2013. *Emotion and Decision Making Explained*. Oxford: Oxford University Press.

Tononi, G. and Koch, C., 2015. Consciousness: Here, there and everywhere? *Philosophical Transactions of the Royal Society B: Biological Sciences*, 370(1668), pp.1–15.

Varela, F.J., Thompson, E., and Rosch, E., 1991. *The Embodied Mind: Cognitive Science and Human Experience*. Cambridge, MA: MIT Press.

Chapter 18: Recursive Realism and the Patterns of Social Interaction

Recursion in Communication and Social Dynamics

Human communication is inherently recursive, involving feedback loops where messages are exchanged, interpreted, and responded to, creating a shared understanding that evolves through interaction. Recursive Realism suggests that social dynamics are shaped by recursive processes, where individual actions influence group behaviour, and group norms in turn shape individual responses. This section explores how recursion manifests in language, cultural evolution, and the formation of social networks, revealing the self-organising nature of human societies.

Language as a Recursive System

Language is one of the most complex tools of human cognition, allowing for the expression of thoughts, emotions, and ideas through symbols and rules. Recursive Realism suggests that language is a recursive system, where grammar, syntax, and meaning are shaped by feedback loops between speakers and listeners, creating a shared framework for communication.

Grammar and syntax allow for the construction of complex sentences, where phrases can be nested within other phrases to create layers of meaning. For example, in the sentence, "The man who wrote the book was praised," the relative clause "who wrote the book" is embedded within the larger sentence, creating a recursive structure. Recursive Realism suggests that language is fundamentally recursive, as it allows for the recombination of elements into new forms, enabling infinite expression from a finite set of rules. This perspective allows for a deeper understanding of how thought and communication are intertwined through linguistic recursion, where words and concepts are continually reconfigured to convey complex ideas.

Pragmatics refers to the use of language in social contexts, where meaning is shaped by cultural norms, intentions, and interpretations. This process is inherently recursive, as speakers adjust their language based on the responses of listeners, creating a dynamic dialogue. For example, during a conversation, a speaker might clarify or rephrase a statement based on the listener's reaction, creating a feedback loop that enhances mutual understanding. Recursive Realism suggests that pragmatic communication is a self-regulating process, where social cues guide the refinement of language use. This perspective highlights how meaning is not fixed but emerges through the interaction of speakers and listeners, creating a shared context through recursive exchange.

Metaphors allow for the extension of meaning by linking concepts from different domains, creating new ways of understanding experience. This process is recursive, as the interpretation of a metaphor involves revisiting and reinterpreting familiar ideas in novel contexts. For example, the metaphor "Time is a river" invites a comparison between the flow of time and the movement of water, creating a layered understanding of temporal experience. Recursive

Realism suggests that metaphor is a self-referential mechanism, where language draws on existing patterns to generate new insights, creating a recursive loop between conceptual domains. This perspective allows for a richer appreciation of how language can expand consciousness, where metaphorical thinking reveals the depths of human creativity.

Cultural Evolution and the Recursion of Norms

Culture is a dynamic system that evolves through interactions between individuals, traditions, and social structures. Recursive Realism suggests that cultural evolution is shaped by recursive processes, where shared beliefs and practices are transmitted, modified, and reintegrated into the collective consciousness, creating a self-organising system that adapts over time.

Memes, units of cultural information that are transmitted from person to person, play a role in shaping cultural norms through imitation and adaptation. This process is inherently recursive, as memes are modified through reinterpretation and adaptation before being passed on. For example, a popular song might inspire remixes and parodies, each of which reshapes the original in a new context. Recursive Realism suggests that memes operate as cultural feedback loops, where ideas evolve through cycles of imitation and innovation, creating a dynamic tapestry of shared meaning. This perspective allows for a deeper understanding of how culture is not static but a living system, where traditions and innovations are interwoven through reciprocal interactions.

Social norms, the shared rules that guide behaviour in a community, are shaped through reciprocal interactions, where individual actions influence group expectations, and group norms shape individual behaviour. This process is recursive, as norms evolve in response to feedback from individuals who accept, challenge, or reinterpret them. For example, shifts in social attitudes toward gender roles have emerged through ongoing dialogue between individual advocates and cultural institutions, leading to changes in laws, media representations, and everyday practices. Recursive Realism suggests that social norms are self-organising patterns, where collective expectations adapt through feedback loops between personal choices and cultural frameworks. This perspective highlights how cultural change is not merely top-down but involves a recursive interplay between individual agency and collective identity.

Rituals play a central role in cultural life, providing symbolic actions that reinforce and transmit meaning across generations. This process is recursive, as rituals are re-enacted with each performance, creating a continuity of meaning that connects the past, present, and future. For example, a wedding ceremony involves repeating symbolic gestures that represent commitment and union, creating a shared experience that is both timeless and personalised. Recursive Realism suggests that rituals are self-referential practices, where the meaning of each act is deepened through repetition and reinterpretation. This perspective allows for a richer understanding of how culture creates a sense of continuity, where rituals serve as anchors that integrate personal experience with collective identity.

Social Networks and the Dynamics of Connectivity

Social networks, the webs of relationships that connect individuals, are fundamental to the structure of human societies, shaping how information, influence, and resources flow within a community. Recursive Realism suggests that social networks operate through recursive interactions, where individual connections create emergent patterns that influence collective behaviour.

Network theory studies how nodes (individuals) are connected through edges (relationships), revealing the structure of social systems. This connectivity is inherently recursive, as relationships influence networks, and networks in turn shape individual relationships. For example, influencers on social media can amplify trends, creating feedback loops that affect the opinions and behaviours of followers. Recursive Realism suggests that social networks are self-organising systems, where patterns of influence emerge through reciprocal interactions between individuals and groups. This perspective allows for a deeper understanding of how ideas, values, and behaviours spread through communities, creating dynamic networks of shared meaning.

Reciprocity, the exchange of favourable actions between individuals, is a key mechanism for building trust and cooperation in social groups. This process is recursive, as cooperative behaviours are reinforced through mutual benefit and reciprocal exchanges. For example, in small communities, acts of generosity can create a culture of sharing, where individuals are motivated to help others because they anticipate reciprocity. Recursive Realism suggests that cooperation evolves through feedback loops, where positive interactions create a self-reinforcing cycle that strengthens social bonds. This perspective highlights how societies can maintain cohesion through mutual respect and reciprocal relationships, creating a sense of solidarity that is both adaptive and resilient.

Leadership in social networks often emerges through a process where individuals gain influence based on their ability to articulate shared values, organise collective efforts, or innovate solutions. This process is recursive, as leaders shape the behaviour of followers, and followers in turn shape the leader's role through feedback and support. For example, a social movement leader might gain influence by voicing concerns that resonate with community members, creating a reciprocal relationship where collective action amplifies the leader's impact. Recursive Realism suggests that leadership is a self-organising phenomenon, where social influence is distributed through feedback loops that connect individual charisma with collective aspirations. This perspective allows for a richer understanding of how leadership evolves in dynamic environments, where reciprocal interactions shape the direction of collective efforts.

Social interactions are shaped by recursive processes of communication, cultural evolution, and network connectivity, creating a dynamic web of relationships that adapts over time. Recursive Realism offers a framework for understanding how language, norms, and social structures emerge through reciprocal feedback loops, revealing how individual actions influence collective behaviour, and group dynamics shape personal experiences. By recognising the recursive nature of social life, we gain insight into how societies evolve through self-organising patterns, where communication and connection create a shared reality.

Recursive Realism and the Evolution of Collective Consciousness

The concept of collective consciousness refers to the shared beliefs, values, and attitudes that shape how a community or society perceives the world. Recursive Realism suggests that collective consciousness is formed through recursive feedback loops between individual thought and group norms, where ideas are amplified, challenged, and integrated into a cohesive worldview. This section explores how recursion drives the evolution of cultural identity, the spread of social movements, and the formation of societal values, offering insights into the dynamic relationship between individuals and societies.

The Formation of Collective Beliefs and Ideologies

Beliefs and ideologies are central to the identity of communities, shaping how societies understand their place in the world and how they organise their values. Recursive Realism suggests that the formation of collective beliefs is shaped by recursive interactions, where individual perspectives are synthesised into shared narratives that create a cohesive social identity.

Echo chambers occur when individuals surround themselves with like-minded perspectives, creating a feedback loop where beliefs are reinforced without external challenge. This process is inherently recursive, as confirmation bias drives the repetition and amplification of specific ideas within a group. For example, online communities that share a common political ideology might amplify their views through discussion forums, creating a shared belief system that becomes increasingly rigid. Recursive Realism suggests that echo chambers reflect the self-reinforcing nature of collective thought, where ideas gain momentum through social validation. This perspective allows for a deeper understanding of how collective beliefs can become entrenched, creating cultural divides that are resistant to new information.

Myths play a central role in shaping cultural identity, offering stories that explain the origins, values, and aspirations of a society. This process is recursive, as myths are reinterpreted and retold across generations, creating a narrative loop that connects the past to the present. For example, the mythology of nationhood often involves stories of heroes, struggles, and founding moments that are passed down through literature, rituals, and monuments, creating a shared sense of identity. Recursive Realism suggests that myth-making is a self-referential process, where stories evolve through dialogue between tradition and innovation, creating a dynamic narrative that adapts to changing contexts. This perspective highlights how cultural identity is not static but a living narrative that evolves through collective reflection and reinterpretation.

Ideological movements, whether political, religious, or social, often arise when shared ideas coalesce into a cohesive vision that seeks to reshape society. This process is recursive, as individuals are inspired by the movement's narrative, while the movement is shaped by the actions and expressions of its followers. For example, the civil rights movement in the United States involved a reciprocal dialogue between leaders who articulated a vision of equality and communities who organised protests, sit-ins, and advocacy efforts to advance that vision. Recursive Realism suggests that movements are self-organising systems, where narratives evolve through feedback loops between collective ideals and individual actions. This perspective allows for a richer understanding of how social change is driven by reciprocal dynamics, where ideas become embodied in actions, and actions reshape ideas.

The Dynamics of Social Movements and Collective Action

Social movements involve coordinated efforts by groups to address shared concerns and advocate for change. Recursive Realism suggests that social movements are shaped by recursive feedback loops, where collective action and shared goals are reinforced through dialogue, protests, and community organising.

Symbolic actions, such as protests, marches, and public demonstrations, play a key role in social movements, serving as visible expressions of shared grievances and demands for change. These actions are recursive, as they amplify the movement's message while also energising participants through a sense of solidarity. For example, the Women's March in 2017 involved millions of participants around the world, creating a powerful statement of resistance that was amplified through media coverage and social media. Recursive Realism suggests that symbolic actions are self-reinforcing loops, where collective energy and public visibility create a cycle of empowerment that sustains the movement's momentum. This perspective allows for a deeper understanding of how movements use symbolic actions to create visibility and influence, where reciprocal interactions between action and awareness drive social change.

Collective identity is the sense of shared identity that binds individuals together in a movement, creating a feeling of belonging to a larger cause. This identity is recursive, as it is shaped by group interactions and reflected in the symbols, slogans, and rituals that define the movement's culture. For example, the LGBTQ+ rights movement has developed a shared identity through symbols like the rainbow flag, pride parades, and inclusive language, creating a space where individuals feel connected to a broader community. Recursive Realism suggests that collective identity is a self-referential process, where individual experiences are integrated into a larger narrative that creates a sense of purpose and solidarity. This perspective highlights how movements can sustain themselves through collective identity, where shared values create a resilient foundation for organising and advocacy.

Decentralised movements, such as the Occupy movement or environmental activism, often operate through networked structures where local actions are connected through digital platforms and grassroots organising. This structure is recursive, as local initiatives influence the movement's overall direction, while the movement's ethos shapes the approaches taken by local groups. For example, the climate change movement involves activists in different countries who coordinate actions through social media, creating a global network that adapts to regional contexts. Recursive Realism suggests that decentralised movements are self-organising systems, where actions and ideas flow through reciprocal feedback loops that create a collective sense of urgency and purpose. This perspective allows for a richer understanding of how movements can mobilise without central leadership, where networked recursion creates a fluid structure that is adaptive and resilient.

Collective Consciousness and the Emergence of Social Values

Social values are the principles and norms that guide behaviour within a community, shaping how individuals interact with each other and with the world. Recursive Realism suggests that social values emerge through recursive processes, where cultural narratives, ethical debates, and shared experiences create a framework for understanding what is right, just, and important.

Narrative ethics suggests that stories play a central role in shaping moral understanding, as they offer examples and parables that illustrate ethical dilemmas and values. This process is recursive, as stories are interpreted and reinterpreted through discussion, reflection, and cultural shifts. For example, literature such as George Orwell's *1984* or Harper Lee's *To Kill a Mockingbird* has shaped generations' understanding of freedom and justice, creating shared reference points for debating social issues. Recursive Realism suggests that narrative ethics operates through a loop between storytelling and moral reflection, where individuals and communities engage in ongoing dialogue about what values are most meaningful. This perspective allows for a deeper appreciation of how societies use stories to transmit and evolve ethical principles, creating a shared language for moral discourse.

Social Movements as Laboratories of New Values: Social movements often serve as laboratories for the development of new values, challenging established norms and proposing alternative visions for society. This process is recursive, as movements engage in internal debates about strategy and ethics, while external pressure and public reaction shape the movement's evolution. For example, the feminist movement has continually revised its goals and strategies in response to shifts in society, from suffrage to workplace equality to intersectional feminism. Recursive Realism suggests that social movements are dynamic systems, where new values emerge through dialogue and experiment, creating feedback loops that transform society over time. This perspective highlights how movements can innovate in the realm of values, offering alternative frameworks for understanding justice, equality, and freedom.

Societies often struggle to balance respect for tradition with the need for change, creating a recursive tension between continuity and innovation. This tension is resolved through recursive dialogue, where cultural practices are reinterpreted in light of new realities. For example, religious communities might adapt doctrines to address modern issues like climate change or technology, creating a synthesis that respects tradition while embracing relevance. Recursive Realism suggests that cultural evolution is a self-referential process, where societies continuously reintegrate old ideas into new contexts, creating a dynamic balance between roots and branches. This perspective allows for a richer understanding of how societies evolve, where tradition and innovation are woven together through reciprocal reflection.

Collective consciousness emerges through recursive processes that integrate individual thoughts and shared narratives, creating a cohesive social identity that evolves over time. Recursive Realism offers a framework for understanding how beliefs, movements, and social values are shaped by feedback loops between individual experiences and cultural frameworks, revealing how societies adapt through self-organising interactions. By recognising the recursive nature of collective thought, we gain insight into how movements inspire change, how values are redefined, and how societies continually reinvent themselves through reflection and dialogue.

In this chapter, we have explored how Recursive Realism offers a framework for understanding the patterns of social interaction, revealing how communication, collective action, and cultural evolution are shaped by recursive feedback loops. By examining the self-organising dynamics of language, social movements, and shared beliefs, Recursive Realism allows us to see societies as dynamic systems, where individual agency and collective identity are intertwined through reciprocal interactions. This perspective provides a deeper understanding of how social change is not a linear process but a complex dance between tradition and innovation, where collective consciousness evolves through self-reflection and adaptation.

References

Aarts, H. and Dijksterhuis, A., 2003. The silence of the library: Environment, situational norm, and social behavior. *Journal of Personality and Social Psychology*, 84(1), pp.18–28.

Anderson, B., 2006. *Imagined Communities: Reflections on the Origin and Spread of Nationalism*. Revised edition. London: Verso Books.

Bakshy, E., Messing, S., and Adamic, L.A., 2015. Exposure to ideologically diverse news and opinion on Facebook. *Science*, 348(6239), pp.1130–1132.

Bandura, A., 1977. Self-efficacy: Toward a unifying theory of behavioral change. *Psychological Review*, 84(2), pp.191–215.

Barabási, A.-L., 2002. *Linked: How Everything Is Connected to Everything Else and What It Means for Business, Science, and Everyday Life*. New York: Basic Books.

Boyd, D., 2014. *It's Complicated: The Social Lives of Networked Teens*. New Haven: Yale University Press.

Cialdini, R.B. and Goldstein, N.J., 2004. Social influence: Compliance and conformity. *Annual Review of Psychology*, 55, pp.591–621.

Dawkins, R., 1976. *The Selfish Gene*. Oxford: Oxford University Press.

Eckert, P. and McConnell-Ginet, S., 2003. *Language and Gender*. Cambridge: Cambridge University Press.

Foucault, M., 1977. *Discipline and Punish: The Birth of the Prison*. Translated by A. Sheridan. New York: Pantheon Books.

Gladwell, M., 2000. *The Tipping Point: How Little Things Can Make a Big Difference*. Boston: Little, Brown.

Granovetter, M.S., 1973. The strength of weak ties. *American Journal of Sociology*, 78(6), pp.1360–1380.

Habermas, J., 1984. *The Theory of Communicative Action*. Translated by T. McCarthy. Boston: Beacon Press.

Lakoff, G. and Johnson, M., 1980. *Metaphors We Live By*. Chicago: University of Chicago Press.

Putnam, R.D., 2000. *Bowling Alone: The Collapse and Revival of American Community*. New York: Simon & Schuster.

Rheingold, H., 2002. *Smart Mobs: The Next Social Revolution*. Cambridge, MA: Basic Books.

Scott, J., 2017. *Social Network Analysis*. 4th edition. London: SAGE Publications.

Shirky, C., 2008. *Here Comes Everybody: The Power of Organizing Without Organizations*. New York: Penguin Books.

Tomasello, M., 2008. *Origins of Human Communication*. Cambridge, MA: MIT Press.

Watts, D.J., 2004. *Six Degrees: The Science of a Connected Age*. New York: W.W. Norton & Company.

Chapter 19: Recursive Realism and the Philosophy of Time

Time as a Recursive Structure in Mind and Nature

Time has long been a central enigma in philosophy, science, and consciousness studies, with differing views on whether it is a fundamental aspect of the universe or a construct of the human mind. Recursive Realism suggests that time is best understood as a recursive structure, where the flow of moments is shaped through feedback loops that connect past, present, and future into a cohesive experience. This section explores how recursion manifests in the perception of time, the nature of memory, and the physical universe, revealing a layered understanding of temporal reality.

The Perception of Time and the Recursive Mind

Human perception of time is not a neutral recording of events but a dynamic process that is shaped by attention, memory, and expectation. Recursive Realism suggests that time perception involves self-referential loops, where the mind integrates past experiences with present awareness and future anticipations to create a continuous sense of time.

The brain integrates sensory information into a coherent temporal sequence, creating the illusion of a smooth flow of time. This process is inherently recursive, as temporal binding involves linking moments together through memory traces and anticipatory models. For example, when listening to a melody, the mind connects each note to the one before it, creating a sense of rhythm that unifies the sequence. Recursive Realism suggests that the sense of continuity is a self-referential construct, where past and present are woven into a temporal thread through feedback loops that maintain awareness of duration and change. This perspective allows for a deeper understanding of how consciousness creates a sense of time, where the present moment is always contextualised within a larger narrative.

Memory as a Time-Binding Mechanism: Memory plays a key role in shaping our experience of time, allowing us to recall past events and anticipate future possibilities. This process is recursive, as memories are not just stored but are continually reinterpreted and integrated into current thoughts and expectations. For example, the act of remembering a birthday might involve layers of emotion, imagery, and meaning that shift with each recollection, creating a dynamic narrative of the past. Recursive Realism suggests that memory functions as a time-binding mechanism, where the mind creates loops between recalled experiences and present states, allowing for a sense of personal history and identity. This perspective highlights how memory is not a passive archive but an active process that shapes our understanding of time, creating a continuity between who we were and who we are becoming.

Anticipation involves projecting current knowledge into the future, allowing us to plan, hope, and prepare for upcoming events. This process is recursive, as the mind draws on past experiences to predict what might happen next, while new experiences continuously reshape our expectations. For example, a weather forecast combines historical data with real-time observations to predict conditions, while each forecast is adjusted based on new patterns that emerge. Recursive Realism suggests that anticipation is a self-referential process, where future possibilities are constructed through feedback loops between memory and perception. This perspective allows for a richer understanding of how time is experienced as a flow, where the future is not fixed but shaped by reciprocal interactions between imagination and reality.

Time in the Physical Universe: Recursion and Entropy

In physics, time is often understood through the concept of entropy, where the arrow of time points toward increasing disorder. Recursive Realism suggests that time's nature in the physical universe is shaped by recursive processes, where energy exchanges, information flows, and feedback mechanisms create the temporal patterns that govern cosmic evolution.

Thermodynamics describes how energy is transferred and transformed within systems, with the second law of thermodynamics stating that entropy tends to increase in isolated systems. This irreversibility gives rise to the arrow of time, where events move from ordered states to disordered states. Recursive Realism suggests that entropy itself involves recursive interactions, where local decreases in disorder (such as the formation of stars or biological life) are balanced by increases in entropy elsewhere in the universe. For example, the energy output of the sun drives photosynthesis on Earth, creating local order while contributing to the overall entropy of the universe. This perspective allows for a more nuanced understanding of time's arrow, where entropy is not just a linear progression but a complex interplay of energy flows that create patterns and structures within the cosmos.

Many cosmological models suggest that the universe may undergo cycles of expansion and contraction, where big bangs and big crunches create recurring epochs of cosmic evolution. This idea introduces a recursive aspect to cosmic time, where the universe's life cycle might be seen as a self-repeating process that operates on vast timescales. For example, some theories propose that black holes could eventually lead to new big bangs, creating a loop of destruction and creation that extends beyond a single lifetime of the universe. Recursive Realism suggests that cosmic cycles represent a recursive structure in time's fabric, where the universe itself may be part of a larger pattern that repeats through feedback loops. This perspective allows for a deeper appreciation of how time's nature might be cyclical, where beginnings and endings are interconnected within a cosmic dance.

In quantum mechanics, time is not absolute but is linked to probabilities and wave functions, where particles exist in superposition until measured. This introduces a recursive relationship between observation and time, where the act of measurement influences the temporal state of a particle. For example, the quantum phenomenon known as the delayed-choice experiment suggests that decisions made in the present can influence the past behaviour of particles, creating a feedback loop that challenges linear time. Recursive Realism suggests that quantum time is a self-referential construct, where the boundaries between past, present, and future become fluid through reciprocal interactions between observer and observed. This perspective offers a more integrated understanding of how time's nature is multi-dimensional, where probability and consciousness shape the experience of temporal events.

Time, Causality, and the Recursive Universe

Causality, the principle that causes lead to effects, is central to our understanding of time's flow. Recursive Realism suggests that causality is not a simple chain of events but a recursive relationship, where effects can loop back to influence their causes, creating a dynamic interplay that shapes the structure of reality.

In complex systems, feedback loops create non-linear relationships between causes and effects, where the outcome of an action can influence its own conditions. For example, in ecological systems, predator-prey dynamics involve reciprocal interactions, where changes in the population of one species affect the other, creating a cyclical pattern of growth and decline. Recursive Realism suggests that causality is shaped by feedback loops, where causes and effects create a self-regulating system that adapts over time. This perspective allows for a more sophisticated understanding of causality, where time is not just a linear flow but a network of interactions that shape outcomes through reciprocal influence.

Time travel, a concept often explored in science fiction, introduces paradoxes such as the grandfather paradox, where changes made in the past could alter the conditions that led to the time traveller's journey. These paradoxes reveal the recursive nature of time, where future actions loop back to influence the past. While physical time travel remains speculative, Recursive Realism suggests that mental time travel, the ability to project oneself into past experiences or future possibilities, operates through self-referential loops that shape our understanding of causality. This perspective allows for a richer exploration of how time can be experienced as a recursive dimension, where the boundaries between cause and effect are fluid and interconnected.

In the physical universe, causality creates the conditions for self-organisation, where simple interactions give rise to complex structures over time. This process is recursive, as each new order creates conditions that allow for further complexity. For example, in the formation of galaxies, gravitational interactions between particles lead to the clustering of stars, creating a feedback loop where gravity pulls matter into new configurations. Recursive Realism suggests that order and complexity emerge through recursive causality, where patterns evolve through cycles of interaction and adaptation. This perspective highlights how time's role in the universe is not just about entropy but also about the emergence of patterns that create structure and cohesion.

Time is a multi-layered phenomenon that is shaped by recursive feedback loops, connecting past, present, and future in a dynamic interplay. Recursive Realism provides a framework for understanding time as both a mental construct and a physical process, where memory, anticipation, and causality create a cohesive sense of temporal reality. By recognising the recursive nature of time, we gain insight into how consciousness and cosmic processes are intertwined, revealing a deeper understanding of how the flow of time shapes the universe and our experience within it.

Recursive Realism, Temporality, and the Human Experience

Time is not just a scientific phenomenon but a core element of human experience, shaping how we understand life, change, and mortality. Recursive Realism suggests that time is experienced through recursive interactions between cultural narratives, individual reflection, and existential questioning, creating a complex tapestry that integrates cosmic rhythms with personal stories. This section explores how Recursive Realism offers a new perspective on temporality, helping us navigate the philosophical dilemmas of life's transient nature and the enduring search for meaning.

Cultural Conceptions of Time and Recursive Narratives

Cultural understandings of time vary widely, reflecting different worldviews and philosophical traditions. Recursive Realism suggests that these conceptions of time are shaped through recursive storytelling, where myths, rituals, and historical narratives create a shared temporal framework that connects past experiences with future hopes.

Many ancient cultures, such as those of the Maya, Hinduism, and Buddhism, conceptualise time as cyclical, where events are believed to repeat in endless loops of birth, death, and rebirth. This view reflects a recursive understanding of existence, where life is seen as a cycle that renews itself through cycles of renewal. For example, the Mayan Long Count calendar envisions cosmic cycles that span thousands of years, with each era linked to mythological events that shape human destiny. Recursive Realism suggests that cyclical time represents a self-referential narrative, where past and future are not linear endpoints but interwoven stages in a continuous process. This perspective allows for a richer understanding of how cultural frameworks can offer a sense of continuity, where life's transience is balanced by the reassurance of eternal return.

Western cultures have traditionally adopted a linear view of time, influenced by Judeo-Christian traditions and Enlightenment thinking, where history is seen as a progressive movement toward goals such as salvation, knowledge, or social improvement. This narrative of progress is recursive in nature, as each generation builds upon the achievements and failures of those that came before, creating a feedback loop between past lessons and future aspirations. For example, the scientific revolution involved rethinking ancient ideas about nature, leading to a new understanding of the cosmos that shaped modernity. Recursive Realism suggests that the linear view of time reflects a dynamic relationship between memory and innovation, where historical progress emerges through reciprocal interactions between tradition and exploration. This perspective allows for a deeper appreciation of how cultural narratives can provide a sense of purpose, where time becomes a pathway toward self-realisation and societal change.

In the modern world, many societies experience a hybrid understanding of time, where traditional cycles coexist with linear progress, creating a complex interplay between heritage and innovation. This hybridity is inherently recursive, as individuals navigate multiple temporalities, from the ritual rhythms of religion to the demands of capitalist schedules. For example, a modern individual might observe seasonal festivals while also measuring life through career milestones and technological advancements. Recursive Realism suggests that temporal hybridity reflects the self-referential nature of contemporary identity, where cultural continuity is balanced with adaptation to new temporal frameworks. This perspective allows for a more fluid understanding of time, where individuals create meaning through reciprocal exchanges between past traditions and future-oriented goals.

Existential Time and the Dilemmas of Transience

Time's passage raises existential questions about mortality, change, and the meaning of life's brevity. Recursive Realism suggests that existential reflection on time is inherently recursive, as individuals turn inward to contemplate their lives, creating loops of self-questioning that shape the experience of being.

Awareness of death is a fundamental aspect of human consciousness, prompting reflection on the nature of life and the value of each moment. Recursive Realism suggests that contemplating mortality involves a recursive dialogue between life's transient nature and the search for enduring meaning. For example, the philosophy of existentialism, as seen in the works of Jean-Paul Sartre and Martin Heidegger, emphasises the awareness of death as a catalyst for authentic living, where the realisation of finite time prompts a commitment to personal values. This recursive reflection creates a loop between facing mortality and finding purpose, where the end of life's timeline deepens the significance of choices made in the present. Recursive Realism allows for a deeper understanding of how existential time is a dialogue between limits and possibilities, where the finite becomes a lens through which the infinite is glimpsed.

Philosophers have long grappled with the question of whether time offers a source of meaning or merely reveals life's inherent absurdity. Recursive Realism suggests that the search for meaning in time involves self-referential inquiry, where individuals revisit fundamental questions through different stages of life. For example, the mid-life crisis often involves re-evaluating goals and values as one reflects on the passage of time, creating a recursive loop of self-assessment that reshapes identity. Similarly, spiritual practices, such as meditation or pilgrimage, often involve reflecting on the relationship between time and eternity, seeking to transcend the limits of the temporal. Recursive Realism suggests that time's meaning emerges through a process of self-reflection, where existential questions are not answered once but are revisited as life unfolds. This perspective highlights how time is a mirror for the mind's inquiry, where each moment deepens the mystery of existence.

Concerns about legacy, the desire to leave a lasting impact, are central to how individuals navigate time's passage. Recursive Realism suggests that legacy is shaped by feedback loops between individual actions and their consequences, where decisions made in the present ripple into the future. For example, artists, scientists, and leaders often reflect on how their work might influence future generations, creating a recursive relationship between present efforts and imagined futures. This process of seeking legacy creates a dialogue between personal significance and collective memory, where time's flow is shaped by acts of creation and remembrance. Recursive Realism allows for a more nuanced understanding of how time connects the individual to the collective, where the desire for lasting impact becomes a driving force that shapes human aspirations and creativity.

Recursive Realism and the Integration of Temporal Perspectives

Recursive Realism provides a framework for integrating multiple temporal perspectives, allowing for a holistic view of how time shapes thought, culture, and existence. This section explores how recursion allows for reconciling different views of time, from the cosmic to the personal, offering a way to navigate the complex interplay between timeless truths and the unfolding moment.

Philosophical traditions such as Platonism and mysticism often seek a state of timelessness, where truths are believed to exist beyond time's flow. At the same time, existentialism and process philosophy emphasise the importance of becoming, where meaning is found in the unfolding of events. Recursive Realism suggests that the tension between timelessness and becoming is inherently recursive, as the mind engages in reflective loops that seek to understand the eternal while participating in the temporal. For example, spiritual practices that focus on transcendence often involve a recursive process of returning to the present moment, where the timeless is glimpsed through awareness of each breath. This perspective allows for a more integrated approach to time, where the search for eternal truths and the appreciation of change are seen as complementary aspects of human experience.

Historical Consciousness and the Recursive Interpretation of the Past: Historical consciousness involves the awareness of how the past shapes the present, creating a recursive relationship between memory and identity. Recursive Realism suggests that understanding history is a self-referential process, where societies continually reinterpret the past in light of current values and future goals. For example, historians and philosophers revisit events such as the French Revolution or the Holocaust to explore their impact on modern ideals of freedom and justice, creating a dialogue between past narratives and contemporary debates. This perspective highlights how time is a recursive process of reinterpretation, where each generation redefines its relationship to history through new lenses of understanding.

Mindfulness practices and contemplative traditions often emphasise the flexibility of awareness, where attention shifts between immediate experience and broader reflections on life's arc. Recursive Realism suggests that temporal fluidity involves a recursive process, where consciousness moves between focused attention and expanded awareness, creating a dynamic understanding of time. For example, a meditator might focus on the sensation of breath, while also being aware of the larger context of life's journey. This perspective allows for a richer appreciation of how time can be experienced as both momentary and expansive, where awareness shifts through self-reflective loops that integrate micro and macro perspectives.

Time is a profound element of human life, shaping our understanding of change, continuity, and meaning. Recursive Realism offers a framework for exploring how cultural narratives, existential reflection, and temporal awareness create a complex interplay between past, present, and future, revealing the recursive nature of how we experience time. By recognising the multi-layered nature of temporality, Recursive Realism allows us to appreciate the depths of time, where the unfolding of each moment is connected to the eternal patterns of existence.

In this chapter, we have explored how Recursive Realism reshapes our understanding of time, revealing how temporal experience is shaped by recursive feedback loops between mind, culture, and the cosmos. By examining the interplay between cyclical and linear time, existential reflection, and the search for timeless truths, Recursive Realism provides a deeper understanding of how time is not just a scientific concept but a core aspect of the human condition. This perspective offers a way to navigate the philosophical challenges of transience, meaning, and the infinite, where time is seen as a reciprocal dance between what is fleeting and what endures.

References

Barbour, J., 1999. *The End of Time: The Next Revolution in Physics*. Oxford: Oxford University Press.

Bergson, H., 1910. *Time and Free Will: An Essay on the Immediate Data of Consciousness*. Translated by F.L. Pogson. London: George Allen & Unwin.

Bohr, N., 1935. Can quantum-mechanical description of physical reality be considered complete? *Physical Review*, 48(8), pp.696–702.

Coveney, P. and Highfield, R., 1991. *The Arrow of Time: The Quest to Solve Science's Greatest Mystery*. London: Flamingo.

Einstein, A., 1916. *Relativity: The Special and the General Theory*. Translated by R.W. Lawson. London: Methuen & Co.

Gell-Mann, M., 1994. *The Quark and the Jaguar: Adventures in the Simple and the Complex*. New York: W.H. Freeman.

Gödel, K., 1949. An example of a new type of cosmological solutions of Einstein's field equations of gravitation. *Reviews of Modern Physics*, 21(3), pp.447–450.

Heidegger, M., 1962. *Being and Time*. Translated by J. Macquarrie and E. Robinson. New York: Harper & Row.

Husserl, E., 1964. *The Phenomenology of Internal Time-Consciousness*. Translated by J.S. Churchill. Bloomington: Indiana University Press.

Kuhn, T.S., 1962. *The Structure of Scientific Revolutions*. Chicago: University of Chicago Press.

Prigogine, I., 1997. *The End of Certainty: Time, Chaos, and the New Laws of Nature*. New York: Free Press.

Rovelli, C., 2018. *The Order of Time*. Translated by E. Segre and S. Carnell. London: Allen Lane.

Sartre, J.-P., 1943. *Being and Nothingness*. Translated by H.E. Barnes. New York: Philosophical Library.

Smolin, L., 2013. *Time Reborn: From the Crisis in Physics to the Future of the Universe*. Boston: Houghton Mifflin Harcourt.

Thompson, E., 2007. *Mind in Life: Biology, Phenomenology, and the Sciences of Mind*. Cambridge, MA: Harvard University Press.

Wheeler, J.A. and Ford, K., 1998. *Geons, Black Holes, and Quantum Foam: A Life in Physics*. New York: W.W. Norton & Company.

Zerubavel, E., 1981. *Hidden Rhythms: Schedules and Calendars in Social Life*. Chicago: University of Chicago Press.

Chapter 20: Recursive Realism and the Boundaries of Reality

Exploring the Edges of Knowledge through Recursive Thinking

The boundaries of reality are where knowledge meets mystery, where science grapples with paradoxes, and philosophy contemplates the unknowable. Recursive Realism suggests that recursion provides a unique lens through which to examine these edges, offering a way to understand how complex patterns emerge from self-referential processes and how reality itself might be a product of recursive structures. This section explores how Recursive Realism engages with cosmological questions, the nature of consciousness, and the possibilities of parallel realities.

The Multiverse and Recursive Cosmology

Cosmology, the study of the universe's origins, structure, and evolution, has expanded to include the possibility of a multiverse, where multiple universes coexist with their own laws of physics and distinct realities. Recursive Realism suggests that the multiverse can be understood through recursive principles, where universes may be nested within each other or interact through feedback loops that transcend traditional space-time.

Some multiverse theories, such as eternal inflation, propose that our universe is just one bubble in a vast foam of universes, each with its own physical constants and initial conditions. This concept introduces a recursive element to cosmology, where each universe is self-contained yet emerges from a larger structure that generates infinite variations. Recursive Realism suggests that bubble universes reflect a nested pattern, where cosmic creation operates through self-referential loops that give rise to new realities. This perspective allows for a deeper understanding of how the multiverse might function as a recursive system, where each bubble is a manifestation of a higher-order process that self-organises across scales. This raises the question of whether life and consciousness could exist in other recursive configurations, suggesting a plurality of realities that challenge human imagination.

The many-worlds interpretation of quantum mechanics suggests that every quantum decision leads to a branching of realities, creating a vast array of parallel worlds where all possible outcomes occur. This introduces a recursive structure to quantum reality, where each moment generates new universes in a self referential manner. For example, if an electron takes path A in one universe, it might take path B in another, with both paths coexisting as different branches of the same reality. Recursive Realism suggests that parallel worlds represent a recursive unfolding of possibilities, where each decision creates a loop that redefines the nature of reality. This perspective allows for a richer appreciation of how reality might be a multilayered tapestry, where every moment is a node in an infinite web of interconnected possibilities.

The holographic principle proposes that the information about a volume of space can be encoded on its boundary, suggesting that the universe might be a projection of lower-dimensional data. This concept introduces a recursive element to space-time, where reality is generated through self-referential encoding that shapes three-dimensional experience from two-dimensional information. For example, the surface area of a black hole contains information about its internal structure, creating a feedback loop between surface and volume. Recursive Realism suggests that the holographic nature of reality reflects a recursive structure, where space-time is woven through feedback loops that connect different dimensions. This perspective offers a deeper understanding of how the fabric of reality might be structured through recursive patterns, where boundaries and interiors are interdependent in ways that challenge classical intuition.

Consciousness and the Recursive Nature of Awareness

Consciousness, the subjective experience of awareness, remains one of the most profound mysteries in science and philosophy. Recursive Realism suggests that consciousness itself is shaped by recursive processes, where self-awareness arises from neural feedback loops that create a continuous sense of self. This section explores how Recursive Realism offers new insights into the nature of consciousness, the mind-body problem, and the possibility of artificial intelligence.

The hard problem of consciousness, coined by philosopher David Chalmers, concerns the question of why subjective experiences arise from physical processes in the brain. Recursive Realism suggests that self-awareness is a recursive loop, where the mind reflects upon its own thoughts, creating a feedback mechanism that gives rise to qualia, the raw feel of experience. For example, the feeling of pain involves not only a neural response to damage but also a self-referential awareness of suffering. This recursive model allows for a deeper understanding of how consciousness might be emergent from neural networks that self-reflect, where awareness is a byproduct of the brain's ability to process information about itself.

Panpsychism is the philosophical view that consciousness is a fundamental property of all matter, not just a feature of biological systems. This concept introduces a recursive element to the nature of awareness, suggesting that mind-like qualities might be distributed throughout the universe at different scales. Recursive Realism suggests that panpsychism can be understood as a self-similar pattern, where basic forms of awareness might coalesce into more complex consciousness through feedback loops. For example, a single cell might have a rudimentary sensitivity to its environment, while multicellular organisms develop coordinated senses that lead to higher-order perception. This perspective allows for a more integrated view of consciousness, where awareness is not a binary property but a continuum shaped by recursive interactions at all levels of reality.

The development of artificial intelligence (AI) raises questions about whether machines could ever achieve consciousness through recursive algorithms that mimic human thought. Recursive Realism suggests that true AI consciousness would require the ability to engage in self-referential loops, where a machine could reflect on its own states, intentions, and experiences. For example, an AI system that can monitor its own decision-making processes and adjust its strategies based on experience is engaging in a form of recursion. However, whether this could produce subjective awareness remains an open question. Recursive Realism offers a framework for exploring the conditions under which recursive algorithms might lead to conscious experience, highlighting the differences between computational recursion and the organic feedback loops of the human brain.

Metaphysical Questions and the Recursive Structure of Reality

Metaphysics explores the fundamental nature of reality, addressing questions that lie beyond the reach of empirical science. Recursive Realism suggests that metaphysical questions can be approached through self-referential thinking, where the boundaries of understanding are explored through recursive reflection on existence and being.

Ontology is the study of what exists and the nature of being. Recursive Realism suggests that being itself is a recursive phenomenon, where entities are defined through relationships that reflect back upon themselves. For example, the concept of self is inherently recursive, as awareness of one's existence requires a loop of self-reference. Similarly, mathematical structures such as fractals exhibit self-similarity across scales, where patterns repeat through nested iterations. This perspective allows for a deeper appreciation of how existence might be structured through reciprocal relationships, where being is not isolated but emerges through interactions with the whole.

Philosophers have long pondered the nature of nothingness, asking why there is something rather than nothing. Recursive Realism suggests that nothingness and existence are interrelated through self-referential paradoxes, where the concept of nothingness becomes meaningful only in contrast to somethingness. For example, mathematical concepts such as zero and infinity involve reciprocal definitions, where each is understood in relation to the other. This recursive perspective allows for a richer understanding of how the boundaries of reality are shaped by paradoxical loops, where absence and presence are entangled in a mysterious dance.

Mystical experiences, often described in terms of unity and transcendence, involve a shift in consciousness where the boundaries between self and the universe seem to dissolve. Recursive Realism suggests that mystical states can be understood as a collapse of self-referential loops, where the mind ceases to distinguish between subject and object, creating a sense of oneness. For example, in Zen Buddhism, the concept of no-mind (mushin) involves a state where thought and awareness flow freely, unencumbered by self-reference. This perspective offers a way to understand how spiritual experiences might reveal a deeper layer of reality, where the recursions of ordinary consciousness give way to a direct encounter with the infinite.

Recursive Realism offers a framework for exploring the mysteries that lie at the edges of understanding, revealing how recursion shapes cosmic structures, consciousness, and metaphysical paradoxes. By recognising the self-referential nature of reality, Recursive Realism allows us to approach questions about the multiverse, the nature of mind, and the mystical in a way that integrates science and philosophy, offering a new lens through which to contemplate the unknown.

Recursive Realism and the Horizon of Human Knowledge

Human knowledge is inherently limited, constrained by the scope of observation, conceptual frameworks, and the tools available for exploration. Yet, the process of discovery is shaped by recursive feedback loops, where questions lead to answers that generate new questions, creating an ever-expanding field of understanding. Recursive Realism suggests that the horizon of knowledge is defined by self-referential inquiry, where exploration and reflection create a dynamic interplay between the known and the unknown. This section delves into how

Recursive Realism can illuminate the limits of certainty, the nature of paradox, and the pathway toward deeper insights.

The Recursive Nature of Scientific Discovery

Science is a method of inquiry that seeks to understand reality through empirical observation, experimentation, and theoretical modeling. Recursive Realism suggests that the process of scientific discovery is inherently recursive, involving cycles of hypothesis, observation, and refinement that create a spiral of understanding.

Scientific progress relies on the formulation of hypotheses, which are tested through experiments and observations. The results of these tests often confirm, refute, or modify the original hypothesis, creating a feedback loop that drives further investigation. Recursive Realism suggests that hypothesis testing is a self-correcting process, where knowledge is refined through iterative cycles of learning. For example, Newton's laws of motion provided a framework for understanding classical mechanics, which was later refined by Einstein's theory of relativity to account for extreme conditions of gravity and velocity. This recursive approach allows for a deeper understanding of how science evolves, where each discovery redefines the boundaries of the known universe and opens new frontiers for exploration.

Thomas Kuhn's concept of paradigm shifts highlights how scientific revolutions occur when existing models are replaced by new frameworks that offer a better explanation of phenomena. Recursive Realism suggests that paradigm shifts are recursive transformations, where anomalies in data lead to a re-evaluation of fundamental assumptions, creating a loop that reconfigures scientific understanding. For example, the shift from a geocentric to a heliocentric model of the solar system fundamentally changed how humans understood their place in the universe, leading to a new era of astronomical discovery. This perspective allows for a richer appreciation of how scientific revolutions are self-organising processes, where doubt and discovery create a recursive dialogue that reshapes our vision of reality.

Cosmology faces the challenge of studying regions of the universe that lie beyond the reach of current instruments, such as the event horizon of a black hole or the expanding edge of the observable universe. These limits create reciprocal tension between what can be measured and what remains speculative, highlighting the recursive boundary between empirical science and theoretical imagination. Recursive Realism suggests that scientific inquiry operates within a feedback loop where observations push the limits of knowledge, while theories provide a framework for exploring the unobservable. This perspective allows for a deeper understanding of how science navigates its own boundaries, where the expansion of instrumentation and conceptual models creates a self-renewing quest for deeper truths.

The Philosophy of Paradox and the Limits of Logic

Philosophy has long grappled with paradoxes that challenge the coherence of logic and the nature of existence. Recursive Realism suggests that paradoxes arise from self-referential loops, where concepts turn back upon themselves, creating tensions that reveal the limits of linear thinking. This section explores how Recursive Realism engages with philosophical paradoxes, offering a way to understand the boundaries of reason and the insights that emerge from contradiction.

Self-Reference and the Paradox of the Liar:

The liar paradox, expressed in statements like "This sentence is false", creates a logical contradiction because it refers to itself in a way that undermines its own truth value. Recursive Realism suggests that such paradoxes are natural outcomes of self-referential thinking, where language and logic encounter their own limits. For example, Gödel's incompleteness theorems demonstrated that any sufficiently complex mathematical system contains statements that cannot be proven within the system itself, highlighting the recursive nature of truth. This perspective allows for a richer understanding of how paradoxes can be illuminating, revealing the depths of reality that lie beyond the bounds of formal systems.

Infinity, Zeno's Paradoxes, and the Nature of Continuity:

Zeno's paradoxes, such as the dichotomy paradox, where a runner must traverse an infinite series of half-distances to reach the finish line, challenge our understanding of motion and infinity. Recursive Realism suggests that Zeno's paradoxes reflect the recursive structure of infinite division, where conceptualising continuity requires self-referential thinking about how infinite sequences can sum to finite values. For example, the mathematical concept of a limit in calculus resolves Zeno's paradoxes by showing how infinite series can converge to a definite outcome. This recursive approach allows for a deeper appreciation of how infinity and continuity are intertwined, revealing how logic and mathematics use self-referential models to grapple with the infinite.

The Problem of Free Will and Determinism:

The debate between free will and determinism raises questions about whether human actions are truly autonomous or predetermined by causal chains. Recursive Realism suggests that free will can be understood as a recursive phenomenon, where conscious choices emerge through self-referential loops of reflection and intention. For example, a person might reflect on their desires and values, making a decision that is shaped by past experiences but also reinterpreted through present awareness. This perspective allows for a more nuanced view of agency, where determinism and freedom are not mutually exclusive but are interwoven through reciprocal interactions between self-awareness and external influences.

Recursive Realism and the Future of Inquiry

Recursive Realism offers a vision for the future of knowledge, suggesting that the expansion of understanding involves a continual dialogue between the known and the unknown, where each discovery reveals new layers of mystery. This section explores how recursion can guide the pursuit of truth, offering a framework for integrating scientific discovery, philosophical reflection, and the search for deeper meaning.

Curiosity drives human exploration, pushing us to ask questions that probe the boundaries of understanding. Recursive Realism suggests that curiosity is inherently recursive, as answers often lead to new questions that redefine the scope of inquiry. For example, the discovery of dark matter and dark energy in cosmology has opened up profound questions about the nature of the universe's unseen components, creating a feedback loop between observation and theory. This perspective allows for a richer appreciation of how curiosity is a self-renewing process, where knowledge grows through cycles of discovery that expand the horizon of what is possible.

Complex problems often require interdisciplinary thinking, where ideas from different fields are integrated into a cohesive understanding. Recursive Realism suggests that interdisciplinary inquiry is a recursive process, where concepts from physics, biology, psychology, and philosophy are synthesised to address fundamental questions. For example, the study of consciousness draws on neuroscience, cognitive science, phenomenology, and quantum theory, creating a multi-layered approach that reflects the complexity of the mind. This perspective highlights how recursion can guide the integration of diverse insights, where each discipline contributes to a holistic understanding of reality's layers.

The search for meaning is a fundamental aspect of the human condition, prompting us to explore questions about life's purpose, the nature of existence, and our place in the cosmos. Recursive Realism suggests that meaning is found in the ongoing dialogue between understanding and mystery, where each insight opens new dimensions of contemplation. For example, the question of why the universe exists at all remains one of the most profound mysteries, inviting a recursive reflection on the nature of being and the possibilities of the infinite. This perspective allows for a deeper appreciation of how the pursuit of meaning is not about finding final answers but about engaging with the mysteries that lie beyond comprehension, where recursion becomes a pathway for exploring the unknown.

Knowledge is a dynamic process that is shaped by recursive inquiry, where the pursuit of understanding involves a continual interplay between certainty and doubt, insight and mystery. Recursive Realism offers a framework for navigating the boundaries of knowledge, revealing how scientific discovery, philosophical reflection, and the search for meaning are interconnected through feedback loops that drive human inquiry. By recognising the self-referential nature of thought, Recursive Realism provides a way to explore the edges of reality, where the known meets the infinite potential of the unknown.

In this chapter, we have explored how Recursive Realism engages with the mysteries that lie at the limits of understanding, offering a framework for integrating scientific exploration, metaphysical reflection, and the quest for deeper truths. By examining the recursive nature of cosmic structures, consciousness, and the philosophy of inquiry, Recursive Realism provides a vision for how knowledge can continue to expand, revealing new layers of reality that challenge the limits of perception. This perspective allows us to approach the universe as a recursive system, where each discovery opens the way for new questions, creating an infinite dialogue between mind and cosmos.

References

Barbour, J., 1999. *The End of Time: The Next Revolution in Physics*. Oxford: Oxford University Press.

Bohm, D., 1980. *Wholeness and the Implicate Order*. London: Routledge.

Chalmers, D.J., 1996. *The Conscious Mind: In Search of a Fundamental Theory*. Oxford: Oxford University Press.

Deutsch, D., 1997. *The Fabric of Reality: The Science of Parallel Universes - and Its Implications*. London: Penguin Books.

Everett, H., 1957. "Relative State" Formulation of Quantum Mechanics. *Reviews of Modern Physics*, 29(3), pp.454–462.

Gödel, K., 1949. An Example of a New Type of Cosmological Solutions of Einstein's Field Equations of Gravitation. *Reviews of Modern Physics*, 21(3), pp.447–450.

Greene, B., 2003. *The Elegant Universe: Superstrings, Hidden Dimensions, and the Quest for the Ultimate Theory*. New York: Vintage Books.

Hawking, S. and Mlodinow, L., 2010. *The Grand Design*. London: Bantam Press.

Kuhn, T.S., 1962. *The Structure of Scientific Revolutions*. Chicago: University of Chicago Press.

Penrose, R., 1989. *The Emperor's New Mind: Concerning Computers, Minds and the Laws of Physics*. Oxford: Oxford University Press.

Prigogine, I., 1997. *The End of Certainty: Time, Chaos, and the New Laws of Nature*. New York: Free Press.

Rovelli, C., 2018. *The Order of Time*. Translated by E. Segre and S. Carnell. London: Allen Lane.

Sartre, J.-P., 1943. *Being and Nothingness*. Translated by H.E. Barnes. New York: Philosophical Library.

Smolin, L., 2013. *Time Reborn: From the Crisis in Physics to the Future of the Universe*. Boston: Houghton Mifflin Harcourt.

Tegmark, M., 2014. *Our Mathematical Universe: My Quest for the Ultimate Nature of Reality*. New York: Vintage Books.

Wheeler, J.A., 1982. Bohr, Einstein, and the Strange Lesson of the Quantum. *Mind in Nature: Nobel Conference XVII*, pp.1–25.

Zurek, W.H., 2003. Decoherence, Einselection, and the Quantum Origins of the Classical. *Reviews of Modern Physics*, 75(3), pp.715–775.

Chapter 21: Recursive Realism and the Dynamics of Ethics and Morality

The Self-Referential Nature of Moral Reasoning

Ethics and morality have always been central to the human experience, providing a framework for navigating choices that affect the well-being of ourselves and others. Recursive Realism suggests that moral reasoning is inherently self-referential, involving feedback loops where actions, intentions, and outcomes are continually evaluated and revised. This section explores how Recursive Realism can illuminate the complex dynamics of moral thought, revealing how ethical principles are shaped through reciprocal interactions between individual reflection and collective values.

The Evolution of Moral Intuition through Recursive Feedback

Moral intuition refers to the immediate sense of right and wrong that guides behaviour without the need for deliberate reasoning. Recursive Realism suggests that moral intuition is shaped by feedback loops between emotional responses and cognitive assessments, where experiences of empathy, guilt, and fairness influence ethical judgements.

Empathy, the ability to understand and share the feelings of others, plays a crucial role in moral development, allowing individuals to recognise the suffering and needs of others. This process is inherently recursive, as empathic responses create a cycle of reflection, where awareness of another's experience prompts internal emotional reactions that guide behaviour. For example, witnessing another person's pain can lead to a sense of compassion that motivates helping behaviour, creating a feedback loop between perception and action. Recursive Realism suggests that empathy operates through self-referential awareness, where emotional resonance with others deepens the moral sense of interconnectedness. This perspective allows for a richer understanding of how moral intuition is not a fixed trait but a dynamic process that evolves through social interactions.

Guilt and shame are moral emotions that play a key role in regulating behaviour, serving as internal signals that indicate deviation from social norms or personal values. These emotions are recursive in nature, as they involve reflecting on one's own actions and their impact on others, creating a cycle of self-assessment. For example, a person who feels guilty about lying might revisit the incident in their mind, imagining how their behaviour might have hurt others, and vowing to act differently in the future. Recursive Realism suggests that guilt and shame function as self-correcting mechanisms, where internal reflection creates a loop between past actions and future intentions. This perspective highlights how moral emotions guide ethical growth, creating a self-regulating system that aligns behaviour with social values.

Reciprocity, the principle of mutual exchange, is a universal aspect of moral reasoning, shaping how individuals evaluate fairness and justice in interpersonal relationships. This process is inherently recursive, as reciprocal interactions create a feedback loop where acts of kindness or injustice are mirrored in social responses. For example, the golden rule, "Treat others as you would like to be treated", is a self-referential principle that guides moral behaviour through reciprocal reflection on one's own preferences and the well-being of others. Recursive Realism suggests that reciprocity is a self-organising pattern in moral life, where fairness norms emerge from cycles of mutual respect and cooperative interactions. This perspective allows for a deeper appreciation of how ethical principles are internalised through reciprocal relationships, creating a shared sense of justice.

Moral Dilemmas and the Complexity of Ethical Decision-Making

Moral dilemmas arise when individuals must choose between competing values or outcomes, creating situations where no option is entirely satisfactory. Recursive Realism suggests that ethical decision-making involves self-referential loops, where individuals weigh intentions, consequences, and principles through a process of reflection that integrates multiple perspectives.

The trolley problem, a famous ethical thought experiment, asks whether it is morally acceptable to divert a runaway trolley, sacrificing one person to save five others. This dilemma reveals the complexity of utilitarian reasoning and the tension between outcome-based ethics and principled decision-making. Recursive Realism suggests that the trolley problem illustrates a self-referential process, where individuals must reflect on their values and the implications of their choices, creating a loop between internal principles and external consequences. For example, a person might initially favour the utilitarian choice of minimising harm but reconsider based on the moral principle that intentionally causing harm is wrong. This perspective allows for a richer understanding of how ethical decisions are shaped by recursive reflection, where reasoning and intuition interact in a dynamic process.

Deontological ethics, as articulated by philosophers like Immanuel Kant, emphasises the importance of moral rules and duties that should be followed regardless of the outcomes. Recursive Realism suggests that moral rules involve self-referential loops, where principles such as truth-telling or respect for autonomy are applied across varied situations and reinterpreted through reflection. For example, a person might follow the rule of honesty but reflect on whether truth-telling is appropriate in a situation where it might cause harm. Recursive Realism allows for a deeper appreciation of how moral rules are not static but are reinterpreted through ongoing reflection, where principled ethics interacts with the nuances of life's complexity.

Virtue ethics, inspired by Aristotelian thought, focuses on the development of moral character through the cultivation of virtues such as courage, compassion, and wisdom. Recursive Realism suggests that virtue ethics is inherently self-referential, as the process of becoming a virtuous person involves continuous reflection on one's actions, motives, and growth over time. For example, a person seeking to develop patience might reflect on moments when they reacted impulsively, creating a loop between self-awareness and intentional improvement. This perspective highlights how virtue development is a dynamic process, where moral growth occurs through reciprocal interactions between self-reflection and social practice.

Moral Evolution and the Dynamics of Social Norms

Moral values and social norms evolve over time, reflecting changes in cultural values, technological advances, and collective experiences. Recursive Realism suggests that moral evolution is a self-referential process, where individual moral judgements and collective norms influence each other through feedback loops, creating a dynamic interplay between tradition and innovation.

Cultural relativism acknowledges that moral values vary between cultures, reflecting different histories, traditions, and belief systems. Recursive Realism suggests that ethical diversity emerges through self-referential adaptation, where societies develop moral frameworks that respond to their unique circumstances. For example, norms around hospitality might be more strict in desert cultures where resources are scarce, creating a reciprocal relationship between environment and values. This perspective allows for a richer appreciation of how cultural norms are self-organising systems, where moral frameworks evolve through cycles of reflection on community needs and individual dignity.

Moral progress often involves challenging entrenched norms and advocating for change, as seen in civil rights movements, feminism, and LGBTQ+ advocacy. Recursive Realism suggests that social change is shaped by feedback loops between moral arguments and social attitudes, where activism and reflection create a cycle that redefines norms. For example, the abolitionist movement against slavery involved a moral re-evaluation of human rights, leading to shifts in legal frameworks and cultural attitudes. This perspective highlights how moral progress is not a linear path but a recursive dialogue between values and action, where new moral visions emerge from critical reflection on past practices.

The rise of digital technology has created new ethical challenges, such as privacy rights, AI ethics, and the impact of social media on mental health. Recursive Realism suggests that digital ethics involves self-referential feedback loops, where technological developments shape social norms, and public responses influence the design and regulation of new technologies. For example, concerns about data privacy have led to regulations like the GDPR in the European Union, creating a dialogue between tech companies, governments, and citizens. This perspective allows for a deeper understanding of how ethics adapts to new contexts, where social norms evolve through reciprocal interactions with technological change.

Moral reasoning is a dynamic process that is shaped by self-referential feedback loops, where empathy, reflection, and social interaction create a complex web of ethical thought. Recursive Realism offers a framework for understanding how moral intuition, dilemmas, and social norms evolve through reciprocal interactions, revealing how individuals and societies shape each other through the search for ethical coherence. By recognising the recursive nature of moral life, Recursive Realism allows us to appreciate how ethics is not a static code but a living dialogue that adapts and grows over time.

Recursive Realism and the Ethics of Interconnectedness

In an increasingly globalised world, the ethical responsibilities of individuals and societies extend beyond local communities to encompass the well-being of the entire planet. Recursive Realism suggests that ethical reasoning must account for the self-referential loops that connect human actions with global systems, recognising the interdependence of ecological balance, technological development, and cultural evolution. This section explores how Recursive Realism offers insights into global ethics, providing a framework for addressing environmental crises, advancing human rights, and navigating the ethical dilemmas of rapid technological change.

Environmental Ethics and the Recursion of Ecological Systems

Environmental ethics addresses the relationship between humans and the natural world, exploring how we should treat ecosystems, species, and the planet's resources. Recursive Realism suggests that environmental stewardship is shaped by feedback loops between human actions and ecological responses, where self-referential awareness of environmental impact can drive sustainable practices.

The Gaia Hypothesis, proposed by James Lovelock, suggests that the Earth functions as a self-regulating system, where living organisms interact with atmospheric, oceanic, and geological processes to maintain conditions conducive to life. This concept introduces a recursive element to ecology, where biotic and abiotic components form feedback loops that stabilise climate, atmospheric composition, and nutrient cycles. Recursive Realism suggests that human activity must be understood as part of this self-regulating system, where actions like deforestation or carbon emissions disrupt the delicate balance of Earth's feedback mechanisms. For example, the melting of polar ice caps due to global warming creates a positive feedback loop that accelerates climate change, reducing reflective surfaces and increasing heat absorption. This perspective allows for a deeper understanding of how environmental ethics must account for the self-reinforcing consequences of human actions, advocating for a reciprocal relationship with nature that prioritises sustainability.

Biodiversity, the variety of life in ecosystems, provides critical services such as pollination, water purification, and soil fertility, which are essential for human survival. Recursive Realism suggests that preserving biodiversity involves recognising the recursive loops that link ecosystem health with human well-being, where the loss of species can create cascading effects that disrupt ecological stability. For example, the decline of pollinator populations like bees threatens agricultural productivity, creating a loop between environmental degradation and food security. This perspective highlights how ethical responsibility extends to the protection of ecosystems, recognising that human flourishing is intertwined with the health of the natural world. By adopting a recursive approach to environmental ethics, societies can foster a sense of stewardship that respects the interconnectedness of all living systems.

Climate change represents one of the most pressing ethical challenges of the 21st century, raising questions about intergenerational justice, global equity, and the responsibility to mitigate harm. Recursive Realism suggests that climate action involves a recursive awareness of how present decisions shape future conditions, where policies on carbon emissions, renewable energy, and conservation create feedback loops that influence global temperatures and climate stability. For example, the Paris Agreement represents an attempt to create a self-reinforcing international commitment to reduce greenhouse gases, where each nation's efforts contribute to a collective outcome. Recursive Realism allows for a deeper appreciation of how climate ethics involves recognising the temporal recursion of environmental impacts, where today's actions have long-term consequences for future generations. This perspective emphasises the need for collaborative solutions that integrate scientific understanding with moral reflection, creating a feedback loop between knowledge and policy.

Global Responsibility and the Ethics of Interconnected Societies

Globalisation has created a world where economies, cultures, and political systems are interconnected in ways that transcend national borders. Recursive Realism suggests that global ethics must account for the self-referential dynamics of interconnected societies, where actions in one region can have far-reaching impacts on others, creating a moral obligation to consider the well-being of the global community.

Human rights, the idea that all individuals have inherent dignity and rights regardless of their background, reflect a universal ethical framework that has evolved through reciprocal dialogue between cultures and political movements. Recursive Realism suggests that human rights discourse is inherently recursive, as principles of equality and justice are interpreted and reinterpreted through local practices and international norms. For example, the Universal Declaration of Human Rights provides a baseline for global ethical standards, but its application varies across different legal systems, creating a feedback loop between universal principles and context-specific adaptations. This perspective allows for a richer understanding of how global ethics involves balancing universal ideals with cultural particularity, creating a recursive dialogue that respects diversity while upholding shared values.

Economic inequality between nations and within societies raises questions about fairness, opportunity, and the ethical responsibilities of wealthier countries toward poorer ones. Recursive Realism suggests that economic ethics involves recognising the feedback loops that link global trade, investment, and development with patterns of inequality. For example, the policies of multinational corporations can shape labour conditions in developing countries, creating a loop where global demand affects local economies. This perspective highlights the ethical imperative to consider the reciprocal effects of economic actions, advocating for policies that promote fair trade, poverty reduction, and sustainable development. By adopting a recursive approach to global economics, societies can work toward reducing disparities while fostering a more just world.

Migration and the refugee crisis present ethical challenges that involve balancing national interests with the rights of individuals seeking asylum and opportunity. Recursive Realism suggests that the ethics of migration involves recognising the reciprocal relationship between nations of origin and host countries, where policies on borders and integration shape the well-being of both migrants and receiving communities. For example, the acceptance or rejection of refugees fleeing conflict can create feedback loops that influence global stability and regional tensions. This perspective allows for a deeper understanding of how migration is part of a larger network of human movement, where ethical reasoning must account for the complex interplay of individual rights, community resilience, and global solidarity.

Technological Progress and the Ethics of the Future

Technological advancement has transformed human life, offering unprecedented opportunities for innovation while also creating new ethical dilemmas. Recursive Realism suggests that the ethics of technology must address the self-referential impacts of innovation, where new tools and capabilities reshape human society and challenge traditional norms.

Artificial intelligence (AI), particularly through machine learning, involves recursive algorithms that adapt and improve based on data inputs. Recursive Realism suggests that the ethical implications of AI involve reflecting on the feedback loops between technological capability and social impact, where AI systems influence decision-making, privacy, and power dynamics. For example, algorithmic bias in AI models can create reinforcing patterns of inequality, affecting hiring practices, criminal justice, and healthcare. This perspective highlights the need for ethical frameworks that monitor, evaluate, and adjust AI systems to ensure fairness, transparency, and accountability. By adopting a recursive approach to AI ethics, societies can navigate the complex dynamics of automation and human dignity.

Genetic engineering, including CRISPR technology, offers the potential to edit genes to prevent disease, enhance abilities, or alter traits. Recursive Realism suggests that the ethics of genetic modification involves reflecting on how altering human biology creates self-referential loops that influence future generations. For example, decisions about editing germline cells have implications that extend beyond individuals to their descendants, creating a feedback loop between scientific innovation and ethical responsibility. This perspective allows for a deeper understanding of how genetic engineering raises questions about the limits of human intervention, where enhancement and natural evolution intersect in complex ways. By considering the reciprocal effects of genetic choices, societies can navigate the ethical complexities of biotechnology.

Social media platforms create digital environments that influence communication, identity, and public discourse. Recursive Realism suggests that the ethics of digital connectivity involves recognising the feedback loops between individual behaviour and platform algorithms, where user interactions shape the content that is amplified and prioritised. For example, the spread of misinformation through viral posts can create a self-reinforcing loop that distorts public understanding and influences political outcomes. This perspective highlights the need for ethical considerations in the design and regulation of digital platforms, where the recursive impact of information sharing must be managed to protect truth, privacy, and civic integrity. By adopting a recursive approach to digital ethics, societies can work toward creating online spaces that promote healthy discourse and meaningful connection.

Ethical reasoning in a globalised world involves recognising the self-referential feedback loops that connect individual actions with collective outcomes, shaping the well-being of the planet and its inhabitants. Recursive Realism offers a framework for understanding how environmental stewardship, global responsibility, and technological progress are interconnected, revealing how moral decisions create reciprocal effects that influence the future of humanity. By adopting a recursive perspective on ethics, societies can navigate the complex interplay of local actions and global impacts, fostering a sense of shared responsibility that respects the interdependence of all life forms.

In this chapter, we have explored how Recursive Realism provides a framework for understanding ethics and morality, revealing how moral intuition, ethical decision-making, and social responsibility are shaped by self-referential feedback loops. By examining the reciprocal relationships between individual reflection, cultural norms, and global challenges, Recursive Realism offers a deeper understanding of how ethical systems evolve, adapting to the complexities of human existence. This perspective allows us to appreciate how moral reasoning is not a static code but a living dialogue that grows through interaction, reflection, and responsibility, creating a dynamic framework for navigating the ethical dimensions of life in a complex world.

References

Aristotle, 350 BCE. *Nicomachean Ethics.* Translated by W.D. Ross. Oxford: Oxford University Press.

Darwall, S., 2003. *The Second-Person Standpoint: Morality, Respect, and Accountability.* Cambridge, MA: Harvard University Press.

Foot, P., 1978. *Virtues and Vices and Other Essays in Moral Philosophy.* Oxford: Oxford University Press.

Frankl, V.E., 1984. *Man's Search for Meaning.* New York: Pocket Books.

Held, V., 2006. *The Ethics of Care: Personal, Political, and Global.* Oxford: Oxford University Press.

Hume, D., 1739. *A Treatise of Human Nature.* London: John Noon.

Kant, I., 1785. *Groundwork of the Metaphysics of Morals.* Translated by M. Gregor. Cambridge: Cambridge University Press.

Lovelock, J., 1979. *Gaia: A New Look at Life on Earth.* Oxford: Oxford University Press.

MacIntyre, A., 1981. *After Virtue: A Study in Moral Theory.* Notre Dame, IN: University of Notre Dame Press.

Noddings, N., 1984. *Caring: A Feminine Approach to Ethics and Moral Education.* Berkeley, CA: University of California Press.

Rawls, J., 1971. *A Theory of Justice.* Cambridge, MA: Harvard University Press.

Singer, P., 1975. *Animal Liberation: A New Ethics for Our Treatment of Animals.* New York: HarperCollins.

Slote, M., 2007. *The Ethics of Care and Empathy.* London: Routledge.

Sandel, M.J., 2020. *The Tyranny of Merit: What's Become of the Common Good?* London: Penguin Books.

Taylor, C., 1989. *Sources of the Self: The Making of the Modern Identity.* Cambridge, MA: Harvard University Press.

Williams, B., 1985. *Ethics and the Limits of Philosophy.* Cambridge, MA: Harvard University Press.

Chapter 22: Recursive Realism and the Nature of Consciousness

The Recursive Structure of Consciousness

Consciousness, the experience of being aware, remains one of the most challenging puzzles in philosophy, neuroscience, and psychology. Recursive Realism suggests that consciousness is inherently self-referential, arising from feedback loops that integrate sensory inputs, cognitive processing, and self-awareness. This section explores how Recursive Realism can illuminate the structure of consciousness, offering insights into the nature of awareness, the emergence of the self, and the relationship between the mind and environment.

Awareness as a Recursive Process

Awareness involves the ability to perceive and respond to the environment, integrating information from multiple sensory channels into a coherent experience. Recursive Realism suggests that awareness is a recursive process, where sensory inputs are continuously monitored, evaluated, and reintegrated into a dynamic representation of the world.

Conscious awareness involves the integration of diverse sensory inputs, such as sight, sound, touch, and proprioception, into a cohesive experience. This process is recursive in nature, as the brain constantly updates its internal model of the environment based on incoming data and past experiences. For example, when navigating a busy street, the mind integrates visual information about the movement of cars, auditory cues from honking horns, and proprioceptive feedback from the body's position, creating a feedback loop that allows for adaptive decision-making. Recursive Realism suggests that awareness emerges from the self-referential integration of sensory information, where each moment is contextualised within a larger narrative of experience.

Attention involves the selective focus on certain aspects of the environment while filtering out others, allowing for detailed processing of relevant information. Recursive Realism suggests that attention operates through a feedback loop, where the focus of awareness is continuously adjusted based on shifts in external stimuli and internal goals. For example, when reading a book, attention may shift from one paragraph to the next, depending on the content's relevance and interest, while distracting noises may redirect focus momentarily. This perspective highlights how attention is a dynamic process of recursive selection, where the mind constantly evaluates and re-evaluates what demands awareness in a given moment.

Recursive Awareness and the Emergence of Qualia

Qualia refer to the subjective qualities of conscious experience, such as the redness of a rose or the bitterness of coffee, which are inherently difficult to quantify. Recursive Realism suggests that qualia emerge from the self-referential nature of awareness, where neural feedback loops create a rich, first-person experience that is unique to each individual. For example, the experience of colour involves not only the activation of photoreceptors in the eye but also the integration of this sensory information with memories, emotional associations, and cultural meanings. This recursive model allows for a deeper understanding of how subjective experience is shaped by the interplay of sensory perception and cognitive reflection, where awareness becomes a multi-layered phenomenon that cannot be reduced to simple neural activity.

The Emergence of the Self through Recursive Reflection

The self is a central aspect of consciousness, involving the sense of personal identity that persists over time. Recursive Realism suggests that the sense of self emerges through recursive reflection, where the mind continuously evaluates its own thoughts, memories, and intentions, creating a cohesive narrative of who we are.

Self-reflection involves thinking about oneself, one's actions, and one's experiences, creating a feedback loop that shapes personal identity. Recursive Realism suggests that self-reflection is a self-referential process, where the mind turns inward to examine its own thoughts, creating a continuous dialogue that shapes the experience of being. For example, reflecting on a past decision, such as choosing a career path, involves considering the motivations, emotions, and consequences associated with that choice, creating a narrative that contributes to one's identity. This perspective highlights how the self is not a static entity but a dynamic process of ongoing reflection, where identity is shaped through recursive interactions between memory, intention, and awareness.

Narrative Identity

Narrative identity refers to the story that individuals create about their lives, integrating past experiences, present circumstances, and future aspirations into a coherent narrative. Recursive Realism suggests that narrative identity is a recursive construct, where the mind continually revisits and reinterprets the events of one's life, creating a loop between memory and imagination. For example, a person might reinterpret a challenging event, such as a failed relationship, as a learning experience that contributed to personal growth, creating a narrative that integrates suffering with resilience. This perspective allows for a deeper understanding of how identity is shaped by self-referential storytelling, where the self is constructed through an ongoing process of reinterpreting the past and envisioning the future.

The continuity of the self, the sense that we are the same person over time, is a foundational aspect of consciousness. Recursive Realism suggests that continuity is a self-referential construct, where the mind creates a cohesive narrative by linking moments of awareness into a seamless whole. For example, the memory of a childhood event is integrated with current experiences to create a sense of ongoing identity, even though thoughts, emotions, and circumstances have changed significantly over time. This recursive process allows for a richer appreciation of how the self is an illusion of continuity, where awareness weaves disparate moments into a unified experience that feels coherent and consistent.

The Reciprocal Interaction between Mind, Body, and Environment

Consciousness does not exist in isolation but is shaped by the reciprocal interactions between the mind, body, and environment. Recursive Realism suggests that consciousness emerges from self-referential loops that link internal states with external stimuli, creating a dynamic interplay that shapes experience.

Embodied cognition is the view that cognition is shaped by the body's interactions with the environment, where sensory inputs and motor outputs create a reciprocal feedback loop that influences thought and awareness. Recursive Realism suggests that consciousness is deeply rooted in these sensory-motor loops, where the mind continuously adapts to the changing environment through bodily movement and perception. For example, reaching for a cup of coffee involves not only the perception of the cup's location but also the adjustment of muscle movements to ensure a smooth grasp, creating a feedback loop between perception and action. This perspective highlights how awareness is not a disembodied phenomenon but a product of the recursive interplay between body and environment, where cognition emerges from physical interactions with the world.

The extended mind hypothesis suggests that cognitive processes can extend beyond the brain to include external tools, technologies, and social interactions that support thinking. Recursive Realism suggests that the use of tools creates a self-referential loop between the mind and external artefacts, where thought is extended through recursive interaction with the environment. For example, using a smartphone to store information creates a loop where the mind relies on an external device to enhance memory, creating a dynamic interplay between biological cognition and technological extension. This perspective allows for a deeper appreciation of how consciousness is shaped by reciprocal relationships with tools and technologies, where the boundaries of the self are fluid and extend into the environment.

Social cognition, the ability to understand and interpret the thoughts, emotions, and intentions of others, is a key aspect of human consciousness. Recursive Realism suggests that social awareness involves reciprocal feedback loops, where individuals engage in self-referential reflection on how they are perceived by others and adjust their behaviour accordingly. For example, in a conversation, a person might notice a friend's facial expression, interpret it as confusion, and adjust their explanation to clarify a point, creating a loop between perception, interpretation, and response. This recursive process allows for a richer understanding of how consciousness is inherently social, where awareness of others shapes self-awareness, creating a dynamic interplay that forms the foundation of human relationships.

Consciousness is a multi-layered phenomenon that emerges from recursive feedback loops involving sensory integration, self-reflection, and interaction with the environment. Recursive Realism provides a framework for understanding how awareness, identity, and the sense of self are shaped by self-referential processes, revealing how the mind creates a cohesive experience through continuous adaptation, reflection, and integration. By recognising the recursive nature of consciousness, Recursive Realism offers a deeper appreciation of the dynamic interplay between the mind, body, and environment, where awareness is not a static state but a process of ongoing creation.

Recursive Realism and the Boundaries of Consciousness

The boundaries of consciousness, the limits of awareness, the unconscious mind, and altered states, represent some of the most intriguing aspects of human experience. Recursive Realism suggests that consciousness is shaped by self-referential processes that create layers of awareness, where different states of mind are shaped by the interplay between perception, reflection, and environmental influences. This section explores how Recursive Realism can illuminate the boundaries of consciousness, examining altered states, the unconscious mind, and the potential for expanded awareness through self-referential exploration.

Altered States of Consciousness and the Recursive Nature of Awareness

Altered states of consciousness, such as dreams, meditative states, and psychedelic experiences, involve shifts in perception, thought, and emotion that differ from ordinary waking consciousness. Recursive Realism suggests that altered states involve modifications to the feedback loops that create conscious experience, leading to new configurations of awareness.

Dreams provide a window into the unconscious mind, where sensory input is minimised, and internal thoughts, memories, and emotions become the primary content of experience. Recursive Realism suggests that dreaming involves a self-referential loop, where the mind creates narratives and imagery from reprocessed memories and emotional states, leading to new associations and insights. For example, a recurring dream might involve the revisiting of a traumatic event in different forms, reflecting a recursive attempt to process and understand the emotional significance of that experience. This perspective allows for a richer appreciation of how dreams are a recursive exploration of the mind, where the self interacts with its own memories and emotions to create a dynamic experience of symbolic meaning.

Meditation involves focused attention and self-observation, leading to a state of heightened awareness that differs from ordinary consciousness. Recursive Realism suggests that meditative states involve a recursive feedback loop, where attention is turned inward to observe thoughts, sensations, and emotions without reacting to them. For example, in mindfulness meditation, the practitioner observes their thoughts as they arise, creating a loop between awareness and detachment that allows for a deeper understanding of the mind's contents. This recursive process enables a shift in perspective, where thoughts are seen as transient events rather than defining characteristics of the self. This perspective highlights how meditation offers a way to alter the recursive loops of awareness, leading to a state of expanded consciousness where self-awareness becomes a tool for transcendence.

Psychedelic substances, such as LSD, psilocybin, or ayahuasca, can lead to profound alterations in consciousness, including enhanced perception, ego dissolution, and a sense of interconnectedness with the universe. Recursive Realism suggests that psychedelic experiences involve a disruption of the normal feedback loops that create conscious awareness, allowing for new connections between different parts of the mind. For example, the experience of ego dissolution, where the boundaries between self and environment dissolve, can be understood as a collapse of the recursive loops that maintain the sense of individuality, leading to a state of unified awareness. This perspective allows for a deeper understanding of how psychedelics offer a way to explore the boundaries of consciousness, revealing the plasticity of the mind and the potential for expanded states of awareness.

The Unconscious Mind and the Recursive Dynamics of Hidden Awareness

The unconscious mind refers to the aspects of thought, emotion, and memory that lie outside of conscious awareness but still influence behaviour and experience. Recursive Realism suggests that the unconscious operates through self-referential loops that shape patterns of thought and behaviour without direct awareness, creating a layered structure of consciousness.

Freud's Model

Sigmund Freud described the unconscious as a repository of repressed desires, conflicts, and memories that shape conscious behaviour through symbolic expression. Recursive Realism suggests that the unconscious operates through self-referential loops, where conflicting desires create internal tensions that are expressed through dreams, slips of the tongue, or neurotic symptoms. For example, a repressed desire for success might manifest as self-sabotage, creating a loop where the unconscious motivation influences conscious decisions without direct awareness. This recursive model allows for a deeper understanding of how unconscious processes shape conscious experience, where layers of awareness interact through feedback loops that create complex behavioural patterns.

Jung's Collective Unconscious

Carl Jung introduced the concept of the collective unconscious, a shared layer of the unconscious mind that contains universal archetypes, symbolic patterns that recur across cultures and individuals. Recursive Realism suggests that archetypes operate through self-referential patterns, where universal themes are expressed through individual experience, creating a loop between the personal and the collective. For example, the hero archetype, representing courage, struggle, and transformation, may emerge in an individual's dreams or creative works, reflecting the recursive interplay between individual identity and universal human experiences. This perspective highlights how the unconscious is shaped by recursive dynamics, where individual awareness is influenced by deeply embedded patterns that link the self to the collective psyche.

Implicit Memory

Implicit memory refers to memories that are not consciously accessible but still influence behaviour, such as skills, habits, or conditioned responses. Recursive Realism suggests that implicit memory involves self-referential loops, where past experiences create patterns of behaviour that are reinforced through repetition. For example, the ability to ride a bicycle is an implicit memory that involves procedural knowledge and muscle coordination, creating a loop between sensory feedback and motor control. This recursive model allows for a deeper appreciation of how the unconscious influences conscious behaviour through automatic processes, where learned patterns of thought and action operate below the level of direct awareness.

Expanded Awareness and the Potential for Recursive Growth

Expanded awareness refers to the state of heightened consciousness where insights, creativity, and self-transcendence become possible. Recursive Realism suggests that expanded awareness involves the enhancement of self-referential loops, where reflection on one's thoughts, emotions, and connections to the world creates a state of greater insight and understanding.

Flow states, as described by Mihaly Csikszentmihalyi, involve complete immersion in an activity, where attention, action, and awareness are synchronised in a state of effortless engagement. Recursive Realism suggests that flow involves the recursive alignment of cognitive processes, where the mind creates a loop of focused attention and feedback that enhances performance and creativity. For example, an artist painting a masterpiece might enter a flow state, where the brushstrokes, colours, and emotional intent are integrated into a seamless experience of creation. This perspective highlights how expanded awareness can be achieved through the refinement of recursive processes, where focused engagement leads to a state of optimal functioning.

Mystical experiences, often described in terms of unity, transcendence, and oneness, involve a shift in awareness where the boundaries between self and the universe dissolve. Recursive Realism suggests that mystical states involve a collapse or transformation of the recursive loops that maintain the sense of individuality, leading to a state of expanded consciousness. For example, during a mystical experience, an individual may feel an overwhelming sense of interconnectedness with all living beings, as the distinction between self and other becomes blurred. This perspective allows for a deeper understanding of how expanded awareness can be achieved through the alteration of self-referential dynamics, leading to a profound shift in how the self is experienced.

Self-inquiry, the process of questioning one's own beliefs, assumptions, and values, involves a recursive exploration of the mind, where reflection leads to new insights and understanding. Recursive Realism suggests that self-inquiry creates a feedback loop between awareness and self-concept, where exploring one's own thoughts leads to greater clarity and expanded awareness. For example, questioning a long-held belief about success might lead to a shift in values, where material wealth is re-evaluated in favour of meaningful relationships and personal growth. This recursive process allows for a deeper appreciation of how expanded awareness is achieved through ongoing reflection, where self-awareness becomes a tool for growth and transformation.

The boundaries of consciousness, whether altered states, the unconscious, or expanded awareness, are shaped by recursive feedback loops that create layers of awareness, hidden motivations, and potential for growth. Recursive Realism provides a framework for understanding how different states of consciousness are shaped by self-referential processes, revealing how altered awareness can lead to new insights, creative breakthroughs, and a deeper connection with the universe. By recognising the recursive dynamics of the mind, Recursive Realism offers a way to explore the limits and possibilities of consciousness, revealing the profound complexity of human awareness and the potential for transformation.

We have explored how Recursive Realism offers a framework for understanding consciousness, revealing how self-awareness, altered states, and expanded awareness emerge from recursive feedback loops. By examining the structure of consciousness, the boundaries of awareness, and the dynamic interplay between the mind, body, and environment, Recursive Realism provides a deeper understanding of how awareness is not a static state but a process of ongoing creation and adaptation. This perspective allows us to appreciate the complexity of the human mind, where layers of consciousness interact through self-referential processes that shape thought, emotion, and identity, offering a rich landscape for exploration and growth.

References

Baars, B.J., 1988. *A Cognitive Theory of Consciousness*. Cambridge: Cambridge University Press.

Block, N., 1995. On a confusion about a function of consciousness. *Behavioral and Brain Sciences*, 18(2), pp. 227–247.

Chalmers, D.J., 1996. *The Conscious Mind: In Search of a Fundamental Theory*. Oxford: Oxford University Press.

Csikszentmihalyi, M., 1990. *Flow: The Psychology of Optimal Experience*. New York: Harper & Row.

Damasio, A.R., 1999. *The Feeling of What Happens: Body and Emotion in the Making of Consciousness*. New York: Harcourt Brace.

Dennett, D.C., 1991. *Consciousness Explained*. Boston: Little, Brown and Co.

Edelman, G.M. and Tononi, G., 2000. *A Universe of Consciousness: How Matter Becomes Imagination*. New York: Basic Books.

Freud, S., 1915. The unconscious. In: *The Standard Edition of the Complete Psychological Works of Sigmund Freud*, Vol. 14, pp. 159–215. London: Hogarth Press.

Jung, C.G., 1968. *Archetypes and the Collective Unconscious*. Princeton: Princeton University Press.

Koch, C., 2004. *The Quest for Consciousness: A Neurobiological Approach*. Englewood, CO: Roberts & Company.

Llinás, R. and Ribary, U., 2001. Consciousness and the brain: The thalamocortical dialogue in health and disease. *Annals of the New York Academy of Sciences*, 929(1), pp. 166–175.

Nagel, T., 1974. What is it like to be a bat? *Philosophical Review*, 83(4), pp. 435–450.

Tononi, G., 2004. An information integration theory of consciousness. *BMC Neuroscience*, 5(1), pp. 1–22.

Varela, F.J., Thompson, E. and Rosch, E., 1991. *The Embodied Mind: Cognitive Science and Human Experience*. Cambridge, MA: MIT Press.

Winkielman, P., Zajonc, R.B. and Schwarz, N., 1997. Subliminal affective priming resists attributional interventions. *Cognition and Emotion*, 11(4), pp. 433–465.

Chapter 23: Recursive Realism and the Nature of Reality

The Mind, Perception, and the Construction of Reality

Reality, as experienced by humans, is a combination of objective physical phenomena and subjective interpretation. Recursive Realism suggests that reality is inherently self-referential, shaped by the recursive feedback loops between perception, consciousness, and the material world. This section explores how the mind plays a role in constructing reality, focusing on the relationship between perception, cognition, and the physical world.

Perception as a Recursive Construct

Perception involves the process of interpreting sensory information to create a mental representation of the environment. Recursive Realism suggests that perception is inherently recursive, involving a continuous feedback loop between sensory inputs, cognitive models, and expectations.

Perception involves both bottom-up processing (the analysis of sensory input) and top-down processing (the application of prior knowledge and expectations). Recursive Realism suggests that perception emerges from the dynamic interaction between these two processes, creating a loop where sensory information is filtered through existing cognitive models and then refined through further experience. For example, when walking through a forest, the visual perception of trees is shaped by both the sensory details of light and texture and the cognitive expectation of what a forest should look like. This recursive feedback between bottom-up and top-down processing allows for a coherent representation of reality that is both informed by the environment and shaped by internal cognitive models.

The Bayesian brain hypothesis suggests that the mind creates predictive models of the environment and updates these models based on sensory feedback, using a process similar to Bayesian inference. Recursive Realism suggests that this predictive coding is inherently recursive, as predictions create expectations that influence sensory perception, which in turn leads to updates to the internal model. For example, when expecting to hear a specific word in a conversation, the brain may use this prediction to interpret auditory input more efficiently. This recursive process creates a loop where reality is constructed through continuous interactions between expectation and sensory experience, allowing the mind to create a coherent model of an ever-changing world.

Perceptual illusions provide evidence of the recursive nature of perception, as they reveal how the mind creates interpretations of sensory information that do not always align with objective reality. For example, the Müller-Lyer illusion, where two lines of the same length appear to be different due to the presence of arrow-like ends, illustrates how prior assumptions and cognitive processing create a recursive loop that influences how we see. Recursive Realism suggests that illusions arise from the mismatch between sensory input and the internal model, revealing how perception is not a direct reflection of the external world but a construct shaped by self-referential processing.

The Role of Consciousness in Reality Construction

Consciousness plays a central role in the construction of reality, as it provides the subjective experience through which the world is understood. Recursive Realism suggests that consciousness interacts with perception and cognition in a self-referential loop that shapes the experience of reality.

Conscious awareness allows for the integration of sensory information, cognitive processes, and emotions into a cohesive experience of reality. Recursive Realism suggests that consciousness involves a recursive process, where awareness is continuously updated based on sensory input and internal thoughts, creating a dynamic interplay that shapes how reality is experienced. For example, the experience of a sunset is shaped not only by the visual perception of colour but also by emotional associations and memories, creating a multi-layered experience that is constructed recursively. This perspective highlights how reality is not a static entity but a process shaped by the recursive interplay between consciousness and the environment.

The Observer Effect

The observer effect, the idea that the act of observation can influence the outcome, is well known in quantum mechanics but also has implications for conscious experience. Recursive Realism suggests that attention acts as a recursive filter that shapes reality by selectively focusing on certain aspects of the environment while ignoring others. For example, in a crowded room, paying attention to a particular conversation creates a loop where the content of the conversation becomes the dominant reality, while other stimuli fade into the background. This self-referential process reveals how reality is shaped by what we choose to focus on, highlighting the power of attention in constructing our subjective world.

Meaning is central to human experience, as it shapes how we interpret reality and engage with the world. Recursive Realism suggests that meaning is constructed through a recursive process of reflection, where experiences are interpreted and reinterpreted in the context of the self. For example, reflecting on a challenging experience, such as losing a job, can lead to a reinterpretation of that event as an opportunity for growth, creating a loop where the mind constructs new meaning from past events. This recursive process allows for a deeper appreciation of how reality is shaped by meaning-making, where the self continuously redefines its relationship with the world through reflection and reinterpretation.

The Material World and the Self-Referential Universe

Reality is not only shaped by perception and consciousness but also involves the material world and the laws of physics that govern it. Recursive Realism suggests that the universe itself may be self-referential, with cosmic structures and natural laws reflecting recursive patterns that shape the emergence of complexity.

Fractals are mathematical structures that exhibit self-similarity across different scales, meaning that their patterns repeat at smaller and larger scales. Recursive Realism suggests that fractal patterns are a manifestation of the self-referential nature of the universe, where complex forms emerge through repeated iterations. For example, the branching patterns of trees, rivers, and blood vessels all exhibit fractal geometry, reflecting a recursive process that creates complex structures from simple rules. This perspective highlights how the material world is shaped by recursive dynamics, where natural laws create patterns that repeat across different levels of reality.

The universe itself may be understood as a recursive system, where simple physical laws give rise to complex phenomena through repeated interactions. Recursive Realism suggests that cosmic structures, such as galaxies, solar systems, and life itself, emerge from self-referential feedback loops that create order from chaos. For example, the formation of stars involves the collapse of gas clouds under gravity, creating a feedback loop where gravity and nuclear fusion shape the evolution of celestial bodies. This perspective allows for a deeper understanding of how the universe evolves through reciprocal interactions, where the laws of physics create complex systems that exhibit self-referential properties.

The anthropic principle suggests that the universe must be such that it allows for the existence of conscious observers, otherwise, we would not be here to observe it. Recursive Realism suggests that the act of observation is itself a recursive process, where the universe creates observers who, in turn, reflect on and interpret the universe. For example, the laws of physics are finely tuned to allow for the formation of atoms, molecules, and life, creating a feedback loop where the universe becomes aware of itself through conscious beings. This self-referential perspective highlights the interconnectedness of the observer and the observed, suggesting that reality is not a passive backdrop but a dynamic interplay between consciousness and the cosmos.

Reality, as experienced by humans, is not a fixed entity but a construct shaped by recursive interactions between perception, consciousness, and the material world. Recursive Realism offers a framework for understanding how the mind plays a central role in constructing reality, where sensory input, cognitive models, and subjective experience create a dynamic interplay that shapes how we understand the world. By recognising the self-referential nature of reality, Recursive Realism provides insights into how the universe itself may be a recursive system, where complexity emerges from simple laws through feedback loops that shape the cosmic order.

Recursive Realism and the Nature of Cosmic Reality

The cosmos is a vast and complex system governed by physical laws that shape the emergence of matter, energy, and life. Recursive Realism suggests that cosmic reality is inherently self-referential, involving feedback loops that create patterns across space, time, and consciousness. This section explores how recursion provides insights into the nature of the universe, examining the origin of cosmic structures, the nature of time, and the role of consciousness in shaping reality.

The Origin of the Universe and the Emergence of Complexity

The origin of the universe is one of the most profound mysteries of science and philosophy. Recursive Realism suggests that the cosmos itself may be understood as a self-referential system, where complexity emerges from simple principles through recursive processes.

The Big Bang theory describes the beginning of the universe as an event where space, time, and energy emerged from an initial singularity. Recursive Realism suggests that the Big Bang can be understood as the initial act of recursion, where the universe began with a singular state and expanded into a system of increasing complexity. For example, the emergence of fundamental particles, atoms, and galaxies can be seen as the result of recursive processes that built upon the initial conditions set in motion by the Big Bang. This perspective highlights how the universe evolves through reciprocal interactions, where simple physical laws create complex structures through a continuous process of iteration and self-organisation.

Cosmic inflation refers to the rapid expansion of space that occurred shortly after the Big Bang, leading to the uniformity and flatness of the observable universe. Recursive Realism suggests that cosmic inflation is an example of a self-referential process, where the expansion of space created a feedback loop that set the stage for the formation of galaxies and large-scale structures. For example, the exponential growth of space allowed for quantum fluctuations to be stretched into cosmic structures, creating a loop where small-scale variations led to the formation of galaxies and clusters. This perspective highlights how cosmic reality is shaped by recursive dynamics, where early events in the universe influenced the formation of complex systems through a process of iteration.

The universe exhibits patterns of order at multiple levels, from the formation of galaxies to the emergence of life. Recursive Realism suggests that cosmic order emerges through self-organisation, a recursive process where local interactions give rise to global patterns. For example, the gravitational attraction between particles led to the formation of stars, which in turn led to the creation of heavier elements through nuclear fusion, elements that would later become the building blocks of planets and life. This recursive process reveals how the universe is a self-organising system, where complexity emerges from simple principles through repeated interactions that create higher levels of structure.

The Nature of Time and the Recursion of Temporal Reality

Time is a fundamental aspect of reality, shaping the evolution of the universe and the experience of consciousness. Recursive Realism suggests that time is inherently recursive, involving feedback loops that create patterns of change and continuity.

Many cultures and philosophical traditions have conceptualised time as cyclical, where events repeat in endless cycles. Recursive Realism suggests that cyclic time reflects the recursive nature of cosmic processes, where patterns of creation and destruction repeat across different scales. For example, the life cycle of stars, from birth in nebulae to death as supernovae, creates a loop where the materials from one generation of stars are recycled into the next, creating a cycle of cosmic renewal. This perspective highlights how the universe is shaped by recurring cycles, where temporal patterns emerge from recursive processes that create order and continuity across time.

The arrow of time refers to the directionality of time, from past to future, often associated with the increase of entropy as described by the second law of thermodynamics. Recursive Realism suggests that entropy is a self-referential process, where the increase of disorder creates new opportunities for complexity to emerge. For example, the formation of stars and planets involves a local decrease in entropy through self-organisation, even as the overall entropy of the universe continues to increase. This perspective allows for a deeper understanding of how the arrow of time is shaped by recursive dynamics, where local systems create order while contributing to the global increase in disorder, reflecting the self-referential interplay between complexity and entropy.

Time is also a subjective experience, shaped by the mind's perception of change and continuity. Recursive Realism suggests that the perception of time is inherently recursive, involving the self-referential integration of memory, anticipation, and present awareness. For example, the experience of waiting can feel longer or shorter depending on the level of attention and engagement, creating a loop between conscious focus and the perception of time. This perspective highlights how time is not only an objective phenomenon but also a construct of consciousness, shaped by the recursive dynamics of awareness and reflection.

Consciousness and the Self-Referential Universe

Consciousness plays a unique role in the cosmos, providing a means for the universe to become aware of itself. Recursive Realism suggests that consciousness and the universe are part of a self-referential system, where the act of observation creates a feedback loop that shapes reality.

The Participatory Universe and the Role of Consciousness

The participatory universe hypothesis, as suggested by physicist John Wheeler, proposes that conscious observers play a role in bringing the universe into existence through the act of observation. Recursive Realism suggests that the universe is inherently self-referential, where consciousness creates a loop between the observer and the observed, shaping the nature of reality. For example, in quantum mechanics, the measurement problem suggests that the act of observation determines the state of a quantum system, highlighting the interplay between consciousness and physical reality. This perspective allows for a deeper understanding of how consciousness is an integral part of the cosmic system, creating a reciprocal relationship where the universe becomes aware of itself through the minds of conscious beings.

Panpsychism and the Recursion of Mind in Matter

Panpsychism is the view that consciousness is a fundamental aspect of matter, present at all levels of the universe. Recursive Realism suggests that consciousness may emerge from recursive interactions within matter, where self-referential processes at the quantum or subatomic level give rise to awareness. For example, the neural networks in the brain create a complex web of recursive connections that lead to the emergence of consciousness, suggesting that similar processes may occur at different levels of cosmic reality. This perspective highlights the possibility that the universe itself may be conscious in some form, with mind and matter intertwined through recursive dynamics that create layers of awareness.

Cosmic Consciousness and the Recursion of Awareness

The concept of cosmic consciousness suggests that the universe is a living entity with an awareness that transcends individual minds. Recursive Realism suggests that cosmic consciousness may be understood as a self-referential system, where individual consciousness is a reflection of a greater cosmic awareness. For example, the experience of mystical unity, where individuals feel a sense of oneness with the universe, can be seen as a moment when the recursive boundaries between individual and cosmic consciousness dissolve, revealing a deeper connection between the self and the cosmos. This perspective allows for a richer appreciation of how consciousness and the universe are part of a single self-referential system, where awareness is not limited to individual beings but is a fundamental aspect of cosmic reality.

Cosmic reality is a self-referential system, shaped by recursive processes that create patterns of order, complexity, and awareness. Recursive Realism offers a framework for understanding how the universe evolves through reciprocal interactions, where simple principles give rise to complex structures through iteration and self-organisation. By examining the origin of the universe, the nature of time, and the role of consciousness, Recursive Realism provides a deeper understanding of how the cosmos and consciousness are interconnected, revealing a dynamic interplay where the universe becomes aware of itself through the recursive dynamics of observation and reflection.

In Chapter 23, we explored how Recursive Realism offers a framework for understanding the nature of reality, revealing how perception, consciousness, and the material world are interconnected through recursive processes. By examining the construction of reality through perception, the self-referential nature of cosmic structures, and the role of consciousness in shaping cosmic reality, Recursive Realism provides a vision of the universe as a living, dynamic system that evolves through reciprocal interactions. This perspective allows us to appreciate how the universe and consciousness are part of a single self-referential system, where awareness and reality are co-created through the interplay of the mind, matter, and the laws of nature.

References

Bohm, D. (1980) *Wholeness and the Implicate Order*. London: Routledge.

Chalmers, D.J. (1996) *The Conscious Mind: In Search of a Fundamental Theory*. Oxford: Oxford University Press.

Csikszentmihalyi, M. (1990) *Flow: The Psychology of Optimal Experience*. New York: Harper & Row.

Dennett, D.C. (1991) *Consciousness Explained*. Boston: Little, Brown.

DeWitt, B. and Graham, N. (eds.) (1973) *The Many-Worlds Interpretation of Quantum Mechanics*. Princeton: Princeton University Press.

Feynman, R.P., Leighton, R.B. and Sands, M. (1964) *The Feynman Lectures on Physics: Volume 1*. Reading, MA: Addison-Wesley.

Freeman, W.J. (2000) *How Brains Make Up Their Minds*. London: Phoenix.

Jung, C.G. (1968) *The Archetypes and the Collective Unconscious*. 2nd edn. Princeton: Princeton University Press.

Lovelock, J.E. (1979) *Gaia: A New Look at Life on Earth*. Oxford: Oxford University Press.

Penrose, R. (1994) *Shadows of the Mind: A Search for the Missing Science of Consciousness*. Oxford: Oxford University Press.

Sagan, C. (1980) *Cosmos*. New York: Random House.

Smolin, L. (1997) *The Life of the Cosmos*. Oxford: Oxford University Press.

Tegmark, M. (2014) *Our Mathematical Universe: My Quest for the Ultimate Nature of Reality*. New York: Alfred A. Knopf.

Wheeler, J.A. (1988) 'World as system self-synthesized by quantum networking', *IBM Journal of Research and Development,* 32(1), pp. 4–15.

Zeki, S. (1993) *A Vision of the Brain*. Oxford: Blackwell Science.

Chapter 24: Recursive Realism and the Concept of Free Will

Free Will, Determinism, and the Recursive Nature of Choice

The question of free will, whether humans are truly free to make choices, or whether all actions are determined by prior causes, has been a central debate in philosophy for centuries. Recursive Realism suggests that free will is inherently self-referential, involving feedback loops where the mind navigates between deterministic influences and autonomous decisions. This section explores how Recursive Realism can offer insights into the nature of free will, examining how choice emerges from the recursive interplay between determinism, conscious intention, and self-awareness.

The Deterministic Universe and the Illusion of Autonomy

Determinism is the view that all events in the universe, including human actions, are the result of prior causes governed by natural laws. This perspective suggests that free will is an illusion, as choices are ultimately determined by genetics, environment, and physical processes. Recursive Realism, however, offers a nuanced view of how deterministic influences interact with the mind's recursive processes.

Causal Chains and the Recursiveness of Decision-Making: In a deterministic universe, every event is part of a causal chain, where each action is determined by previous events. Recursive Realism suggests that decision-making involves recursion within these causal chains, where the mind reflects on past experiences and potential outcomes to shape actions. For example, choosing to study a new language may be influenced by cultural background, previous exposure to languages, and personal interests, creating a loop between deterministic factors and reflective choice. This perspective allows for a deeper understanding of how autonomy operates within a deterministic framework, where the mind uses self-referential reflection to navigate the constraints of causality.

Neuroscience and the Recurrence of Decision Processes: Neuroscientific studies have suggested that brain activity associated with a decision occurs milliseconds before conscious awareness of the choice, leading some to argue that free will is an illusion. Recursive Realism suggests that neural processes involve feedback loops where automatic brain functions and conscious intention interact to shape behaviour. For example, the readiness potential observed in neuroscience, a signal that occurs before a voluntary action, may reflect a recursive dialogue between unconscious processing and conscious reflection, where awareness becomes part of the loop that shapes action. This perspective highlights how consciousness plays a reciprocal role in decision-making, where awareness is both influenced by and influences the brain's processes.

The Illusion of Control and the Recursive Construction of Agency: The illusion of control refers to the tendency to overestimate one's influence over events, such as believing that one's thoughts can change outcomes beyond one's direct actions. Recursive Realism suggests that the sense of agency is constructed through self-referential loops, where the mind creates a narrative that links intentions with outcomes. For example, a person might feel responsible for success in a team project, even though the outcome was influenced by multiple factors beyond their control. This recursive model allows for a deeper appreciation of how agency is shaped by the mind's interpretation of causal relationships, creating a sense of autonomy even within a deterministic framework.

Conscious Intention and the Dynamics of Choice

While determinism emphasises the role of causes, free will focuses on the experience of choice and the ability to act according to intentions. Recursive Realism suggests that conscious intention is shaped by self-referential processes, where the mind reflects on possibilities, values, and desires to guide actions.

Intentionality as a Recursive Loop: Intentionality refers to the directedness of thoughts and actions toward particular goals or outcomes. Recursive Realism suggests that intentionality involves a feedback loop where the mind continually evaluates its own goals and adjusts actions to achieve them. For example, the decision to exercise regularly may involve a reciprocal process where short-term desires (e.g., avoiding discomfort) are weighed against long-term goals (e.g., improving health), creating a loop of internal dialogue that shapes the final choice. This perspective allows for a richer understanding of how intentions are not static but are refined through reflection, where free will emerges as a dynamic process of self-directed reasoning.

The Role of Reflection in Ethical Decisions:

Ethical decisions often involve weighing conflicting values and anticipating the consequences of actions, making them a prime example of self-referential decision-making. Recursive Realism suggests that ethical reflection involves a recursive loop, where the mind considers the implications of choices for oneself and others, creating a feedback process that guides moral actions. For example, deciding whether to donate to charity might involve reflecting on the potential impact of the donation, the importance of self-sacrifice, and the balance between personal needs and altruistic values. This perspective highlights how free will operates through recursive ethical reasoning, where the mind uses self-awareness to navigate between competing values and create meaning in moral choices.

Goal Setting and the Feedback Loop of Autonomy:

Goal setting is a key aspect of self-determination, where individuals establish objectives and pursue them through deliberate actions. Recursive Realism suggests that goal-directed behaviour is inherently recursive, involving a cycle of intention, action, evaluation, and adjustment. For example, pursuing a career goal might involve repeatedly adjusting plans based on feedback from successes and setbacks, creating a loop where the mind refines its strategies based on changing circumstances. This perspective allows for a deeper understanding of how autonomy is shaped by the recursive nature of goal pursuit, where free will is not about absolute independence from influences but about navigating them in a self-directed manner.

The Role of Uncertainty and the Space for Free Will

While determinism emphasises predictable outcomes, free will involves navigating uncertainty and making choices in the face of unknowns. Recursive Realism suggests that uncertainty creates a space where free will can operate, as the mind uses self-referential reflection to explore possibilities and choose among alternatives.

Quantum Uncertainty and the Recursiveness of Probability:

Quantum mechanics introduces a level of uncertainty at the subatomic level, where particles exist in probabilistic states until measured. Recursive Realism suggests that the mind operates within a world that is not fully determined but includes probabilistic elements, creating room for choice. For example, everyday decisions, such as choosing a meal or selecting a book, might be influenced by small uncertainties in internal states or external conditions, creating a loop where self-reflection interacts with uncertain outcomes. This perspective highlights how free will is not about defying natural laws but about navigating the uncertainties that arise in a complex world.

Creative Problem Solving and the Exploration of Possibilities:

Creativity involves generating new ideas and solutions that transcend conventional thinking, making it a key example of free will in action. Recursive Realism suggests that creativity is inherently recursive, involving a loop of exploration, evaluation, and refinement. For example, an artist creating a painting may experiment with different colours and techniques, continuously adjusting based on the emerging composition until the final artwork takes shape. This perspective highlights how free will operates through creative processes, where the mind uses self-referential exploration to navigate the unknown and create novel outcomes.

The Experience of Choice and the Paradox of Free Will:

Free will involves a paradox, where the experience of choice feels genuine, even though every decision is shaped by internal and external factors. Recursive Realism suggests that this paradox is resolved through self-referential awareness, where the mind creates a narrative that integrates influences with autonomous intentions. For example, choosing to move to a new city may involve considering career opportunities, family needs, and personal aspirations, creating a loop where freedom is experienced through the process of balancing influences with intentional direction. This perspective allows for a deeper appreciation of how free will is not about absolute freedom but about the self-referential construction of agency within a structured world.

The concept of free will is complex, involving a dynamic interplay between deterministic influences and self-directed choices. Recursive Realism provides a framework for understanding how free will operates through self-referential feedback loops, where the mind navigates between causal constraints and intentional reflection. By recognising the recursive nature of decision-making, intentionality, and creativity, Recursive Realism offers a nuanced perspective on how autonomy emerges within a deterministic universe, revealing how free will is a process of self-reflection and self-direction rather than an absolute state.

Recursive Realism and the Ethics of Free Will

The question of free will is not only a matter of philosophical inquiry but also has profound ethical implications. Responsibility, moral accountability, and the possibility of change are all tied to the idea that individuals have the capacity to choose their actions. Recursive Realism suggests that ethics involves a recursive process where autonomy and responsibility are continuously re-evaluated through self-reflection and social feedback. This section explores how Recursive Realism can provide insights into the moral dimensions of free will, focusing on the relationship between individual choice, social norms, and the evolution of ethical values.

Moral Responsibility and the Recursive Nature of Accountability

Moral responsibility involves the idea that individuals are accountable for their actions and can be praised or blamed based on their choices. Recursive Realism suggests that responsibility is inherently self-referential, involving a loop where individuals reflect on their intentions and actions within the context of social norms and moral principles.

Guilt and regret are emotional responses that arise when individuals perceive that they have acted wrongly or failed to meet moral standards. Recursive Realism suggests that these emotions involve a self-referential loop, where the mind reflects on past actions and compares them to moral expectations, leading to feelings of remorse. For example, feeling guilty for breaking a promise involves reflecting on the harm caused and re-evaluating one's values, creating a loop that can lead to moral growth. This perspective highlights how moral responsibility is shaped by the mind's capacity for self-reflection, where guilt and regret become mechanisms for aligning behaviour with moral principles.

Accountability is often enforced through social interactions, where communities hold individuals responsible for their actions through praise, criticism, or punishment. Recursive Realism suggests that accountability involves a recursive dialogue between individuals and society, where actions are judged based on social norms, and feedback influences future behaviour. For example, legal systems provide a formal structure for holding individuals accountable for crimes, creating a loop where laws and punishments shape social behaviour. This perspective allows for a deeper understanding of how free will is embedded within a social context, where individual autonomy is balanced by the expectations and values of the community.

Intention plays a crucial role in moral judgements, as actions that are intended to cause harm are often seen as more blameworthy than those that result from accidental causes. Recursive Realism suggests that intention involves a self-referential process, where individuals consider the motivations behind their actions and evaluate their alignment with moral values. For example, a person who accidentally causes harm might feel less responsible than someone who deliberately intends to hurt another, reflecting the role of intention in moral reasoning. This perspective highlights how moral responsibility is shaped by the mind's ability to reflect on intentions, where ethical judgement becomes a recursive process that integrates inner motivations with external consequences.

The Ethics of Autonomy and the Freedom to Change

Autonomy, the ability to direct one's life according to one's values, is a central aspect of free will and has significant ethical implications. Recursive Realism suggests that autonomy involves a recursive process, where individuals continually re-evaluate their beliefs, desires, and goals in light of new experiences and insights.

The ability to change, to grow, learn, and transform, is a key aspect of human autonomy. Recursive Realism suggests that change involves a self-referential feedback loop, where individuals reflect on their past behaviours and adjust their future actions based on new insights. For example, a person who recognises a destructive habit, such as procrastination, might engage in self-reflection to understand the underlying causes and develop strategies for improvement, creating a loop of self-directed change. This perspective allows for a deeper appreciation of how free will is not only about making choices but also about the capacity to reshape oneself, where autonomy is exercised through recursive self-improvement.

Consent is a fundamental principle in ethics, reflecting the right of individuals to make decisions about their own lives. Recursive Realism suggests that consent is inherently recursive, as it involves self-reflection on one's desires and boundaries, allowing for informed decision-making. For example, the act of consenting to a medical treatment involves reflecting on the risks, benefits, and personal values, creating a loop where autonomy is expressed through deliberate choice. This perspective highlights how ethical autonomy involves the capacity for self-directed decision-making, where individual freedom is balanced with consideration of the consequences for oneself and others.

Moral Growth and the Recursive Development of Character: Character, the set of traits and dispositions that shape moral behaviour, is not fixed but evolves through experiences and choices. Recursive Realism suggests that moral growth involves a self-referential loop, where individuals reflect on their actions, learn from mistakes, and cultivate virtues through ongoing practice. For example, developing the virtue of patience might involve recognising impulsive tendencies and intentionally practising restraint in challenging situations, creating a loop where reflection and action reinforce each other. This perspective allows for a deeper understanding of how free will is tied to moral development, where autonomy is used to shape character through recursive engagement with ethical ideals.

Free Will, Society, and the Balance of Rights and Duties

Free will is not only a personal capacity but also has implications for societal structures, where the balance of individual rights and social duties is crucial for coexistence. Recursive Realism suggests that society functions as a self-referential system, where individual freedoms interact with collective responsibilities through feedback loops that shape social norms.

The social contract, the idea that individuals agree to follow rules in exchange for protection and benefits, reflects a recursive relationship between individual autonomy and collective well-being. Recursive Realism suggests that rights and responsibilities are negotiated through a feedback loop where individual actions influence social norms, and societal expectations shape individual choices. For example, freedom of speech allows individuals to express their opinions, but it is balanced by laws that protect against harm, creating a dynamic interaction between personal freedom and public interest. This perspective highlights how free will is exercised within a social context, where autonomy is shaped by the reciprocal relationship between individuals and the community.

Justice systems are designed to hold individuals accountable for their actions, balancing punishment with rehabilitation. Recursive Realism suggests that justice involves a recursive process of evaluation, where the intent, impact, and context of an action are considered in determining consequences. For example, the decision to offer leniency to a first-time offender might reflect a recognition of their potential for change, creating a loop where societal judgement interacts with individual growth. This perspective allows for a richer understanding of how free will is considered within ethical and legal systems, where the goal is not only punishment but also encouraging positive change through a reciprocal process of reflection and reform.

Collective responsibility involves recognising the shared impact of individual actions on the well-being of society and the environment. Recursive Realism suggests that collective ethics involves a self-referential loop, where societies reflect on the consequences of collective actions and adjust norms to promote the common good. For example, addressing climate change involves recognising the impact of individual behaviours (such as energy use and consumption patterns) on global systems, creating a loop where social norms and policies encourage sustainable practices. This perspective highlights how free will is intertwined with collective responsibilities, where autonomy is used to shape a shared future through ethical collaboration.

The ethics of free will involves recognising the reciprocal relationship between individual autonomy and collective responsibility, where moral choices are shaped by self-reflection, social feedback, and the dynamic interplay between rights and duties. Recursive Realism offers a framework for understanding how moral responsibility, accountability, and the potential for change are inherently self-referential, where individuals use autonomy to navigate the complexities of ethical life. By recognising the recursive nature of free will, Recursive Realism provides a way to balance personal freedom with the needs of the community, offering a vision of ethical living that respects both individual agency and the interconnectedness of all life.

In Chapter 24, we have explored how Recursive Realism provides a framework for understanding free will, revealing how autonomy is shaped by self-referential feedback loops that integrate deterministic influences with conscious intention. By examining the dynamics of choice, the ethics of responsibility, and the balance between freedom and duty, Recursive Realism offers a nuanced perspective on the nature of free will in a deterministic universe. This perspective allows us to appreciate how individuals exercise autonomy through a process of ongoing reflection, where moral decisions are shaped by the interplay between self-awareness, social context, and the potential for growth.

References

Aristotle (2009) *Nicomachean Ethics*. Translated by W.D. Ross. Oxford: Oxford University Press.

Baumeister, R.F. (2008) *Free Will and Consciousness: How Might They Work?* Oxford: Oxford University Press.

Chalmers, D.J. (1996) *The Conscious Mind: In Search of a Fundamental Theory*. Oxford: Oxford University Press.

Dennett, D.C. (1984) *Elbow Room: The Varieties of Free Will Worth Wanting*. Cambridge, MA: MIT Press.

Frankfurt, H. (1971) 'Freedom of the Will and the Concept of a Person', *Journal of Philosophy,* 68(1), pp. 5–20.

Gazzaniga, M.S. (2011) *Who's in Charge? Free Will and the Science of the Brain.* New York: Ecco Press.

Heisenberg, W. (1958) *Physics and Philosophy: The Revolution in Modern Science.* New York: Harper.

Hume, D. (2007) *An Enquiry Concerning Human Understanding.* Edited by P. Millican. Oxford: Oxford University Press.

Kane, R. (1996) *The Significance of Free Will.* Oxford: Oxford University Press.

Libet, B. (1985) 'Unconscious Cerebral Initiative and the Role of Conscious Will in Voluntary Action', *Behavioral and Brain Sciences,* 8(4), pp. 529–566.

Nagel, T. (1986) *The View from Nowhere.* New York: Oxford University Press.

Popper, K.R. (1994) *The Myth of the Framework: In Defence of Science and Rationality.* London: Routledge.

Sartre, J.-P. (1943) *Being and Nothingness.* Translated by H.E. Barnes. New York: Washington Square Press.

Skinner, B.F. (1971) *Beyond Freedom and Dignity.* Indianapolis: Hackett Publishing Company.

Smolin, L. (1997) *The Life of the Cosmos.* Oxford: Oxford University Press.

Strawson, G. (1994) *Freedom and Belief.* Oxford: Oxford University Press.

Chapter 25: Recursive Realism and the Evolution of Knowledge

The Self-Referential Nature of Learning and Discovery

Knowledge is not a static entity but a dynamic process that evolves through inquiry, reflection, and cumulative insight. Recursive Realism suggests that the development of knowledge is inherently self-referential, involving feedback loops where ideas are continuously re-evaluated and refined through experience and dialogue. This section explores how Recursive Realism can provide a deeper understanding of learning, scientific discovery, and the evolution of ideas.

The Recursive Nature of Conceptual Understanding

Conceptual understanding involves the ability to see connections between different ideas and apply knowledge in new contexts. Recursive Realism suggests that conceptual learning is inherently self-referential, involving a loop where new information is integrated with existing knowledge to create a more refined understanding.

Learning as a Recursive Process: Learning involves a cycle of acquiring information, integrating it with prior knowledge, and revisiting concepts to achieve a deeper understanding. Recursive Realism suggests that learning is a self-referential process, where the mind uses previous experiences to interpret new data, creating a loop that enhances understanding. For example, a student learning physics might use their understanding of Newtonian mechanics to grasp more complex concepts like relativity, creating a reciprocal relationship between new knowledge and established ideas. This perspective highlights how learning is not a linear accumulation of facts but a recursive process where understanding deepens through reflection and re-interpretation.

Metacognition, the ability to think about one's own thinking, is a key aspect of effective learning. Recursive Realism suggests that metacognition involves a self-referential loop, where learners monitor their thought processes, evaluate their understanding, and adjust their strategies. For example, a scientist conducting research might reflect on their assumptions, identify gaps in their knowledge, and design new experiments to address uncertainties, creating a loop of self-directed inquiry. This perspective allows for a richer understanding of how self-awareness enhances learning, where the mind uses recursive reflection to refine its approach and improve understanding.

Problem solving involves the ability to navigate uncertainty, explore possibilities, and generate solutions. Recursive Realism suggests that problem solving is inherently recursive, involving a loop of hypothesis generation, testing, and revision. For example, a mathematician solving a complex equation might try different approaches, evaluate their results, and adjust their strategy until they arrive at a solution, creating a recursive cycle of exploration and correction. This perspective highlights how problem solving is a dynamic process where knowledge evolves through continuous iteration, allowing ideas to become more precise and solutions more elegant.

The Cumulative Evolution of Scientific Knowledge

Science represents one of the most systematic approaches to understanding the world, involving experimentation, theory-building, and peer review. Recursive Realism suggests that scientific progress involves self-referential feedback loops, where theories are continually tested, refined, and revised through empirical evidence.

The scientific method involves a cycle of observation, hypothesis formation, experimentation, and conclusion, where results are used to revise theories and generate new questions. Recursive Realism suggests that science operates through a self-referential loop, where hypotheses are tested against data, and theories evolve through ongoing inquiry. For example, the shift from Newtonian physics to Einstein's theory of relativity involved revisiting assumptions about space and time, creating a loop of reflection and empirical testing that led to a new understanding of the cosmos. This perspective highlights how scientific knowledge evolves through recursive processes, where the mind uses self-correction and adaptation to achieve greater accuracy in understanding reality.

Falsifiability, the idea that a theory must be testable and refutable, is a cornerstone of scientific inquiry. Recursive Realism suggests that the process of falsification involves a recursive loop, where theories are continually challenged and revised based on new evidence. For example, the theory of evolution has been refined through genetic research, which has provided new insights into the mechanisms of natural selection and mutation, creating a loop where biological theories evolve in response to empirical findings. This perspective allows for a deeper appreciation of how scientific progress is not about achieving final truths but about continuously improving our understanding through self-correcting inquiry.

Thomas Kuhn's concept of paradigm shifts describes how scientific progress involves periods of normal science interrupted by revolutions that change theoretical frameworks. Recursive Realism suggests that paradigm shifts are self-referential events, where the scientific community re-evaluates its core assumptions and reconstructs its worldview. For example, the transition from the geocentric to the heliocentric model of the solar system involved challenging entrenched beliefs and revising theories about the universe, creating a loop of debate, discovery, and redefinition. This perspective highlights how scientific revolutions are part of a recursive process, where knowledge systems evolve through cycles of innovation and reflection.

The Role of Culture in the Evolution of Knowledge

Knowledge is not only a personal endeavour but also a cultural phenomenon, shaped by social interactions, traditions, and shared understanding. Recursive Realism suggests that cultural knowledge evolves through self-referential loops, where communities reflect on their practices, stories, and values to build collective understanding.

Oral traditions, the stories, myths, and wisdom passed down through generations, represent a form of self-referential knowledge that evolves through repetition and interpretation. Recursive Realism suggests that oral traditions involve a loop where stories are retold and reinterpreted to adapt to changing social needs, creating a dynamic form of cultural memory. For example, myths about creation or moral lessons may evolve over time to reflect the values of new generations, while maintaining a connection to the past. This perspective highlights how cultural knowledge is self-referential, where communities use storytelling to preserve and reinterpret their identity through time.

Education systems represent a formal structure for transmitting knowledge across generations, involving a recursive relationship between teachers, students, and societal expectations. Recursive Realism suggests that education involves a feedback loop, where knowledge is transferred and then reinterpreted by students in new contexts. For example, learning history involves understanding past events while reflecting on their relevance to current issues, creating a loop of interpretation that shapes cultural understanding. This perspective allows for a richer appreciation of how education is not about imparting static knowledge but about fostering the ability to think critically and engage with ideas in a dynamic way.

Technology, from the printing press to the internet, has transformed how knowledge is created, shared, and refined. Recursive Realism suggests that technological advancements create self-referential loops, where new tools change the way information is disseminated, leading to shifts in how knowledge is structured. For example, the digital age has created a loop where online platforms allow for instant feedback, collaboration, and revisions of ideas, shaping scientific discourse and cultural debates. This perspective highlights how technology acts as a catalyst for recursive learning, where the flow of information evolves through new forms of interaction and communication.

Knowledge evolves through self-referential processes, where learning, discovery, and cultural exchange create feedback loops that shape understanding. Recursive Realism provides a framework for understanding how conceptual understanding, scientific progress, and cultural knowledge are inherently dynamic, involving a cycle of reflection, adaptation, and refinement. By recognising the recursive nature of learning, Recursive Realism reveals how ideas are not static truths but living constructs that evolve through ongoing engagement with experience and dialogue.

Recursive Realism and the Interplay of Theory and Experience

The development of knowledge involves a dynamic interaction between theory, the abstract frameworks that help us explain the world, and experience, which provides the empirical evidence that tests and challenges those frameworks. Recursive Realism suggests that theories and experience form a self-referential loop, where ideas are refined through real-world feedback and adapted to new discoveries. This section explores how Recursive Realism offers a deeper understanding of the recursive nature of theory-building, the role of experimentation, and the dynamic interplay between ideas and reality.

Theory as a Self-Referential Construct

Theories are conceptual models that provide explanations for natural phenomena, guiding research and shaping understanding. Recursive Realism suggests that theories are inherently self-referential, as they involve hypothetical constructs that are continually tested and refined through feedback from empirical data.

Hypothesis formation is a central part of the scientific method, involving the creation of proposals that can be tested against observations. Recursive Realism suggests that hypotheses are self-referential constructs, where the mind uses existing knowledge to predict outcomes and then adjusts those predictions based on new evidence. For example, the theory of natural selection began as a hypothesis about how species evolve, which was then tested through observations of variation in populations and adaptation to environments. This recursive process allowed the theory to become more refined over time, as new evidence from genetics and ecology provided deeper insights. This perspective highlights how theories are dynamic structures, shaped by a continuous loop between prediction and evidence.

Theories often involve a degree of abstraction, where complex phenomena are simplified into models that capture their essential features. Recursive Realism suggests that abstraction is a self-referential process, where the mind extracts patterns from observations and generalises them into theoretical constructs. For example, the laws of thermodynamics abstract the behaviour of heat and energy into general principles that apply across varied contexts, from steam engines to biological metabolism. This perspective allows for a deeper understanding of how abstraction helps to organise knowledge, creating theories that can be applied to new situations while remaining open to revision through empirical feedback.

The self-correcting nature of science involves theories that are continually challenged and revised based on new data. Recursive Realism suggests that self-correction is a recursive loop, where discrepancies between theory and observation lead to adjustments in conceptual models. For example, the discovery of the expansion of the universe led to a revision of the steady-state model in favour of the Big Bang theory, which better explained observational evidence like cosmic background radiation. This perspective highlights how theories are adaptable, evolving through a cycle of refinement that ensures greater alignment with reality.

Experimentation and the Feedback Loop of Empirical Inquiry

Experimentation is a key method for testing theories, providing controlled conditions where variables can be manipulated and outcomes measured. Recursive Realism suggests that experimentation is inherently self-referential, as it involves designing tests that challenge assumptions and then revisiting theories based on results.

Designing an experiment involves hypothesising about causal relationships, manipulating variables, and observing outcomes. Recursive Realism suggests that the experimental method is a feedback loop, where data is used to confirm or refute hypotheses, leading to revisions in theoretical understanding. For example, experiments in quantum physics, such as the double-slit experiment, have challenged classical assumptions about particles and waves, leading to the development of quantum mechanics. This recursive process reveals how experiments act as reflections on theory, providing insights that guide conceptual refinement.

Iteration is central to scientific inquiry, where experiments are repeated and adjusted to achieve more precise results. Recursive Realism suggests that iteration involves a self-referential loop, where each round of testing builds on the previous one, creating a cycle of increasing accuracy. For example, the development of vaccines involves multiple stages of testing in clinical trials, where dosage and delivery methods are refined based on feedback from earlier trials. This perspective highlights how knowledge evolves through iterative refinement, where the mind uses experimental results to adjust theories and improve outcomes.

Anomalies, observations that do not fit with existing theories, often act as catalysts for scientific breakthroughs. Recursive Realism suggests that anomalies create a loop of re-evaluation, where theories are challenged and expanded to accommodate new data. For example, the discovery of the orbit of Mercury deviating from predictions made by Newtonian mechanics led to Einstein's theory of general relativity, which provided a more accurate explanation of gravitational effects. This perspective allows for a richer appreciation of how anomalies drive the evolution of knowledge, where the recognition of limits in current theories leads to innovative shifts in understanding.

Theory, Experience, and the Recursive Growth of Understanding

The relationship between theory and experience is central to the growth of knowledge, as ideas are shaped by real-world feedback and adjusted through reflection. Recursive Realism suggests that understanding evolves through a loop between abstract models and practical experiences, where the mind integrates theoretical insights with empirical data to create a cohesive worldview.

Pragmatism, the philosophy that ideas should be tested based on their practical outcomes, reflects the recursive nature of knowledge-building. Recursive Realism suggests that pragmatic testing involves a feedback loop, where theories are applied to real-world situations and then revised based on their effectiveness. For example, the theory of democracy has evolved through political practice, where different models of governance have been tested and adapted based on social outcomes. This perspective highlights how understanding is shaped by the recursive interaction between ideas and experiences, where theories are validated through their ability to work in practical contexts.

Intuition, the ability to understand or know something without conscious reasoning, often plays a role in theoretical insight and discovery. Recursive Realism suggests that intuition is a self-referential process, where the mind draws on deeply internalised knowledge to generate new hypotheses. For example, Einstein's thought experiments about light and relativity relied on intuitive insights that were later confirmed through mathematical formulation and empirical testing. This recursive interplay allows for a deeper appreciation of how intuition and formal theory are interconnected, where abstract insights are refined through empirical validation.

Worldviews, the collective understanding that shapes how cultures interpret reality, evolve through a recursive process of reflection and adaptation. Recursive Realism suggests that cultural worldviews involve a feedback loop, where shared theories about existence, morality, and nature are revised through generational experience. For example, the shift from a geocentric to a heliocentric worldview changed how societies understood their place in the cosmos, leading to shifts in religious beliefs, philosophical ideas, and scientific inquiry. This perspective highlights how knowledge systems are not fixed but dynamic, evolving through a continuous process of theory revision and cultural adaptation.

The evolution of knowledge is driven by the dynamic interplay between theory and experience, where abstract ideas are continually tested, refined, and adapted through real-world feedback. Recursive Realism provides a framework for understanding how theories are living constructs, shaped by self-referential loops that integrate empirical data with conceptual models. By recognising the recursive nature of theoretical development, Recursive Realism reveals how understanding is not linear but iterative, involving a cycle of hypothesis testing, experimental feedback, and the refinement of ideas that drives the growth of knowledge.

In Chapter 25, we have explored how Recursive Realism offers a framework for understanding the development of knowledge, revealing how learning, scientific inquiry, and cultural evolution are shaped by self-referential feedback loops. By examining the nature of conceptual understanding, the role of experimentation, and the dynamic interaction between theory and experience, Recursive Realism provides a vision of knowledge-building as a recursive process that evolves through reflection, adaptation, and empirical testing. This perspective allows us to appreciate how theories are not static truths but adaptive frameworks that are continually refined through engagement with the world, leading to a more nuanced understanding of reality.

References

Bachelard, G. (1984) *The New Scientific Spirit.* Boston: Beacon Press.

Dewey, J. (1916) *Democracy and Education: An Introduction to the Philosophy of Education.* New York: Macmillan.

Einstein, A. and Infeld, L. (1938) *The Evolution of Physics.* Cambridge: Cambridge University Press.

Feyerabend, P.K. (1975) *Against Method.* London: Verso.

Gopnik, A. and Meltzoff, A.N. (1997) *Words, Thoughts, and Theories.* Cambridge, MA: MIT Press.

Hempel, C.G. (1965) *Aspects of Scientific Explanation and Other Essays in the Philosophy of Science.* New York: Free Press.

Kuhn, T.S. (1962) *The Structure of Scientific Revolutions.* Chicago: University of Chicago Press.

Lakatos, I. (1970) 'Falsification and the Methodology of Scientific Research Programmes', in Lakatos, I. and Musgrave, A. (eds.) *Criticism and the Growth of Knowledge.* Cambridge: Cambridge University Press, pp. 91–196.

Popper, K.R. (1959) *The Logic of Scientific Discovery.* London: Routledge.

Quine, W.V.O. (1969) *Ontological Relativity and Other Essays.* New York: Columbia University Press.

Rescher, N. (1980) *Induction: An Essay on the Justification of Inductive Reasoning.* Pittsburgh: University of Pittsburgh Press.

Vygotsky, L.S. (1978) *Mind in Society: The Development of Higher Psychological Processes.* Cambridge, MA: Harvard University Press.

Whewell, W. (1847) *The Philosophy of the Inductive Sciences, Founded Upon Their History.* London: Parker.

Chapter 26: Recursive Realism and Language

The Evolution of Language as a Recursive Phenomenon

Language is one of the most distinctive features of human cognition, allowing for the expression of thoughts, emotions, and abstract ideas. Recursive Realism suggests that language is inherently self-referential, involving recursive structures that allow for the infinite generation of meaning from finite elements. This section explores how Recursive Realism offers insights into the evolution of language, examining the development of syntax, the relationship between words and concepts, and the role of language in shaping human thought.

Syntax and the Recursiveness of Grammar

Syntax, the rules that govern how words are combined into sentences, is central to the structure of language. Recursive Realism suggests that syntax is inherently recursive, allowing for the creation of complex sentences through the repetition and nesting of grammatical structures.

The Recursive Nature of Sentence Formation: Sentences are built from words and phrases that can be combined in nested structures, creating complex expressions from simple rules. Recursive Realism suggests that syntax operates through a self-referential process, where phrases can contain sub-phrases that follow the same grammatical patterns. For example, in the sentence "The cat that chased the mouse is sleeping," the phrase "that chased the mouse" is embedded within the larger structure, creating a recursive loop that adds meaning to the description. This perspective highlights how language allows for infinite creativity through recursion, where simple grammatical rules can generate an endless variety of expressions.

The Role of Recursion in Language Evolution: The evolution of language may have been driven by the development of recursive grammar, which allowed early humans to express increasingly complex ideas. Recursive Realism suggests that the ability to nest phrases and create hierarchical structures enabled the expansion of communication, from simple commands to stories, myths, and abstract reasoning. For example, the transition from proto-languages, which may have involved simple word combinations, to fully developed languages likely involved the emergence of recursive syntax, allowing for nuanced descriptions of events, relationships, and possibilities. This perspective allows for a deeper understanding of how language evolved as a recursive tool, where complexity in syntax led to greater flexibility in expression.

Chomsky's Universal Grammar and Recursive Structures: Noam Chomsky's theory of universal grammar suggests that the capacity for recursive grammar is an innate feature of the human mind, shared across all languages. Recursive Realism builds on this idea by suggesting that universal grammar reflects a self-referential process, where the mind is structured to recognise and generate recursive patterns in language. For example, the ability to understand relative clauses ("the dog that barked") is found across diverse languages, indicating a shared cognitive capacity for handling nested structures. This perspective highlights how language is deeply intertwined with the brain's recursive abilities, suggesting that the evolution of syntax is a manifestation of the mind's broader capacity for self-referential thinking.

Words, Concepts, and the Recursive Construction of Meaning

Words are the building blocks of language, serving as symbols that represent concepts and ideas. Recursive Realism suggests that the relationship between words and concepts is inherently self-referential, as meanings are shaped by context, experience, and cultural interpretation.

The Dynamic Nature of Word Meanings: Words do not have fixed meanings but gain their significance through context and usage. Recursive Realism suggests that the meaning of a word is a self-referential construct, shaped by the feedback loop between how it is used and how it is understood. For example, the word "light" can mean illumination, weightlessness, or even clarity depending on the context in which it is used, creating a loop where meaning is adjusted through interpretation. This perspective highlights how language is not a static system but a dynamic process where words adapt to changing contexts through self-referential interactions.

Metaphor and the Recursion of Conceptual Mapping: Metaphors allow for the mapping of concepts from one domain to another, creating new ways of understanding abstract ideas. Recursive Realism suggests that metaphors are inherently self-referential, involving a loop where concepts are reinterpreted through figurative language. For example, describing time as a river creates a metaphorical loop where the flow of water helps us understand the passage of time, leading to new insights about change and continuity. This perspective highlights how metaphors are a recursive tool for conceptual innovation, allowing language to expand its expressive capacity by reorganising and redefining meanings.

Semantic Networks and the Self-Referential Nature of Knowledge: Semantic networks, the webs of association between words and concepts, reflect the recursive structure of meaning in language. Recursive Realism suggests that semantic understanding involves a loop where concepts are connected and reconnected through language use, creating networks that organise knowledge. For example, the word "tree" is associated with concepts like forest, nature, and growth, creating a web of meanings that evolves through cultural context and personal experience. This perspective highlights how language serves as a recursive framework for organising reality, where words and concepts form a dynamic network that guides thought and communication.

Language as a Tool for Self-Referential Thought

Language is not only a means of communication but also a tool for thinking, allowing us to reflect on our own thoughts and construct complex ideas. Recursive Realism suggests that language enables self-referential thinking, where the mind can describe, analyse, and question its own processes.

Internal Dialogue and the Recursion of Self-Awareness: Internal dialogue, the silent conversation we have with ourselves, is a key feature of human consciousness. Recursive Realism suggests that internal dialogue is inherently self-referential, involving a loop where the mind uses language to reflect on its thoughts and emotions. For example, when making a difficult decision, we might mentally weigh pros and cons, creating a loop of self-questioning that helps clarify intentions. This perspective highlights how language enables self-reflection, allowing us to explore and modify our understanding of ourselves and the world.

Narrative Identity and the Recursive Construction of the Self: Narratives, the stories we tell about ourselves, play a crucial role in shaping identity. Recursive Realism suggests that narrative identity involves a self-referential loop, where individuals use language to construct a coherent story about their past, present, and future. For example, a person might describe themselves as a "survivor" of hardship, creating a narrative loop where past challenges are integrated into a story that shapes their current self-concept. This perspective allows for a deeper understanding of how language is used to construct meaning in identity, where the self is continuously revised through the stories we create.

The Role of Language in Abstract Thought: Language enables abstract thought, allowing us to conceptualise ideas that are not directly tied to sensory experience. Recursive Realism suggests that abstract reasoning involves a self-referential process, where the mind uses words and symbols to manipulate concepts and explore hypothetical scenarios. For example, discussing mathematics or philosophy involves manipulating symbols and formulating arguments that transcend immediate reality, creating a loop where language allows for the exploration of possibilities beyond the physical world. This perspective highlights how language serves as a recursive tool for extending cognition, where words become a medium for conceptual exploration and the creation of new ideas.

Language is inherently self-referential, allowing for the creation of complex expressions, the development of meaning, and the construction of thought through recursive processes. Recursive Realism provides a framework for understanding how syntax, semantics, and internal dialogue interact to shape the evolution of language, revealing how the mind uses words to organise reality and navigate complex ideas. By recognising the recursive nature of language, Recursive Realism offers insights into how language serves as both a tool for communication and a framework for self-reflection, enabling humans to create, understand, and transform the world through words.

Recursive Realism and the Role of Language in Cultural Evolution

Culture encompasses the shared beliefs, practices, values, and knowledge systems that shape human communities. Language plays a central role in cultural evolution, acting as a medium through which ideas are transmitted, shared, and transformed. Recursive Realism suggests that the relationship between language and culture is inherently self-referential, involving feedback loops where cultural norms shape language use, and language in turn influences the development of cultural identities. This section explores how language facilitates the transmission of knowledge, the formation of social structures, and the evolution of worldviews, offering a deeper understanding of how Recursive Realism can illuminate the dynamic interaction between language and cultural change.

Language as a Vehicle for Cultural Knowledge

Language is a primary tool for transmitting cultural knowledge, allowing communities to preserve their histories, traditions, and wisdom across generations. Recursive Realism suggests that the transmission of knowledge through language involves a self-referential loop, where ideas are passed down, reinterpreted, and adapted to new contexts.

Oral Tradition and the Recursion of Storytelling: Oral traditions, the stories, myths, and songs that are shared verbally, represent a form of cultural memory that evolves through repetition and reinterpretation. Recursive Realism suggests that storytelling is inherently self-referential, as each retelling involves a dialogue between the past and the present, where new meanings are added while core elements are preserved. For example, a myth about creation might be retold across generations, with each storyteller adapting it to address current concerns, creating a loop where cultural values are both maintained and transformed. This perspective highlights how language serves as a medium for cultural continuity, where stories act as recursive structures that carry knowledge through time.

Writing Systems and the Recursion of Recorded History: The development of writing allowed for the preservation of knowledge in a more permanent form, enabling the recording of history, laws, and philosophies. Recursive Realism suggests that writing creates a self-referential loop, where ideas are captured in texts and then revisited by future generations for interpretation and debate. For example, ancient texts like Homer's *Iliad* or religious scriptures have been reinterpreted across different eras, allowing for a continuous dialogue between past wisdom and current understanding. This perspective allows for a deeper appreciation of how written language contributes to cultural evolution, where texts act as anchors for reflecting on tradition while providing a basis for new interpretations.

Language, Ritual, and the Recursion of Cultural Practices: Rituals, the formalised actions that mark important events or social transitions, often rely on language to codify meanings and connect participants to a shared tradition. Recursive Realism suggests that ritual language is a self-referential tool, where words and phrases used in ceremonies create a loop between individuals and collective beliefs. For example, reciting a prayer or chanting a mantra creates a shared experience that reinforces group identity and cultural continuity, while allowing for personal reflection within the collective act. This perspective highlights how language shapes cultural practices through recursion, where ritual language creates a feedback loop between individual meaning and communal identity.

The Role of Language in Shaping Social Structures

Language is not only a means of expressing ideas but also a tool for organising society, influencing social hierarchies, power dynamics, and cultural norms. Recursive Realism suggests that social structures are shaped through self-referential interactions between language use and social behaviour, where language both reflects and reinforces cultural expectations.

Language and the Construction of Social Identity: Language plays a key role in shaping identity, as it provides the terms and concepts through which individuals understand their role within a community. Recursive Realism suggests that social identity involves a recursive loop, where language provides labels and categories that influence how people see themselves and others. For example, the use of pronouns or titles (such as "Mr.", "Ms.", or "Doctor") can shape perceptions of status and identity, creating a loop where language choices reinforce social roles. This perspective allows for a deeper understanding of how language contributes to social cohesion, where the words we use create a shared understanding of who we are and how we relate to others.

Power Dynamics and the Recursion of Discourse: Language is often used to exert power and influence, shaping the way issues are discussed and understood. Recursive Realism suggests that power dynamics in discourse create a self-referential loop, where certain terms and narratives dominate the conversation, influencing public perception and social norms. For example, the language of politics can shape attitudes toward policies and social groups, creating a loop where rhetorical strategies influence voter behaviour and political outcomes. This perspective highlights how language is not a neutral medium but a tool for shaping power relations, where the control of discourse plays a key role in structuring society.

Cultural Norms and the Feedback Loop of Language Use: Cultural norms are often embedded in language, influencing how individuals behave and interact within their community. Recursive Realism suggests that language use creates a loop between cultural expectations and social behaviour, where linguistic conventions shape norms and norms influence language. For example, the use of politeness markers (like "please" and "thank you") in certain cultures reflects values of respect and courtesy, creating a feedback loop where language and social norms reinforce each other. This perspective allows for a deeper appreciation of how language helps to codify and transmit cultural norms, ensuring that cultural values are maintained over time while allowing for gradual adaptation.

The Evolution of Worldviews through Language and Recursion

Language is a central element in the formation of worldviews, as it shapes how people interpret reality, understand their place in the cosmos, and make sense of existence. Recursive Realism suggests that worldviews evolve through self-referential dialogues, where language enables communities to reflect on their beliefs and reorganise their understanding of the world.

Philosophy, the systematic exploration of existence, knowledge, and morality, relies heavily on language to articulate concepts and explore abstract ideas. Recursive Realism suggests that philosophical discourse is inherently recursive, as it involves critique, debate, and re-evaluation of ideas over time. For example, Western philosophy has evolved through a series of dialogues between thinkers like Plato, Aristotle, Descartes, and Kant, creating a loop where new ideas build upon and challenge earlier frameworks. This perspective highlights how language serves as a medium for intellectual evolution, where philosophical concepts are refined through recursive debate and reflection.

Religious beliefs and myths often provide a framework for understanding the universe, morality, and the meaning of life. Recursive Realism suggests that religious language creates a self-referential loop, where sacred texts and rituals are interpreted and reinterpreted to address the spiritual needs of different eras. For example, scriptural interpretations in Judaism, Christianity, or Islam involve ongoing debates about the meaning of texts, creating a loop where traditional beliefs are adapted to modern contexts. This perspective allows for a richer understanding of how language shapes spiritual worldviews, where religious language provides a reciprocal process of tradition and renewal.

Science, the pursuit of knowledge through observation and experimentation, also relies on language to formulate theories, share discoveries, and build a collective understanding of the natural world. Recursive Realism suggests that scientific discourse is a feedback loop, where theories are tested through experimentation and then refined through peer review and academic dialogue. For example, the theory of evolution has been expanded and refined through genetic research and ecological studies, creating a loop where the language of biology evolves alongside new discoveries. This perspective highlights how language allows for the accumulation of scientific knowledge, creating a recursive process where ideas are continually tested and revised through collective inquiry.

Language plays a crucial role in the evolution of culture, shaping how communities transmit knowledge, organise society, and construct worldviews. Recursive Realism provides a framework for understanding how cultural practices, social norms, and intellectual traditions are shaped through self-referential dialogue, where language acts as a bridge between individual minds and collective experience. By recognising the recursive nature of language, Recursive Realism reveals how words and ideas evolve through cultural feedback loops, shaping the growth of understanding and social cohesion over time.

In Chapter 26, we explored how Recursive Realism offers a framework for understanding the nature of language, revealing how syntax, meaning, and cultural expression are shaped by self-referential processes. By examining the evolution of grammar, the role of language in shaping identity and social structures, and the dynamic interplay between words and worldviews, Recursive Realism provides a vision of language as a living, dynamic system that evolves through reflection, adaptation, and cultural interaction. This perspective allows us to appreciate how language is not only a tool for communication but also a framework for understanding reality, enabling humans to navigate and transform their cultural and intellectual landscape through the power of words.

References

Chomsky, N. (1957) *Syntactic Structures*. The Hague: Mouton.

Chomsky, N. (2002) *On Nature and Language*. Cambridge: Cambridge University Press.

Clark, A. and Chalmers, D. (1998) 'The Extended Mind', *Analysis*, 58(1), pp. 7–19.

De Saussure, F. (1983) *Course in General Linguistics*. Translated by R. Harris. London: Duckworth.

Fauconnier, G. and Turner, M. (2002) *The Way We Think: Conceptual Blending and the Mind's Hidden Complexities*. New York: Basic Books.

Hockett, C.F. (1960) 'The Origin of Speech', *Scientific American*, 203(3), pp. 88–96.

Lakoff, G. and Johnson, M. (1980) *Metaphors We Live By*. Chicago: University of Chicago Press.

Pinker, S. (1994) *The Language Instinct: How the Mind Creates Language*. New York: William Morrow.

Tomasello, M. (1999) *The Cultural Origins of Human Cognition*. Cambridge, MA: Harvard University Press.

Vygotsky, L.S. (1986) *Thought and Language*. Cambridge, MA: MIT Press.

Whorf, B.L. (1956) *Language, Thought, and Reality: Selected Writings of Benjamin Lee Whorf*. Edited by J.B. Carroll. Cambridge, MA: MIT Press.

Zipf, G.K. (1949) *Human Behavior and the Principle of Least Effort.* Cambridge, MA: Addison-Wesley.

Chapter 27: Recursive Realism and the Dynamics of Human Emotion

The Recursive Nature of Emotional Experience

Emotions are fundamental to human experience, shaping how we perceive the world, relate to others, and understand ourselves. Recursive Realism suggests that emotions are inherently self-referential, involving feedback loops where feelings interact with thoughts, memories, and social contexts to create a dynamic emotional landscape. This section explores how Recursive Realism can offer insights into the nature of emotions, examining the role of self-reflection in emotional regulation, the relationship between emotions and cognition, and the self-referential dynamics that shape emotional responses.

Emotions as Feedback Loops

Emotions are often described as reactions to external events or internal thoughts, but they also involve a continuous cycle of reflection and adjustment. Recursive Realism suggests that emotional experiences are feedback loops, where feelings are influenced by cognitive appraisals, and in turn shape how we think and interpret situations.

The Interaction Between Emotion and Thought: Cognitive theories of emotion, such as those proposed by Lazarus and Schachter, suggest that emotions arise from interpretations of situations and events. Recursive Realism builds on this by suggesting that thoughts and emotions form a recursive loop, where interpretations influence emotional responses, and emotional states reshape thought patterns. For example, feeling anxious about an upcoming presentation may lead to thoughts that magnify potential risks, which in turn heighten the anxiety, creating a cycle that intensifies the emotional experience. This perspective allows for a deeper understanding of how emotions are not static reactions but dynamic processes shaped by ongoing reflection and cognitive feedback.

Self-Reflection and the Regulation of Emotions: Self-reflection, the process of thinking about one's own feelings, plays a crucial role in emotional regulation. Recursive Realism suggests that self-reflection creates a feedback loop, where awareness of emotions allows for adjustments in behaviour and thoughts. For example, recognising that one is feeling irritable might lead to adjustments in communication style, creating a loop where emotional awareness guides social interactions. This perspective highlights how self-reflection allows for greater control over emotional responses, where the mind can moderate feelings through recursive awareness.

Emotional Feedback in Social Interactions: Emotions are also shaped by social feedback, where the reactions of others influence how we feel about ourselves and our experiences. Recursive Realism suggests that social interactions create recursive loops of emotional exchange, where expressing emotions leads to responses from others, which in turn influence our feelings. For example, expressing sadness might elicit comfort from a friend, creating a loop where the support from the social environment helps to diminish the negative emotion. This perspective allows for a richer understanding of how emotions are shaped within social contexts, where feedback loops between individuals create a dynamic flow of emotional experiences.

The Role of Memory in Emotional Recursion

Memories play a central role in shaping emotions, as they provide context for understanding current experiences and anticipating future events. Recursive Realism suggests that emotional memories are inherently self-referential, involving feedback loops where past feelings influence present emotions, and current experiences reshape emotional recollections.

Emotional Memory and the Recurrence of Past Feelings: Memories of emotional events, such as loss, joy, or fear, can resurface in similar situations, creating a loop where past experiences shape present feelings. Recursive Realism suggests that emotional memories create a feedback process, where recalling a past event triggers similar feelings, which in turn reinforce the memory. For example, recalling a comforting moment with a loved one may evoke feelings of warmth, creating a loop where memory and emotion sustain each other. This perspective highlights how emotions are deeply intertwined with memory, where the mind revisits past experiences to make sense of present feelings.

Anticipation and the Recursiveness of Emotional Forecasting: Anticipation of future events often involves emotional forecasting, where the mind predicts how a situation will feel based on past experiences. Recursive Realism suggests that anticipation is a self-referential process, where memories of past feelings are projected onto future scenarios, creating a loop that influences decision-making. For example, anticipating the stress of a challenging task might evoke memories of past difficulties, which in turn shape the emotional state leading up to the event. This perspective allows for a deeper understanding of how emotions guide behaviour, where the mind uses emotional memories to navigate potential outcomes.

Trauma and the Feedback Loop of Emotional Triggers: Traumatic memories can create intense feedback loops, where re-experiencing a traumatic event through triggers leads to heightened emotional responses. Recursive Realism suggests that trauma involves a self-referential loop, where triggers in the environment evoke vivid memories that reignite past feelings, reinforcing the emotional impact. For example, a sound or scent associated with a traumatic event might trigger a flood of memories and emotional distress, creating a loop where the past reasserts itself in the present moment. This perspective highlights how emotions are not isolated from memory but are part of a complex network of self-referential processes that shape emotional health.

Emotional Complexity and the Recursion of Self-Concept

Self-concept, the understanding of who we are, is closely tied to our emotional life, as feelings about ourselves influence our sense of identity. Recursive Realism suggests that emotional self-concept is inherently self-referential, involving feedback loops where the way we feel about ourselves shapes our behaviour, which in turn reinforces or challenges our self-image.

Self-Esteem and the Recursion of Positive and Negative Emotions: Self-esteem, the value we place on ourselves, is often influenced by recurring emotional patterns, where positive experiences reinforce a sense of worth, and negative experiences diminish it. Recursive Realism suggests that self-esteem is shaped by a feedback loop, where emotional reactions to success or failure influence self-perception. For example, feeling proud of an achievement might boost self-confidence, which in turn encourages further positive actions, creating a loop of self-reinforcement. Conversely, experiencing repeated setbacks might lead to negative self-talk, creating a downward spiral of self-doubt. This perspective highlights how emotional self-concept is not static but evolves through reciprocal interactions between feelings and self-reflection.

Emotional Ambivalence and the Complexity of Mixed Feelings: Emotional ambivalence, the experience of holding conflicting feelings about a person, situation, or decision, reveals the complexity of human emotion. Recursive Realism suggests that ambivalence involves a self-referential loop, where opposing feelings create a cycle of inner conflict and reflection. For example, feeling both excitement and fear about a new opportunity might create a loop where the mind oscillates between positive and negative appraisals, leading to a richer understanding of the situation's complexities. This perspective allows for a deeper appreciation of how emotions are multidimensional, where the mind engages in recursive processes to navigate the nuances of feeling.

Empathy and the Recursion of Emotional Resonance: Empathy, the ability to understand and share the feelings of others, involves a self-referential process, where the mind uses its own experiences to mirror and interpret the emotions of another person. Recursive Realism suggests that empathy creates a loop of emotional resonance, where feeling the emotions of others influences our own state, and our response shapes the emotional dynamic of interpersonal relationships. For example, feeling sympathy for a friend's loss might involve recalling a similar experience and reflecting on the shared aspects of grief, creating a loop where connection deepens through mutual emotional understanding. This perspective highlights how empathy is not a passive experience but an active, recursive process that builds emotional bridges between individuals.

Emotions are inherently self-referential, involving complex feedback loops where feelings, thoughts, and social interactions interact to shape emotional experiences. Recursive Realism provides a framework for understanding how self-reflection, memory, and social feedback shape the dynamics of emotion, revealing how feelings are not fixed reactions but dynamic processes that evolve through self-awareness and contextual interpretation. By recognising the recursive nature of emotions, Recursive Realism offers insights into the rich complexity of emotional life, where the mind navigates a constantly changing landscape of internal states.

Recursive Realism and the Role of Emotions in Social Dynamics

Emotions play a fundamental role in shaping social behaviour, influencing how individuals interact, build relationships, and create social norms. Recursive Realism suggests that emotions are inherently self-referential in a social context, involving feedback loops where individual feelings resonate with group sentiments, shaping shared emotional experiences. This section explores how Recursive Realism offers insights into the dynamics of emotional communication, the formation of collective emotions, and the role of empathy and conflict in social life.

Emotional Contagion and the Recursion of Shared Feelings

Emotional contagion refers to the phenomenon where emotions spread from one individual to others, creating shared feelings within groups. Recursive Realism suggests that emotional contagion involves a self-referential process, where the expression of emotions triggers reciprocal reactions in others, creating a loop of mutual influence.

The Dynamics of Group Emotion: Group emotions emerge when individual emotional states become synchronised, creating a shared mood that influences group behaviour. Recursive Realism suggests that group emotions are self-reinforcing, involving feedback loops where the expression of emotion by one member influences the emotional state of others, creating a collective experience. For example, in a celebratory gathering, enthusiasm and joy can spread through expressions like laughter and cheering, creating a loop where individual excitement amplifies the group's overall joy. This perspective highlights how emotional contagion plays a key role in building social cohesion, where shared feelings create a sense of unity among group members.

Empathy and the Feedback Loop of Emotional Resonance: Empathy, the ability to understand and share the feelings of others, is crucial for emotional contagion and social bonding. Recursive Realism suggests that empathy creates a recursive loop where perceiving the emotions of another influences one's own emotional state, which in turn shapes how one responds. For example, witnessing a friend's sadness may evoke feelings of compassion, leading to comforting actions, which then reinforce the bond between friends. This perspective allows for a deeper understanding of how empathy facilitates social connection, where reciprocal emotional responses build trust and mutual understanding.

The Role of Nonverbal Communication in Emotional Feedback: Nonverbal cues, such as facial expressions, tone of voice, and body language, are key elements in emotional contagion. Recursive Realism suggests that nonverbal communication creates a loop of emotional feedback, where subtle cues shape how feelings are perceived and responded to. For example, a smile can signal warmth and openness, encouraging positive interactions that create a loop of mutual friendliness. Conversely, defensive body language might trigger tension in others, creating a cycle of distance. This perspective highlights how nonverbal communication is a self-referential aspect of emotional dynamics, where unspoken signals shape the emotional tone of interactions.

The Formation of Collective Emotions and Social Identity

Collective emotions, the shared feelings that emerge within groups, play a crucial role in shaping social identity and group cohesion. Recursive Realism suggests that collective emotions are inherently self-referential, involving feedback loops where individual feelings are amplified and organised into group-level experiences.

Shared Emotional Experiences and the Construction of Group Identity: Emotional experiences, such as celebrations, mourning rituals, or social movements, often become foundations for group identity, creating a sense of belonging among members. Recursive Realism suggests that collective emotions involve a loop where participating in shared rituals reinforces group solidarity, while the group's collective energy shapes individual emotional states. For example, participating in a cultural festival might evoke pride and joy, creating a feedback loop where individual emotions contribute to a shared sense of cultural heritage. This perspective allows for a richer appreciation of how emotions contribute to social cohesion, where shared feelings become anchors for group identity.

Social Movements and the Recursion of Emotional Energy: Social movements often harness collective emotions like anger, hope, or solidarity to mobilise action and challenge existing structures. Recursive Realism suggests that social movements create self-referential loops where emotional appeals energise participants, leading to actions that reinforce the movement's goals. For example, protests against social injustice might evoke anger and determination, creating a cycle where participation reinforces commitment to change. This perspective highlights how collective emotions can drive social transformation, where shared feelings act as catalysts for coordinated action.

National Identity and the Emotional Dynamics of Patriotism: Patriotism, a sense of pride and loyalty to one's nation, often involves collective emotions that shape national identity. Recursive Realism suggests that patriotic emotions are self-referential, involving feedback loops where symbols like flags, anthems, and historical narratives evoke feelings of unity, which in turn strengthen national cohesion. For example, the emotional resonance of national celebrations like Independence Day can create a shared experience that reaffirms a sense of national belonging. This perspective highlights how emotions are central to forming collective identities, where shared feelings about history and values shape how communities see themselves in relation to the larger world.

Conflict, Cooperation, and the Role of Emotions in Social Dynamics

Emotions play a key role in both cooperation and conflict, influencing how groups interact, resolve disputes, and build alliances. Recursive Realism suggests that social dynamics involve self-referential loops where emotional responses shape group behaviour, creating cycles of collaboration or division.

Trust and the Feedback Loop of Cooperative Emotions: Trust is fundamental to cooperation, as it allows individuals to work together toward common goals. Recursive Realism suggests that trust involves a feedback loop, where acts of trust evoke positive emotional responses like gratitude and loyalty, which in turn reinforce the relationship. For example, collaborating on a project might build trust as partners demonstrate reliability, creating a loop where emotional bonds strengthen cooperative behaviour. This perspective allows for a deeper understanding of how positive emotions like trust and respect are central to social cohesion, where emotional reinforcement creates a foundation for collaboration.

Conflict and the Escalation of Negative Emotions: Conflict often involves negative emotions like anger, resentment, and fear, which can create cycles of escalation. Recursive Realism suggests that conflict involves self-referential loops where emotional responses to perceived threats or injustices fuel aggressive actions, leading to retaliation and further negative emotions. For example, a dispute between groups over resources might lead to hostility, creating a loop where each side's actions reinforce the other's fear and anger. This perspective highlights how emotions can contribute to social division, where the failure to break negative feedback loops can lead to prolonged conflicts.

Reconciliation and the Recursion of Forgiveness: Reconciliation, the process of repairing relationships after conflict, often requires emotional shifts like forgiveness and empathy. Recursive Realism suggests that forgiveness involves a self-referential loop, where letting go of resentment leads to positive emotional responses, which in turn open the possibility for renewed trust. For example, a reconciliation process between divided communities might involve acknowledging past hurts, leading to a cycle where forgiveness allows for emotional healing. This perspective highlights how emotions play a key role in restoring social harmony, where breaking negative feedback loops can pave the way for new cycles of cooperation.

Emotions are central to shaping social dynamics, influencing how individuals form bonds, navigate conflicts, and build collective identities. Recursive Realism provides a framework for understanding how emotions are inherently self-referential, involving feedback loops that shape the flow of feelings within communities. By recognising the reciprocal nature of emotional exchanges, Recursive Realism reveals how emotions serve as catalysts for social cohesion or division, offering a vision of social life where individual feelings and collective experiences are interconnected through self-referential dynamics.

In this chapter, we have explored how Recursive Realism offers a framework for understanding the nature of emotions, revealing how feelings interact with thoughts, memories, and social contexts through self-referential feedback loops. By examining the dynamics of emotional experience, the role of shared feelings in group identity, and the interplay between emotions and social dynamics, Recursive Realism provides a nuanced perspective on the complex nature of human emotional life. This perspective allows us to appreciate how emotions are not just biological reactions but dynamic processes that shape how individuals and communities understand and navigate the world.

References

Barrett, L.F. (2017) *How Emotions Are Made: The Secret Life of the Brain*. Boston: Houghton Mifflin Harcourt.

Ekman, P. (1992) 'An Argument for Basic Emotions', *Cognition and Emotion*, 6(3–4), pp. 169–200.

Fredrickson, B.L. (2001) 'The Role of Positive Emotions in Positive Psychology: The Broaden-and-Build Theory of Positive Emotions', *American Psychologist*, 56(3), pp. 218–226.

Goleman, D. (1995) *Emotional Intelligence: Why It Can Matter More Than IQ*. New York: Bantam Books.

Gross, J.J. (1998) 'The Emerging Field of Emotion Regulation: An Integrative Review', *Review of General Psychology*, 2(3), pp. 271–299.

Izard, C.E. (2009) 'Emotion Theory and Research: Highlights, Unanswered Questions, and Emerging Issues', *Annual Review of Psychology*, 60, pp. 1–25.

Lazarus, R.S. (1991) *Emotion and Adaptation*. New York: Oxford University Press.

Panksepp, J. (1998) *Affective Neuroscience: The Foundations of Human and Animal Emotions*. New York: Oxford University Press.

Schachter, S. and Singer, J.E. (1962) 'Cognitive, Social, and Physiological Determinants of Emotional State', *Psychological Review*, 69(5), pp. 379–399.

Thompson, R.A. (1994) 'Emotion Regulation: A Theme in Search of Definition', *Monographs of the Society for Research in Child Development*, 59(2–3), pp. 25–52.

Tracy, J.L. and Robins, R.W. (2007) 'Emerging Insights into the Nature and Function of Pride', *Current Directions in Psychological Science*, 16(3), pp. 147–150.

Chapter 28: Recursive Realism and the Perception of Time

Memory, Anticipation, and the Self-Referential Nature of Time

Time is a central aspect of human experience, shaping how we understand change, process events, and imagine possibilities. Recursive Realism suggests that the perception of time is inherently self-referential, involving feedback loops where memory and anticipation interact with present awareness to create a cohesive sense of temporal flow. This section explores how Recursive Realism offers insights into the dynamics of time perception, examining the role of memory in shaping the past, the importance of anticipation in imagining the future, and the recursive integration of past, present, and future in the experience of time.

Memory as a Window into the Past

Memory allows us to retain and recall past experiences, providing a sense of continuity that connects who we are now with who we have been. Recursive Realism suggests that memory is a self-referential process, involving loops where past experiences are revisited, reinterpreted, and integrated into the present moment.

The Recursiveness of Episodic Memory: Episodic memory, the ability to remember specific events in our personal history, is central to the perception of time. Recursive Realism suggests that episodic memory involves a feedback loop, where the act of remembering brings the past into the present, allowing for reflection and re-evaluation. For example, recalling a childhood memory might evoke vivid imagery, creating a loop where the past event is relived in the present mind, influencing current emotions and understanding. This perspective highlights how memory creates a bridge between past and present, where revisiting memories allows for new insights and emotional growth.

Autobiographical Memory and the Construction of a Personal Timeline: Autobiographical memory, the narrative we create about our own life, involves organising events into a coherent story that provides a sense of self over time. Recursive Realism suggests that autobiographical memory is inherently self-referential, as the mind continually revisits and reorganises past experiences to make sense of who we are. For example, reflecting on key life events like graduation or moving to a new city allows us to understand how those moments have shaped our identity, creating a loop where the past is continually integrated into our current self-concept. This perspective allows for a deeper understanding of how memory contributes to temporal continuity, where self-reflection allows individuals to make sense of their journey through time.

The Role of Memory in Temporal Orientation: Memory not only connects us to the past but also helps us orient ourselves in the present and anticipate the future. Recursive Realism suggests that temporal orientation involves a feedback loop, where memories of similar events guide our interpretation of current experiences and shape our expectations. For example, remembering previous winters might influence how we prepare for the coming season, creating a loop where the past informs our present actions and future planning. This perspective highlights how memory is a dynamic process, where revisiting past experiences helps us navigate the flow of time.

Anticipation and the Projection of the Future

Anticipation allows us to imagine future possibilities, providing a sense of direction that guides decision-making and goal-setting. Recursive Realism suggests that anticipation is a self-referential process, where the mind uses past experiences to predict future outcomes, creating a loop between expectation and preparation.

The Recursion of Imagining Future Scenarios: Imagining the future involves visualising potential events based on past knowledge and current desires. Recursive Realism suggests that anticipation involves a feedback loop, where the mind constructs scenarios of what might happen, allowing for adjustments in plans and expectations. For example, planning a career move might involve envisioning possible challenges and opportunities, creating a loop where anticipated outcomes influence current decisions. This perspective highlights how anticipation is not about predicting the future with certainty but about exploring possibilities through recursive reflection.

Hope, Fear, and the Emotional Dimensions of Anticipation: Emotions like hope and fear are closely tied to anticipation, as they shape how we feel about the future. Recursive Realism suggests that emotional anticipation involves a self-referential loop, where expectations about positive or negative outcomes influence how we prepare and respond. For example, hoping for success might lead to optimistic behaviour, while fearing failure might result in defensive actions, creating a loop where emotional states shape our approach to future events. This perspective allows for a deeper understanding of how emotions interact with temporal perception, where anticipation creates a cycle of feeling and preparation that guides our journey through time.

Planning and the Recursive Organisation of Time: Planning involves organising actions in a sequence to achieve future goals, allowing us to structure our time effectively. Recursive Realism suggests that planning is a self-referential process, where the mind creates timelines for future actions and adjusts them based on new information. For example, preparing for a project deadline might involve breaking down tasks, adjusting priorities, and revisiting timelines as circumstances change, creating a loop where anticipation shapes the structure of time. This perspective highlights how planning is an adaptive process, where the mind uses recursive thinking to navigate the complexities of temporal constraints.

The Integration of Past, Present, and Future

The experience of time involves integrating past experiences, present awareness, and future possibilities into a cohesive whole. Recursive Realism suggests that this integration is a self-referential process, where the mind creates a continuous narrative that connects the flow of time.

Present Awareness as a Temporal Anchor: Present awareness provides a moment-to-moment anchor that allows us to stay grounded in the current experience, even as we recall the past or imagine the future. Recursive Realism suggests that present awareness creates a loop where the mind balances attention between past memories, current sensations, and future thoughts. For example, practising mindfulness might involve observing thoughts about upcoming events while remaining aware of present sensations, creating a cycle where awareness connects past and future through the immediacy of the now. This perspective highlights how the present serves as a pivot point in temporal perception, allowing us to navigate time through a balance of memory, awareness, and anticipation.

Narrative Time and the Self-Referential Story of Life: Human beings often understand their lives as stories, where events are organised into a coherent narrative that provides meaning and direction. Recursive Realism suggests that narrative time is a self-referential construct, where the mind creates a story that integrates past experiences, present challenges, and future aspirations. For example, reflecting on life milestones like relationships, career achievements, or personal growth allows us to construct a narrative that explains how we became who we are, creating a loop where our understanding of time shapes our sense of identity. This perspective allows for a richer understanding of how stories help us make sense of time, where the self is seen as a character in a larger unfolding narrative.

The Temporal Loop of Growth and Change: Personal growth involves reflecting on past experiences, understanding current challenges, and aspiring toward future goals. Recursive Realism suggests that growth is a recursive process, where each stage of life is informed by previous stages and shapes the path ahead. For example, learning from past mistakes might guide present choices, while envisioning future potential creates motivation to overcome current obstacles, creating a loop of self-improvement. This perspective highlights how time is not just a sequence of events but a self-referential journey where the mind integrates change into a continuous story of development.

Time perception involves a dynamic interplay between memory, anticipation, and present awareness, where the mind creates a cohesive understanding of past, present, and future. Recursive Realism provides a framework for understanding how memory brings the past into the present, anticipation projects future possibilities, and awareness anchors the experience of now, creating a self-referential narrative that shapes our understanding of time. This perspective allows us to appreciate how time is not a linear flow but a recursive process, where the mind navigates temporal complexity through reflection, imagination, and self-awareness.

Recursive Realism and the Interplay Between Objective Time and Subjective Experience

Time has long been a subject of scientific investigation and philosophical inquiry, with objective time representing the measurable flow of seconds, minutes, and years, while subjective time reflects individual experiences of how time feels. Recursive Realism suggests that the perception of time is inherently self-referential, involving a loop where objective measures and subjective experiences influence each other, creating a complex understanding of temporal reality. This section explores how Recursive Realism provides insights into the relationship between time as a physical dimension and time as a lived experience, examining how the mind reconciles the contrasts between measured time and the elasticity of subjective time.

Time as a Physical Dimension: The Objective Perspective

Physics views time as a dimension that is measured and quantified, providing a framework for understanding the sequence of events and the laws of nature. Recursive Realism suggests that objective time is structured through self-referential models, where the mind uses measurement to make sense of the passage of time.

Newtonian Time and the Concept of Absolute Flow:

Isaac Newton described time as an absolute flow, moving uniformly regardless of external influences. This view treats time as a linear dimension, where events unfold in a fixed sequence. Recursive Realism suggests that the concept of absolute time reflects a self-referential attempt to define time in measurable terms, creating a model that allows for consistent calculations of motion and change. For example, clocks and calendars provide a framework for organising activities, creating a loop where measured time shapes how we structure daily life. This perspective highlights how objective time provides a shared reference point that enables coordination and scientific exploration.

Relativity and the Elasticity of Time:

Albert Einstein's theory of relativity introduced the idea that time is not absolute but relative, influenced by gravity and motion. Relativity suggests that time can stretch or contract depending on the observer's perspective, challenging the notion of a fixed temporal flow. Recursive Realism suggests that relativity reflects a recursive model where time is understood through the interplay between measurement and observation, creating a loop that allows for multiple experiences of the same event. For example, time dilation, the slowing down of time near a massive object, illustrates how the structure of time changes in response to context. This perspective allows for a richer appreciation of how physical theories of time are dynamic models, where reality is shaped by the self-referential interaction between the observer and the observed.

Quantum Mechanics and the Uncertainty of Time:

In quantum mechanics, time is understood as part of the probabilistic nature of reality, where events unfold with a degree of uncertainty. Quantum theories challenge classical notions of time by suggesting that the sequence of events is not always predictable. Recursive Realism suggests that quantum time represents a self-referential approach to understanding uncertainty, where probabilistic models provide a framework for navigating the complexities of temporal change. For example, the uncertainty principle implies that the exact time of quantum events cannot be determined with precision, creating a loop where time is both a measure and a dynamic potential. This perspective highlights how objective time is multifaceted, involving self-referential models that adapt to different scales of reality.

The Subjective Experience of Time: The Elasticity of Perception

While science seeks to measure time, individuals experience time in a way that can feel stretchable or compressed, influenced by attention, emotion, and context. Recursive Realism suggests that subjective time is shaped by self-referential processes, where perception and cognition create a loop that influences how time is felt.

The Elasticity of Time in Everyday Experience: Subjective time can feel like it speeds up or slows down depending on circumstances. For example, time may seem to drag during moments of boredom or anxiety but fly by during engaging activities. Recursive Realism suggests that this elasticity involves a feedback loop, where the mind adjusts its perception of time based on focus and emotional engagement. For example, during a flow state, where attention is fully absorbed in a task, time seems to disappear, creating a loop where deep concentration alters awareness of duration. This perspective allows for a deeper understanding of how time perception is subjective, shaped by recursive interactions between thought and experience.

Memory, Attention, and the Subjective Construction of Time: The perception of time is closely tied to memory and attention, as the mind uses recollections to organise the past and focus to mark the passage of time. Recursive Realism suggests that subjective time involves a self-referential process, where memories of events create a sense of time's length and attention determines how time is experienced in the moment. For example, a day filled with novel experiences might feel longer in retrospect, as each moment is richly encoded in memory, creating a loop where time perception is influenced by the density of memories. Conversely, repetitive routines may make time feel shorter because fewer distinct memories are formed. This perspective highlights how subjective time is not uniform but context-dependent, shaped by self-referential reflections on experience.

Emotion and the Temporal Distortion of Experience: Emotions like fear, joy, or anticipation can distort the perception of time, making moments feel longer or shorter. Recursive Realism suggests that emotional time involves a feedback loop, where feelings influence how time is perceived, and perceptions of time shape emotional responses. For example, fear might intensify focus, making seconds feel like minutes, while joy can make hours feel like moments. This perspective allows for a deeper appreciation of how subjective time is intertwined with emotional states, where the mind uses recursive awareness to calibrate the passage of time based on internal feelings.

Bridging the Gap Between Objective and Subjective Time

Time can be understood as both an objective measurement and a subjective experience, creating a tension between scientific models and personal perception. Recursive Realism suggests that the relationship between objective and subjective time involves self-referential loops, where scientific understanding of temporal flow is informed by human experience, and subjective perceptions are shaped by cultural constructs of time.

Cultural Conceptions of Time and Temporal Norms: Different cultures understand time in unique ways, shaping how individuals perceive and measure temporal flow. Recursive Realism suggests that cultural norms about time create a feedback loop where social expectations influence individual experiences and collective values shape temporal habits. For example, cultures with a linear view of time may emphasise progress and future planning, while cyclical conceptions of time might focus on repetition and renewal. This perspective allows for a deeper understanding of how culture mediates the relationship between objective measures and subjective experiences, creating a loop where shared beliefs influence personal time perception.

Philosophy and the Recursion of Temporal Understanding: Philosophical inquiry has long explored the nature of time, questioning whether time exists independently or is a construct of human experience. Recursive Realism suggests that philosophical reflections on time are inherently self-referential, involving a loop where theories of time are tested against lived experience and then refined through reflection. For example, Henri Bergson's concept of duration emphasises the qualitative flow of time as felt through consciousness, contrasting with scientific models that treat time as measurable intervals. This perspective highlights how philosophical ideas create a dialogue between objective and subjective time, using recursive thought to bridge the gap between abstract models and personal experience.

The Mind's Temporal Loop: Integrating Science and Subjectivity: Recursive Realism proposes that the mind creates a temporal loop where objective time and subjective time are interwoven, allowing for a holistic understanding of reality. For example, the awareness of clock time can structure daily routines, while subjective feelings about time's flow influence how we experience those routines. This perspective suggests that time perception is a recursive construct, where measurable aspects of time interact with the fluidity of human experience, creating a dynamic relationship that shapes our understanding of existence.

Time is both a scientific concept and a personal experience, shaped by self-referential processes where the mind integrates measurable intervals with lived sensations. Recursive Realism provides a framework for understanding how objective models of time interact with subjective perceptions, creating a dynamic loop that influences how humans understand the flow of time. By recognising the recursive nature of time perception, Recursive Realism offers a vision of temporal experience where science and consciousness are interconnected, allowing for a deeper appreciation of time as both a physical reality and a psychological construct.

In Chapter 28, we have explored how Recursive Realism offers a framework for understanding the perception of time, revealing how memory, anticipation, and scientific models create a self-referential understanding of temporal reality. By examining the role of memory in shaping the past, the nature of anticipation in imagining the future, and the relationship between objective and subjective time, Recursive Realism provides a nuanced perspective on how humans navigate time. This perspective allows us to appreciate time as a dynamic interplay between physical measurements and personal experiences, offering a vision of reality that is both scientifically grounded and deeply human.

References

Bergson, H. (1910) *Time and Free Will: An Essay on the Immediate Data of Consciousness*. London: George Allen & Company.

Chomsky, N. (1965) *Aspects of the Theory of Syntax*. Cambridge, MA: MIT Press.

Einstein, A. (1916) *Relativity: The Special and General Theory*. Translated by R.W. Lawson. New York: Henry Holt and Company.

Lazarus, R.S. (1991) *Emotion and Adaptation*. New York: Oxford University Press.

Newton, I. (1687) *Philosophiæ Naturalis Principia Mathematica*. London: Royal Society.

Schachter, S. and Singer, J.E. (1962) 'Cognitive, Social, and Physiological Determinants of Emotional State', *Psychological Review*, 69(5), pp. 379–399.

Suddendorf, T. and Corballis, M.C. (2007) 'The Evolution of Foresight: What Is Mental Time Travel, and Is It Unique to Humans?', *Behavioral and Brain Sciences*, 30(3), pp. 299–313.

Thompson, E. (2007) *Mind in Life: Biology, Phenomenology, and the Sciences of Mind*. Cambridge, MA: Harvard University Press.

Tulving, E. (1983) *Elements of Episodic Memory*. New York: Oxford University Press.

Wittmann, M. (2018) *Felt Time: The Science of How We Experience Time*. Cambridge, MA: MIT Press.

Zimbardo, P.G. and Boyd, J.N. (1999) 'Putting Time in Perspective: A Valid, Reliable Individual-Differences Metric', *Journal of Personality and Social Psychology*, 77(6), pp. 1271–1288.

Chapter 29: Recursive Realism and the Emergence of Complexity in Nature

Self-Organisation and Recursive Patterns in Natural Systems

Nature is characterised by astonishing complexity, where simple elements combine to form intricate structures and dynamic systems. Recursive Realism suggests that this complexity emerges through self-referential processes, where patterns repeat, adapt, and evolve through recursive interactions. This section explores how Recursive Realism provides insights into the self-organising principles that shape biological life, ecosystems, and natural structures, examining how feedback loops and recursive interactions contribute to the emergence of order in nature.

Recursion in the Formation of Life

The emergence of life from simple molecules represents one of the most profound examples of complexity arising from simple rules. Recursive Realism suggests that the origin of life is characterised by self-referential processes, where molecular interactions create feedback loops that lead to increasing complexity.

Autocatalysis and the Recursion of Chemical Reactions: Autocatalytic reactions, where chemical processes catalyse themselves, are thought to be crucial to the origin of life. Recursive Realism suggests that autocatalysis involves a self-referential loop, where molecules interact in ways that accelerate their own formation, leading to complex networks of chemical reactions. For example, in prebiotic chemistry, simple molecules like amino acids and nucleotides might have self-organised into larger structures through repetitive reactions, creating a loop that drove increasing molecular complexity. This perspective highlights how life's origin can be understood as a recursive process, where the repetition of self-catalysing reactions created the building blocks of biology.

DNA and the Recursion of Genetic Information: DNA is a self-replicating molecule, capable of encoding and transmitting information across generations. Recursive Realism suggests that genetic information is inherently self-referential, as DNA sequences contain instructions for building the proteins that copy and repair the DNA itself. For example, the process of DNA replication involves a cycle where DNA serves as a template for its own duplication, creating a feedback loop that ensures genetic continuity. This perspective allows for a deeper understanding of how life is built on recursive patterns, where the structure of DNA embodies self-reference in its capacity to replicate and transmit information.

Metabolic Pathways and the Recursive Organisation of Biochemistry: Metabolism, the network of chemical reactions that sustains life, is organised through feedback loops that regulate and maintain balance within cells. Recursive Realism suggests that metabolic pathways are self-referential networks, where the products of one reaction serve as inputs for another, creating cycles that ensure biochemical stability. For example, the Krebs cycle is a series of reactions that generates energy for cellular functions, creating a loop where molecules are recycled through a sequence of interconnected steps. This perspective highlights how biological complexity emerges through recursively structured processes, where metabolism provides a foundation for the sustained organisation of life.

Recursive Patterns in Ecosystems

Ecosystems are complex networks of organisms, environmental factors, and interactions that create dynamic webs of life. Recursive Realism suggests that ecosystem dynamics are characterised by self-referential feedback loops, where species interactions and environmental processes shape the emergence of ecological order.

Food Webs and the Recursion of Energy Flow: Food webs represent the flow of energy through ecosystems, where producers, consumers, and decomposers interact in complex cycles. Recursive Realism suggests that food webs are self-referential structures, where energy transfer is organised through feedback loops that balance populations and resource availability. For example, an increase in herbivores might lead to overgrazing, reducing plant populations, which in turn influences predator numbers, creating a cycle where species dynamics adjust to resource changes. This perspective allows for a deeper understanding of how ecosystems maintain stability through recursive interactions, where each level of the food web influences and responds to the others.

Symbiosis and the Recursion of Mutual Benefit: Symbiotic relationships, where different species interact for mutual benefit, represent a form of biological recursion, where organisms create feedback loops that enhance survival. Recursive Realism suggests that symbiosis involves self-referential interactions, where the benefits provided by one species shape the behaviour and evolution of the other. For example, bees and flowering plants engage in mutualistic relationships, where bees pollinate flowers in exchange for nectar, creating a cycle that supports both species' reproduction. This perspective highlights how ecosystem complexity is shaped by recursively structured relationships, where mutual dependencies drive adaptation and co-evolution.

Ecological Succession and the Feedback Loop of Habitat Change: Ecological succession, the process of habitat change over time, illustrates how ecosystems evolve through feedback loops that drive community transformation. Recursive Realism suggests that succession involves a self-referential process, where initial conditions influence species colonisation, which in turn alters the environment, creating a new context for future species. For example, pioneer plants that colonise bare soil create organic matter that allows for more complex vegetation, creating a loop where each stage of succession sets the conditions for the next. This perspective highlights how ecosystems evolve through recursively structured change, where feedback between organisms and their environment drives the emergence of ecological complexity.

Self-Similarity and Fractals in Natural Patterns

Nature is filled with patterns that exhibit self-similarity, where smaller structures resemble larger forms, creating repeating motifs across different scales. Recursive Realism suggests that self-similar patterns in nature are manifestations of recursive processes, where simple rules create complex forms through iterative repetition.

Fractals and the Recursion of Geometric Patterns: Fractals are geometric shapes that display self-similarity, where each part of the shape is a scaled-down version of the whole. Recursive Realism suggests that fractal patterns reflect self-referential processes in nature, where patterns repeat through recursive iteration to create complex structures. For example, the branching of trees, the structure of snowflakes, and the shape of river networks all exhibit fractal-like patterns, creating a loop where growth processes replicate at different scales. This perspective highlights how nature's complexity emerges through recursive geometry, where self-similar structures reflect underlying rules of pattern formation.

Symmetry and the Recursion of Biological Forms: Symmetry is a common feature in biological forms, from the radial symmetry of jellyfish to the bilateral symmetry of mammals. Recursive Realism suggests that symmetry is a self-referential property, where the repetition of forms across an axis or plane creates a balanced structure. For example, the symmetrical arrangement of leaves on a stem or the structure of DNA's double helix reflects self-similar patterns that are recursively organised. This perspective allows for a deeper appreciation of how symmetry is a manifestation of recursive processes, where nature's forms are shaped by repeating motifs that create harmonious structures.

The Recursive Organisation of Tissues and Organs: Biological structures like organs and tissues often exhibit repeating patterns that create functional complexity. Recursive Realism suggests that the organisation of tissues is self-referential, where cellular patterns repeat to form larger structures with specific functions. For example, the alveoli in the lungs form a branching pattern that maximises surface area for gas exchange, creating a loop where simple structures build a complex organ. This perspective highlights how biological organisation is shaped by recursively structured processes, where repetition and variation create the intricate architecture of living systems.

Nature's complexity emerges from self-referential processes, where simple rules create intricate patterns through recursive repetition and feedback loops. Recursive Realism provides a framework for understanding how life, ecosystems, and geometric forms evolve through self-organisation, revealing how the natural world is structured through patterns that repeat, adapt, and integrate across different scales. By recognising the recursive nature of natural systems, Recursive Realism offers a vision of nature's complexity as a dynamic interplay between simplicity and emergence, where self-referential interactions shape the evolution of life.

Recursive Realism and the Interplay Between Simplicity and Complexity

The natural world is a mosaic of simple rules giving rise to complex phenomena, where order and chaos interact to create systems that are both stable and adaptive. Recursive Realism suggests that this interplay between simplicity and complexity is driven by self-referential feedback loops, where simple patterns repeat, adapt, and integrate into higher-order structures. This section explores how Recursive Realism offers insights into the mechanisms that allow for the emergence of complexity, examining the role of fractal structures, chaos theory, and evolutionary processes in shaping natural systems.

Emergence: How Simple Rules Create Complex Systems

Emergence refers to the phenomenon where simple components interact to create complex behaviours and structures that cannot be predicted from the properties of individual parts.

Recursive Realism suggests that emergent complexity is inherently self-referential, as simple interactions generate patterns that organise themselves into higher levels of order.

Fractal Geometry and the Recursion of Pattern Formation: Fractal geometry is a prime example of how simple rules can create complex forms through recursive iteration. Recursive Realism suggests that fractals represent the self-referential nature of pattern formation, where a basic rule is applied repeatedly to generate intricate structures. For example, the Mandelbrot set is generated through a simple mathematical formula, yet it produces infinitely complex shapes that reveal new details at every level of magnification. This perspective highlights how nature's complexity can arise from recursive processes, where self-similarity and iteration create rich patterns from fundamental principles.

Chaos Theory and the Emergence of Order from Disorder: Chaos theory explores how dynamic systems can exhibit unpredictable behaviour while still following deterministic rules. Recursive Realism suggests that chaotic systems are self-referential, where small changes in initial conditions can lead to vastly different outcomes, creating a loop between predictability and randomness. For example, the butterfly effect, where a small change in a weather system can result in significant shifts over time, illustrates how chaotic processes can lead to emergent patterns. This perspective allows for a deeper understanding of how order arises from chaos, where recursive interactions between deterministic rules and sensitive responses create complex systems that are both stable and adaptive.

Cellular Automata and the Recursion of Simple Rules: Cellular automata are mathematical models that use simple rules to simulate the evolution of patterns over time. Recursive Realism suggests that cellular automata demonstrate the power of recursion in creating complexity from basic interactions. For example, Conway's Game of Life uses a few simple rules to determine how cells live, die, or multiply, yet these rules can produce unexpectedly intricate patterns, from oscillating structures to self-replicating entities. This perspective highlights how nature uses recursive processes to build complex forms, where local interactions give rise to global behaviours that transcend the simplicity of their origins.

Stability, Adaptation, and the Role of Feedback Loops

Natural systems are characterised by their ability to maintain stability while adapting to changes in their environment. Recursive Realism suggests that this balance is achieved through feedback loops, where interactions between components allow for dynamic adjustment and self-regulation.

Homeostasis and the Recursion of Regulatory Mechanisms: Homeostasis, the ability of living organisms to maintain stable internal conditions, relies on feedback loops that adjust physiological processes in response to external changes. Recursive Realism suggests that homeostasis is a self-referential process, where the body's systems monitor their own states and initiate responses to restore balance. For example, thermoregulation in mammals involves a loop where temperature sensors trigger sweating or shivering to maintain optimal body temperature. This perspective allows for a deeper understanding of how organisms achieve stability through recursive regulation, where biological feedback ensures adaptation to changing conditions.

Ecological Equilibrium and the Recursion of Predator-Prey Dynamics: Ecosystems maintain balance through feedback loops between populations of predators and prey, creating dynamic cycles that regulate species numbers. Recursive Realism suggests that predator-prey interactions are self-referential, as the population of one species influences the growth or decline of another, creating cyclical patterns. For example, a decline in prey might lead to reduced predator numbers, which in turn allows prey populations to recover, creating a loop of oscillation. This perspective highlights how ecosystems use recursively structured dynamics to achieve long-term stability, where feedback loops between species drive the balance of nature.

Adaptation and the Feedback Loop of Evolutionary Change: Evolution is driven by natural selection, where traits that enhance survival and reproduction become more common in populations over generations. Recursive Realism suggests that evolutionary adaptation involves a self-referential process, where changes in traits create a loop between organisms and their environment, leading to the gradual emergence of complex adaptations. For example, the evolution of camouflage in predators and prey involves a cycle where each adaptation prompts counter-adaptations, creating an arms race that drives the evolution of intricate forms. This perspective allows for a richer understanding of how evolution is shaped by recursive interactions, where traits and behaviours co-evolve through a continuous process of adaptation and feedback.

Creativity, Innovation, and the Emergence of Novelty

Nature is not only complex but also innovative, capable of producing new forms, behaviours, and ecosystem dynamics. Recursive Realism suggests that creativity in nature emerges from self-referential processes, where simple rules and feedback loops create the conditions for novelty to arise.

Genetic Mutation and the Recursion of Variability: Genetic mutations introduce variations in organisms, providing the raw material for evolutionary innovation. Recursive Realism suggests that genetic variability is a self-referential process, where mutations are filtered through natural selection, creating a loop that allows for the exploration of new possibilities. For example, a mutation that enhances an organism's ability to survive in a changing environment may become more common, creating a loop where new traits are tested and refined. This perspective highlights how nature uses recursively structured change to generate novelty, where genetic variation provides the basis for evolutionary creativity.

Ecosystem Resilience and the Recursion of Adaptive Responses: Ecosystems are capable of recovering from disturbances through adaptive responses that maintain function and diversity. Recursive Realism suggests that ecosystem resilience is a self-referential process, where feedback loops allow for adjustments to environmental changes, ensuring long-term stability. For example, after a natural disaster like a wildfire, an ecosystem might undergo successional changes that lead to new species compositions, creating a loop where recovery shapes future resilience. This perspective allows for a deeper appreciation of how nature can innovate through self-referential adaptation, where the capacity for renewal is built into the structure of life.

Symmetry Breaking and the Emergence of Asymmetry: Symmetry breaking is a process where systems that are initially uniform develop asymmetrical structures that allow for diversity and specialisation. Recursive Realism suggests that symmetry breaking involves a self-referential loop, where slight variations create new pathways for development. For example, in embryonic development, a symmetrical cell cluster may undergo differentiation to form specialised tissues, creating a loop where cells communicate to determine distinct roles. This perspective highlights how biological diversity emerges through recursive processes, where symmetry breaking creates opportunities for novel forms and functions.

Nature's complexity arises from simple principles that interact through recursive processes, creating systems that are both predictable and capable of adaptation. Recursive Realism provides a framework for understanding how self-similarity, feedback loops, and emergent dynamics shape the natural world, revealing how simple interactions lead to innovative outcomes. By recognising the self-referential nature of complexity, Recursive Realism offers a vision of nature where order and chaos coexist, and where the rules of the universe are flexible enough to allow for creativity and change.

In Chapter 29, we have explored how Recursive Realism offers a framework for understanding the emergence of complexity in nature, revealing how self-referential processes guide the formation of life, ecosystem dynamics, and natural patterns. By examining the role of feedback loops, fractal structures, and evolutionary adaptation, Recursive Realism provides a nuanced perspective on how nature achieves a balance between stability and novelty, allowing for the emergence of intricate structures and adaptive systems. This perspective allows us to appreciate nature's complexity as a dynamic interplay between simple principles and innovative outcomes, where self-referential interactions drive the evolution of life and the development of natural systems.

References

Bateson, G. (1972) *Steps to an Ecology of Mind*. Chicago: University of Chicago Press.

Darwin, C. (1859) *On the Origin of Species by Means of Natural Selection*. London: John Murray.

Eigen, M. and Schuster, P. (1979) *The Hypercycle: A Principle of Natural Self-Organization*. Berlin: Springer.

Gleick, J. (1987) *Chaos: Making a New Science*. New York: Viking Penguin.

Goodwin, B. (2001) *How the Leopard Changed Its Spots: The Evolution of Complexity*. Princeton, NJ: Princeton University Press.

Kauffman, S.A. (1993) *The Origins of Order: Self-Organization and Selection in Evolution*. Oxford: Oxford University Press.

Lorenz, E.N. (1963) 'Deterministic Nonperiodic Flow', *Journal of the Atmospheric Sciences*, 20(2), pp. 130–141.

Mandelbrot, B.B. (1982) *The Fractal Geometry of Nature*. New York: W.H. Freeman and Company.

Maynard Smith, J. and Szathmáry, E. (1997) *The Major Transitions in Evolution*. Oxford: Oxford University Press.

Prigogine, I. and Stengers, I. (1984) *Order Out of Chaos: Man's New Dialogue with Nature*. New York: Bantam Books.

West, G.B., Brown, J.H., and Enquist, B.J. (1997) 'A General Model for the Origin of Allometric Scaling Laws in Biology', *Science*, 276(5309), pp. 122–126.

Chapter 30: Recursive Realism and the Philosophy of Consciousness

Self-Awareness and the Recursive Structure of Mind

Consciousness is one of the most profound mysteries in philosophy and science, encompassing the ability to experience, perceive, and reflect on one's own existence. Recursive Realism suggests that consciousness is inherently self-referential, where the mind engages in feedback loops of awareness and self-reflection that give rise to a sense of self. This section explores how Recursive Realism offers insights into the emergence of self-awareness, examining the nature of thought, the recursive dynamics of self-reflection, and the relationship between conscious experience and the mind's ability to model itself.

The Nature of Thought as a Recursive Process

Thought is a core aspect of consciousness, involving the ability to process information, form concepts, and make decisions. Recursive Realism suggests that thinking is a self-referential process, where ideas and concepts interact with each other, creating a dynamic loop that allows for complex reasoning.

Inner Dialogue and the Recursion of Thought: Inner dialogue, the conversation we have with ourselves, is a key aspect of conscious thought. Recursive Realism suggests that inner dialogue is inherently self-referential, as the mind generates thoughts and then reflects on those thoughts, creating a loop of self-examination. For example, when we deliberate over a difficult decision, we might consider different perspectives, evaluate options, and respond to our own doubts, creating a recursive cycle of thought refinement. This perspective highlights how self-reflective thinking is not linear but involves a feedback loop where each thought is shaped by its relationship to previous reflections.

Conceptual Integration and the Recursion of Ideas: Human cognition involves the ability to integrate concepts and form new ideas through analogy and synthesis. Recursive Realism suggests that conceptual thinking is a self-referential process, where ideas are compared, merged, and refined through recursive exploration. For example, understanding complex scientific theories often involves integrating multiple models and revisiting assumptions, creating a loop where the mind constructs a more comprehensive understanding over time. This perspective allows for a deeper appreciation of how thought is a recursive construct, where the mind uses self-reference to build increasingly sophisticated concepts.

Attention and the Recursion of Focus: Attention, the ability to focus on specific stimuli or ideas, is crucial to conscious experience. Recursive Realism suggests that attention is a self-referential mechanism, where the mind selects information to focus on and then monitors its own focus, creating a feedback loop that shapes awareness. For example, during meditation, practitioners observe their thoughts and redirect attention, creating a loop where awareness of focus refines the focus itself. This perspective highlights how attention is a dynamic process, where self-monitoring allows for the regulation of conscious awareness.

Self-Reflection and the Construction of Identity

Self-reflection, the ability to think about oneself, is a hallmark of self-awareness and plays a key role in shaping identity. Recursive Realism suggests that self-reflection involves a self-referential process, where the mind uses introspection to construct a sense of self and understand its own nature.

The Mirror of Consciousness: Self-Reflection as a Feedback Loop: Self-reflection can be thought of as the mind's ability to observe itself, creating a loop where thoughts and perceptions are examined and reinterpreted. Recursive Realism suggests that this mirror-like process is self-referential, as awareness turns back onto itself, allowing for the exploration of inner experiences. For example, reflecting on one's emotions might involve questioning why a particular feeling arose, leading to insights about personal values and beliefs, creating a cycle where self-awareness deepens. This perspective highlights how self-reflection is a recursive phenomenon, where awareness is both the observer and the observed, creating a dynamic interplay that shapes our sense of identity.

The Construction of Self-Concept Through Recursive Thought: Self-concept, the understanding of who we are, is shaped by thoughts about past experiences, future aspirations, and current beliefs. Recursive Realism suggests that self-concept is a self-referential construct, where the mind revisits memories and imagines possibilities, creating a loop that integrates different aspects of the self. For example, reflecting on personal achievements and failures allows us to construct a narrative that explains how we have changed, creating a feedback loop where the story of self evolves over time. This perspective allows for a deeper understanding of how identity is not fixed but is continually refined through self-referential thought, where each reflection contributes to a more nuanced understanding of the self.

Self-Identity and the Recursive Nature of Self-Awareness: Self-identity, the stable sense of who we are, is anchored in self-awareness and the ability to see oneself as a consistent entity. Recursive Realism suggests that self-identity is inherently self-referential, as it involves a loop where the mind recognises itself across different moments in time. For example, the recognition that "I am the same person now as I was yesterday" creates a continuity of self that is constructed through self-reflective awareness. This perspective highlights how self-awareness provides a framework for maintaining a coherent identity, where the mind's ability to reflect on its own continuity shapes the experience of being.

The Mind as a Self-Modelling System

Recursive Realism suggests that the mind can be understood as a self-modelling system, where consciousness arises from the mind's ability to represent and understand itself. This section explores how self-modelling allows for the emergence of conscious experience, where the mind constructs internal models that enable awareness of the world and itself.

The Model of Self and the World: The mind creates internal models of the world to navigate reality, integrating sensory information with conceptual understanding. Recursive Realism suggests that self-modelling extends to the self, where the mind creates a model of its own thoughts and emotions, allowing for self-awareness. For example, the ability to recognise one's own thoughts as separate from external events enables introspection and conscious deliberation, creating a loop where the mind reflects on its own representations. This perspective highlights how self-modelling is a self-referential process, where the mind not only perceives the world but also understands its place within it.

The Role of Metacognition in Conscious Experience: Metacognition, thinking about one's own thinking, enables the mind to assess and regulate cognitive processes. Recursive Realism suggests that metacognition is inherently self-referential, as it involves a loop where thoughts are evaluated and adjusted based on self-reflection. For example, recognising that one is struggling with a problem might lead to adjusting strategies, creating a cycle of cognitive improvement. This perspective allows for a deeper understanding of how metacognitive abilities contribute to conscious awareness, where the mind uses self-reflection to enhance its own functioning.

Consciousness as a Recursive Loop of Awareness: Consciousness itself can be seen as a recursive loop where awareness constantly updates and refines itself through self-reference. Recursive Realism suggests that the conscious mind is engaged in a continuous cycle of monitoring and revising its understanding of reality, creating a dynamic experience of being aware. For example, the ability to be aware of one's own awareness, a hallmark of self-consciousness, creates a loop where the mind is both subject and object of perception. This perspective highlights how conscious experience is shaped by recursive processes, where awareness continually turns back upon itself to create a sense of self and world.

Consciousness is deeply self-referential, involving processes of self-awareness, introspection, and self-modelling that create a dynamic experience of being aware. Recursive Realism provides a framework for understanding how thought, identity, and metacognition emerge from recursive feedback loops, allowing for the construction of a coherent self and the ability to reflect on one's own mind. By recognising the recursive nature of self-awareness, Recursive Realism offers a vision of consciousness as an evolving process, where the mind continuously builds and refines its understanding of itself and the world.

Recursive Realism and the Nature of Conscious Experience

Conscious experience encompasses the rich tapestry of perceptions, sensations, and emotions that constitute what it feels like to be aware. Recursive Realism suggests that consciousness is shaped by self-referential loops, where sensory inputs are processed through layers of reflection, creating the depth and texture of subjective experience. This section explores how Recursive Realism offers insights into the nature of perception, the phenomenon of qualia, and the integration of sensory information, examining how self-referential thinking creates the richness of conscious life.

Perception as a Recursive Process

Perception is the mechanism through which sensory information is translated into meaningful experiences. Recursive Realism suggests that perception is inherently self-referential, involving feedback loops where sensory inputs are refined, interpreted, and reinterpreted to create a cohesive picture of reality.

The Recursion of Sensory Processing: Sensory perception, the ability to detect and process stimuli like light, sound, and touch, relies on multiple layers of processing within the brain. Recursive Realism suggests that sensory processing involves a feedback loop, where raw sensory data is filtered and enhanced through recurrent interactions between different brain regions. For example, the visual cortex processes input from the eyes, but higher brain areas continuously send feedback to adjust interpretation based on expectations and context, creating a loop where perception is dynamically refined. This perspective allows for a deeper understanding of how perception is not passive but actively constructed through recursive neural interactions.

Predictive Coding and the Recursion of Expectations: Predictive coding is a theory that suggests the brain uses predictions to interpret sensory input, constantly comparing expectations with incoming data. Recursive Realism suggests that predictive coding is a self-referential process, where the mind generates expectations and updates them based on feedback from the senses. For example, when we walk into a familiar room, our brain predicts what we will see and feel, creating a loop where deviations from expectations are used to adjust perception. This perspective highlights how perception is a recursive dialogue between the known and the unexpected, where awareness is shaped by the interplay of prediction and sensory reality.

Attention and the Recursion of Focused Perception: Attention is crucial for filtering out irrelevant information and enhancing important details, allowing consciousness to focus on what matters most. Recursive Realism suggests that attention is a self-referential loop, where the focus of perception is continually adjusted based on priorities and cognitive goals. For example, when listening to a conversation in a noisy room, our attention helps us isolate the voice we want to hear, creating a loop where focus enhances the clarity of sensory input. This perspective highlights how attention shapes perception through recursive refinement, where conscious awareness directs the flow of sensory processing.

Qualia: The Subjective Texture of Experience

Qualia, the subjective qualities of conscious experience, like the redness of red or the bitterness of coffee, represent one of the most enigmatic aspects of consciousness. Recursive Realism suggests that qualia arise from self-referential processes, where sensory information is enhanced through layers of interpretation, creating the richness of subjective experience.

The Recursion of Sensory Detail: Qualia are often described as intensely specific, the way a particular sound feels, the texture of a memory, the warmth of a sunlit afternoon. Recursive Realism suggests that this specificity arises from feedback loops that allow the mind to focus on subtle variations, creating layers of detail in experience. For example, the richness of tasting a complex flavour might involve noticing the interplay of sweetness, bitterness, and aroma, creating a loop where attention and sensation enhance each other. This perspective allows for a deeper appreciation of how qualia are not merely data but are constructed through recursive awareness, where the mind explores the nuances of experience.

The Mystery of the Hard Problem: The "hard problem of consciousness", as articulated by philosopher David Chalmers, refers to the question of why certain brain processes give rise to subjective experiences. Recursive Realism suggests that the hard problem reflects the self-referential nature of consciousness, where awareness creates a loop that allows information processing to be experienced as phenomenal qualities. For example, the experience of pain is not just the activation of nociceptors but involves self-reflection on the discomfort, creating a loop where physical sensation is amplified by awareness. This perspective highlights how qualia might emerge from the recursive processes of perception and interpretation, where awareness turns sensory data into lived experience.

Self-Awareness and the Qualia of Being: Self-awareness itself can be thought of as a form of qualia, the experience of being aware of being aware. Recursive Realism suggests that the qualia of self-awareness involve a loop where the mind perceives its own presence, creating a distinct quality of conscious experience. For example, the feeling of being "in the moment" during meditation or deep contemplation involves a heightened awareness of one's own thoughts and existence, creating a rich sense of consciousness that is more than the sum of its parts. This perspective allows for a richer understanding of how self-awareness contributes to the depth of conscious experience, where awareness becomes an object of itself.

The Integration of Sensory Information into a Unified Experience

One of the wonders of consciousness is the ability to integrate diverse sensory inputs into a single, cohesive experience, a unified sense of being. Recursive Realism suggests that this integration is achieved through self-referential feedback loops, where sensory modalities are synthesised into a coherent whole.

Multisensory Integration and the Recursion of Perception: Multisensory integration involves combining information from different senses, like sight, sound, and touch, into a unified experience. Recursive Realism suggests that this integration involves a feedback loop, where the brain cross-references sensory data to create a cohesive perception. For example, when eating a meal, we might simultaneously see the colours of the food, taste its flavours, and hear the crunch of a crisp bite, creating a loop where each sense enhances the overall experience. This perspective highlights how conscious perception is greater than individual senses, where self-referential processing allows for the integration of multiple sensory streams.

The Binding Problem and the Unity of Consciousness: The binding problem refers to the question of how the brain combines different sensory inputs into a single conscious experience. Recursive Realism suggests that binding involves recursive loops, where sensory information is synchronised and refined through self-referential processes. For example, the experience of watching a sunset involves the coordination of visual colours, ambient sounds, and the feeling of warmth, creating a unified moment of awareness. This perspective allows for a deeper understanding of how the mind's recursive nature enables the integration of perception, where consciousness synthesises the complexity of sensory input into a seamless flow.

Holistic Awareness and the Recursive Mind: Holistic awareness, the ability to sense the interconnectedness of experiences, reflects the mind's ability to integrate diverse elements into a coherent whole. Recursive Realism suggests that holistic awareness is a self-referential process, where the mind recognises patterns across different moments and contexts, creating a loop of awareness that sees the bigger picture. For example, during moments of insight or creative inspiration, disparate ideas might come together, creating a feeling of clarity where connections become evident. This perspective highlights how conscious experience is shaped by recursive thinking, where awareness integrates varied experiences into a unified understanding.

Consciousness is characterised by rich, layered experiences, where perception, qualia, and sensory integration create a dynamic interplay of awareness. Recursive Realism provides a framework for understanding how self-referential processes shape the texture of conscious life, revealing how the mind's recursive nature allows for depth and coherence in experience. By recognising the recursive nature of perception and awareness, Recursive Realism offers a vision of consciousness as a dialogue between internal reflection and external reality, where the richness of subjective experience arises from the mind's ability to turn inward and connect with itself.

In Chapter 30, we have explored how Recursive Realism offers a framework for understanding the nature of consciousness, revealing how self-awareness, perception, and subjective experience emerge through self-referential loops. By examining the recursive dynamics of thought, the enigma of qualia, and the integration of sensory information, Recursive Realism provides a nuanced perspective on how consciousness functions as a complex, evolving process. This perspective allows us to appreciate consciousness not as a static state but as a self-organising phenomenon, where the mind's recursive nature enables the depth and diversity of conscious life.

References

Baars, B.J. (1988) *A Cognitive Theory of Consciousness*. Cambridge: Cambridge University Press.

Chalmers, D.J. (1996) *The Conscious Mind: In Search of a Fundamental Theory*. Oxford: Oxford University Press.

Dennett, D.C. (1991) *Consciousness Explained*. Boston: Little, Brown.

Damasio, A.R. (1999) *The Feeling of What Happens: Body and Emotion in the Making of Consciousness*. New York: Harcourt Brace.

Dehaene, S. (2014) *Consciousness and the Brain: Deciphering How the Brain Codes Our Thoughts*. New York: Viking.

Edelman, G.M. (1989) *The Remembered Present: A Biological Theory of Consciousness*. New York: Basic Books.

Friston, K. (2010) 'The Free-Energy Principle: A Unified Brain Theory?', *Nature Reviews Neuroscience*, 11(2), pp. 127–138.

Merker, B. (2007) 'Consciousness Without a Cerebral Cortex: A Challenge for Neuroscience and Medicine', *Behavioral and Brain Sciences*, 30(1), pp. 63–81.

Nagel, T. (1974) 'What Is It Like to Be a Bat?', *Philosophical Review*, 83(4), pp. 435–450.

Tononi, G. (2004) 'An Information Integration Theory of Consciousness', *BMC Neuroscience*, 5(1), p. 42.

Varela, F.J., Thompson, E. and Rosch, E. (1991) *The Embodied Mind: Cognitive Science and Human Experience*. Cambridge, MA: MIT Press.

Chapter 31: Recursive Realism and the Origins of Creativity

Imagination and the Recursive Nature of Creative Thought

Creativity is often seen as one of the defining qualities of human intelligence, encompassing the ability to generate new ideas, envision possibilities, and transform concepts into innovative forms. Recursive Realism suggests that creativity is inherently self-referential, involving feedback loops where ideas and imaginative scenarios are explored, revised, and synthesised into novel concepts. This section explores how Recursive Realism provides insights into the nature of imagination, the dynamics of creative synthesis, and the role of self-reflection in artistic and intellectual innovation.

Imagination as a Recursive Process

Imagination, the ability to create mental images and scenarios that do not currently exist, is at the heart of creativity. Recursive Realism suggests that imagination is a self-referential process, where the mind engages in loops of thought that blend memory, experience, and speculation to envision new possibilities.

Mental Simulation and the Recursion of Possibilities: Imagination allows the mind to simulate scenarios, exploring different outcomes and hypothetical situations. Recursive Realism suggests that mental simulation involves feedback loops, where imagined scenarios are adjusted and revised through self-reflective thought. For example, when an architect envisions a new building, they might mentally construct the structure, adjusting details like design and materials, creating a loop where imagination and evaluation interact. This perspective highlights how imagination is a recursive dialogue, where the mind refines its creations through a cycle of revision and exploration.

Daydreaming and the Loop of Creative Play: Daydreaming, the spontaneous flow of thoughts and fantasies, represents a form of imaginative play that can spark creativity. Recursive Realism suggests that daydreaming is a self-referential process, where the mind explores mental landscapes, creating a loop of free association and conceptual blending. For example, a writer daydreaming about a fictional world might visualise characters and settings, creating a narrative that evolves through imaginative engagement. This perspective allows for a deeper appreciation of how daydreaming fosters creative thought, where the mind uses recursion to play with ideas and discover new connections.

Creative Problem-Solving and the Recursion of Thought Experiments: Creative problem-solving often involves thinking beyond conventional frameworks, using imagination to explore unconventional solutions. Recursive Realism suggests that problem-solving is inherently self-referential, where thought experiments create a loop of hypothesis generation and testing. For example, scientists might use thought experiments like Schrödinger's cat to explore the implications of quantum mechanics, creating a recursive dialogue where imagination reveals new perspectives. This perspective highlights how creativity is driven by recursive thinking, where imaginative exploration allows for the discovery of innovative solutions.

Creative Synthesis and the Role of Self-Reflection

Creative synthesis, the process of combining different ideas into a new, unified concept, is central to artistic and intellectual innovation. Recursive Realism suggests that synthesis involves a self-referential loop, where ideas are blended, revisited, and refined through self-reflection.

Analogy and the Recursion of Conceptual Blending: Analogies allow creatives to connect concepts from different domains, revealing hidden similarities and unexpected insights. Recursive Realism suggests that analogy-making is a self-referential process, where the mind creates a loop between familiar ideas and new contexts, allowing for conceptual blending. For example, Leonardo da Vinci's observations of bird flight influenced his designs for flying machines, creating a recursive connection between nature and engineering. This perspective highlights how analogy serves as a bridge in creative thinking, where recursion allows concepts to cross-pollinate and inspire innovation.

Metaphor and the Recursion of Symbolic Thought: Metaphors use symbolic language to convey complex ideas by relating them to more familiar experiences. Recursive Realism suggests that metaphorical thinking involves a loop, where concepts are reinterpreted through symbolic associations, creating new layers of meaning. For example, the metaphor of the mind as a mirror reflects a recursive understanding of consciousness, where awareness turns back upon itself. This perspective allows for a deeper understanding of how metaphor enriches creative expression, where self-referential thought reveals new dimensions of understanding.

Iterative Revision and the Recursion of Artistic Practice: Artists often engage in iterative processes of creation and revision, where works are adjusted and refined through self-reflection. Recursive Realism suggests that artistic creation involves a feedback loop, where the act of creation leads to self-evaluation, guiding further development. For example, a painter might adjust the colours and composition of a piece based on their evolving vision, creating a loop where the artwork grows through cycles of reflection. This perspective highlights how creative synthesis is dynamic, where recursively reworking ideas allows for the emergence of originality.

Innovation and the Recursive Exploration of Boundaries

Innovation, the creation of new methods, products, or paradigms, often involves pushing the boundaries of existing knowledge. Recursive Realism suggests that innovation is inherently self-referential, where the mind engages in loops of questioning and redesigning concepts to explore the edges of possibility.

Disruptive Ideas and the Loop of Conceptual Breakthroughs: Disruptive innovation often occurs when existing models are questioned and replaced with new frameworks. Recursive Realism suggests that breakthroughs involve a loop of challenging assumptions and rebuilding ideas from the ground up. For example, Einstein's theory of relativity challenged the Newtonian conception of absolute space and time, creating a new paradigm that reshaped our understanding of the universe. This perspective allows for a deeper appreciation of how innovation emerges through recursive thinking, where each breakthrough is preceded by cycles of questioning and conceptual reformation.

Creativity in Science and the Recursion of Hypothesis Testing: Scientific creativity often involves developing hypotheses and testing them against empirical evidence, refining theories through recursive experimentation. Recursive Realism suggests that the scientific method is a self-referential process, where hypotheses are tested, revised, and retested, creating a loop that drives the discovery of new knowledge. For example, the development of quantum mechanics involved repeated adjustments to theoretical models in response to new experimental findings, creating a cycle where theories evolved through recursive inquiry. This perspective highlights how scientific innovation is a dynamic process, where self-correcting loops enable the expansion of understanding.

Cultural Innovation and the Recursion of Artistic Movements: Artistic movements often emerge when creators build on past styles and reinterpret traditions, creating a loop where new expressions arise from existing cultural forms. Recursive Realism suggests that cultural evolution involves self-referential loops, where each generation of artists reflects on previous works to create new forms. For example, modernist artists like Picasso reimagined classical forms, creating abstract art that challenged traditional representation, creating a loop between past and future. This perspective highlights how cultural innovation is shaped by recursion, where new movements are born from cycles of reinterpretation and creative exploration.

Creativity is deeply self-referential, involving processes of imagination, synthesis, and innovation that create new possibilities through recursive loops of thought. Recursive Realism provides a framework for understanding how the mind generates novel concepts and artistic expressions, revealing how creativity arises from the interplay between reflection and exploration. By recognising the recursive dynamics of imagination, Recursive Realism offers a vision of creativity as an evolving dialogue between ideas, visions, and the unknown, where each act of creation builds on a cycle of self-reflective thought.

Recursive Realism and the Dynamics of Cultural Expression

Culture is a living entity, constantly shaped and reshaped by artistic expression, intellectual thought, and social evolution. Recursive Realism suggests that cultural expression is inherently self-referential, involving feedback loops where ideas, art forms, and social norms are reinterpreted across generations. This section explores how Recursive Realism provides insights into the evolution of artistic styles, the dynamics of collective imagination, and the interplay between creativity and social transformation, examining how cultures grow through recursive cycles of influence and innovation.

The Evolution of Artistic Styles: A Recursive Dialogue Across Generations

Artistic styles are not static; they evolve as artists engage in a dialogue with the past, reimagining traditions and responding to cultural shifts. Recursive Realism suggests that artistic evolution is a self-referential process, where each new style reflects a reinterpretation of prior influences, creating a cycle that drives cultural growth.

The Recursion of Influence in Artistic Movements: Artistic movements like Romanticism, Impressionism, and Surrealism illustrate how styles evolve through reactions to previous norms. Recursive Realism suggests that each movement creates a loop where artists draw on influences from earlier periods while challenging their limitations, leading to the emergence of new forms. For example, Impressionist painters like Monet and Renoir broke away from realism by focusing on light and colour, creating a recursive dialogue with classical art that redefined painting. This perspective highlights how cultural creativity involves cycles of reinterpretation, where each era reflects both continuity and innovation.

The Role of Tradition and the Recursion of Cultural Memory: Tradition provides a foundation for cultural expression, offering shared motifs, stories, and symbols that unite communities. Recursive Realism suggests that traditions are self-referential constructs, where cultural memory is revisited and renewed through artistic engagement. For example, mythological themes like the hero's journey or the struggle between good and evil appear across cultures, yet each generation adds its own interpretation, creating a loop of cultural continuity. This perspective allows for a deeper appreciation of how tradition and innovation coexist in art, where recursively engaging with cultural memory allows for the creation of new meanings.

Avant-Garde Art and the Loop of Radical Reinterpretation: Avant-garde movements often seek to break from tradition, pushing artistic boundaries to explore new forms of expression. Recursive Realism suggests that avant-garde art is a self-referential challenge to cultural norms, where artists critique existing conventions and redefine the role of art in society. For example, the Dada movement emerged as a reaction to the absurdity of war, using nonsensical imagery and random compositions to challenge conventional aesthetics, creating a recursive loop of cultural critique. This perspective highlights how artistic radicalism drives cultural evolution, where recursion allows artists to deconstruct and rebuild cultural frameworks.

The Role of Collective Imagination in Cultural Creativity

Collective imagination, the shared visions and aspirations that shape societies, is a powerful driver of cultural change. Recursive Realism suggests that collective imagination involves feedback loops between individual creativity and cultural norms, where societal values and shared stories evolve through reciprocal influences.

Mythology and the Recursion of Archetypal Themes: Myths and archetypal stories play a crucial role in shaping cultural imagination, providing narratives that express universal human experiences. Recursive Realism suggests that mythological themes are self-referential constructs, where cultural stories are reinterpreted across eras, reflecting changing social realities. For example, the archetype of the trickster, a figure who challenges norms and defies authority, appears in myths from Loki in Norse mythology to Anansi in African folklore, creating a loop where each retelling adapts the theme to new cultural contexts. This perspective highlights how myths serve as a recursive reservoir of cultural meaning, where societies draw on ancient stories to navigate contemporary challenges.

Cultural Movements and the Feedback Loop of Social Change: Cultural movements like the Renaissance, the Enlightenment, and the Civil Rights movement illustrate how collective imagination can reshape societies through shared visions of progress and justice. Recursive Realism suggests that social movements involve self-referential feedback loops, where new ideas inspire social change, which in turn creates the context for further innovation. For example, the Enlightenment emphasised reason and individual rights, challenging monarchical authority and religious dogma, creating a loop where philosophical ideas reshaped political structures. This perspective allows for a richer understanding of how cultural imagination drives social transformation, where recursively challenging norms allows for the evolution of societal values.

Art as a Mirror and the Recursion of Social Reflection: Art often serves as a mirror for society, reflecting social conditions and critiquing cultural norms. Recursive Realism suggests that artistic expression involves self-referential loops, where artworks are informed by social realities and, in turn, influence public perception. For example, film and literature that address themes like inequality or identity can shape public discourse, creating a loop where art becomes a catalyst for social awareness. This perspective highlights how art functions as a recursive dialogue with society, where cultural expression both reflects and shapes collective consciousness.

Creativity, Innovation, and the Dynamics of Cultural Transformation

Cultural creativity is not just about individual expression but involves the transformation of societies through innovative ideas and artistic breakthroughs. Recursive Realism suggests that cultural change is driven by self-referential processes, where ideas are revisited, challenged, and transformed through cycles of reflection.

Technological Innovation and the Recursion of Cultural Shifts: Technological advances like the printing press, the internet, and artificial intelligence have redefined cultural landscapes, creating new ways for ideas to spread and evolve. Recursive Realism suggests that technology influences culture through feedback loops, where new tools change how societies think, and cultural shifts drive further innovation. For example, the internet has transformed communication and media, allowing for global collaboration and the exchange of ideas, creating a loop where technology and culture co-evolve. This perspective allows for a deeper appreciation of how technological progress shapes cultural creativity, where recursive interactions between tools and societal values drive cultural change.

Globalisation and the Recursion of Cultural Exchange: Globalisation has created a world where cultural ideas and artistic forms are shared and adapted across borders, leading to a dynamic interplay of local traditions and global influences. Recursive Realism suggests that cultural exchange involves self-referential loops, where local cultures are influenced by global trends, and global trends are shaped by local adaptations. For example, musical genres like jazz or hip-hop have spread globally, being reinterpreted in different cultural contexts, creating a loop of cross-cultural creativity. This perspective highlights how cultural creativity is enriched through recursive interactions, where the exchange of ideas fosters innovation and cultural evolution.

The Cycles of Cultural Renaissance: Cultural renaissances, periods of intense artistic and intellectual flourishing, often arise when societies revisit past traditions and reinterpret them in new ways. Recursive Realism suggests that renaissance periods are self-referential phenomena, where cultures look backward to draw inspiration while looking forward to envision new possibilities. For example, the Harlem Renaissance involved African American artists and writers reclaiming cultural heritage and redefining identity, creating a loop where the past became a foundation for future expressions. This perspective highlights how cultural creativity thrives in recursively reflective eras, where societies engage in a dialogue with their own history to create new cultural forms.

Cultural creativity is shaped by self-referential processes, where artistic expression, collective imagination, and social change interact through feedback loops that drive cultural evolution.

Recursive Realism provides a framework for understanding how art and innovation emerge from cycles of reinterpretation, revealing how cultural movements are both informed by tradition and inspired by radical change. By recognising the recursive nature of cultural creativity, Recursive Realism offers a vision of society as a living system, where the interplay of ideas allows for the growth and transformation of cultural life.

In Chapter 31, we have explored how Recursive Realism offers a framework for understanding the origins of creativity, revealing how imagination, artistic expression, and cultural innovation arise from self-referential processes. By examining the recursive dynamics of thought, the evolution of artistic styles, and the role of collective imagination, Recursive Realism provides a nuanced perspective on how creativity functions as a cycle of reflection and exploration. This perspective allows us to appreciate creativity not as a singular act but as an ongoing dialogue, where the mind's recursive nature enables the synthesis of new ideas and the reimagining of cultural norms.

References

Anderson, M.L. (2010) 'Neural Reuse: A Fundamental Organizational Principle of the Brain', *Behavioral and Brain Sciences*, 33(4), pp. 245–313.

Boden, M.A. (1994) *Dimensions of Creativity*. Cambridge, MA: MIT Press.

Carruthers, P. (2002) 'The Cognitive Functions of Language', *Behavioral and Brain Sciences*, 25(6), pp. 657–726.

Csikszentmihalyi, M. (1996) *Creativity: Flow and the Psychology of Discovery and Invention*. New York: HarperCollins.

Dennett, D.C. (2017) *From Bacteria to Bach and Back: The Evolution of Minds*. New York: W.W. Norton & Company.

Koestler, A. (1964) *The Act of Creation*. London: Hutchinson.

Lumsden, C.J. and Wilson, E.O. (1981) *Genes, Mind, and Culture: The Coevolutionary Process*. Cambridge, MA: Harvard University Press.

Runco, M.A. (2014) *Creativity: Theories and Themes: Research, Development, and Practice*. 2nd edn. San Diego: Academic Press.

Sawyer, R.K. (2012) *Explaining Creativity: The Science of Human Innovation*. 2nd edn. Oxford: Oxford University Press.

Simonton, D.K. (2010) 'Creative Thought as Blind Variation and Selective Retention: Combinatorial Models of Exceptional Creativity', *Physics of Life Reviews*, 7(2), pp. 156–179.

Thagard, P. (2012) *The Cognitive Science of Science: Explanation, Discovery, and Conceptual Change*. Cambridge, MA: MIT Press.

Tomasello, M. (1999) *The Cultural Origins of Human Cognition*. Cambridge, MA: Harvard University Press.

Vygotsky, L.S. (2004) *Imagination and Creativity in Childhood.* Journal of Russian and East European Psychology, 42(1), pp. 7–97.

Chapter 32: Recursive Realism and the Ethical Dimensions of Self-Reflection

Moral Reasoning as a Recursive Process

Ethical thinking involves the ability to evaluate actions, consider consequences, and navigate conflicts between values. Recursive Realism suggests that moral reasoning is inherently self-referential, involving feedback loops where principles are reflected upon, questioned, and refined through ongoing deliberation. This section explores how Recursive Realism provides insights into moral reasoning, examining the nature of ethical reflection, the role of empathy, and the dynamics of moral development.

Ethical Reflection and the Recursion of Moral Deliberation

Ethical reflection involves weighing different values and considering the broader implications of decisions. Recursive Realism suggests that moral deliberation is a self-referential process, where individuals engage in loops of evaluation to determine the right course of action.

The Feedback Loop of Ethical Dilemmas: Moral dilemmas, situations where competing values or duties conflict, often require deep reflection to resolve. Recursive Realism suggests that moral reasoning involves a loop where different perspectives are considered, reconsidered, and synthesised into a coherent judgment. For example, in a dilemma where telling the truth may cause harm, an individual might reflect on the importance of honesty versus compassion, creating a cycle of moral evaluation. This perspective allows for a deeper understanding of how ethical reasoning is not linear but involves a recursive dialogue between values and outcomes.

Moral Intuition and the Recursion of Gut Reactions: Moral intuitions, the immediate sense of what is right or wrong, are often shaped by cultural norms and personal experiences. Recursive Realism suggests that moral intuition involves self-referential feedback, where initial reactions are filtered through reflection to determine whether they align with broader values. For example, a person might feel instinctively that helping others is right but then consider the limits of their resources, creating a loop where intuitive responses are reassessed. This perspective highlights how ethical judgments are shaped by recursive interactions between intuition and cognitive reflection.

Self-Reflection and the Recursion of Ethical Principles: Principles like justice, fairness, and autonomy serve as guidelines for ethical decision-making. Recursive Realism suggests that these principles are self-referential constructs, evolving through a process of reflection and testing against real-life scenarios. For example, the principle of fairness might evolve as individuals encounter new situations that challenge their understanding of what is equitable, creating a loop where principles are refined over time. This perspective allows for a richer appreciation of how ethical thinking is dynamic, where self-reflection shapes the development of moral principles.

Empathy, Perspective-Taking, and the Role of Self-Reference

Empathy, the ability to understand and share the feelings of others, is central to moral reasoning. Recursive Realism suggests that empathy involves self-referential thought, where the mind uses its own experiences as a reference point to imagine the experiences of others.

Perspective-Taking and the Recursion of Imagination: Empathy relies on the ability to take the perspective of another, imagining their feelings and thoughts. Recursive Realism suggests that perspective-taking involves a loop, where the mind uses its own experiences as a template to simulate the experiences of others. For example, when seeing someone in distress, we might recall similar feelings of pain or fear, creating a recursive connection that allows for empathic understanding. This perspective highlights how empathy is not merely a reaction but a recursive process of imaginative engagement.

Compassion and the Feedback Loop of Moral Motivation: Compassion, the desire to alleviate suffering, can drive ethical actions by creating a sense of connection to others' well-being. Recursive Realism suggests that compassion involves a self-referential feedback loop, where awareness of another's suffering triggers a reflection on one's own capacity to help. For example, the sight of someone struggling might prompt a cycle of internal questioning about how to offer assistance, creating a loop where empathy leads to action. This perspective allows for a deeper understanding of how compassion arises from recursively engaging with the needs of others and our own sense of responsibility.

Moral Growth and the Recursion of Ethical Self-Awareness: Moral development often involves reflecting on past mistakes and learning from experiences to become more ethical. Recursive Realism suggests that moral growth is a self-referential process, where individuals use past reflections to guide future behaviour. For example, someone might reflect on a time when they acted unfairly and resolve to treat others more kindly, creating a loop where self-awareness fosters moral improvement. This perspective highlights how ethical maturity is shaped by recursive thought, where each reflection contributes to the evolution of character.

The Dynamics of Moral Development in Societies

Moral values are not only personal but are shared across communities, shaping laws, cultural norms, and collective aspirations. Recursive Realism suggests that moral values in societies evolve through feedback loops, where collective reflection influences social norms and laws adapt to changing ethical insights.

Social Norms and the Feedback Loop of Cultural Ethics: Social norms provide shared standards of right and wrong, shaping behaviour in communities. Recursive Realism suggests that social norms are self-referential constructs, evolving through cycles of collective reflection and adaptation to new contexts. For example, norms around gender equality have evolved as societies have reflected on historical inequalities, creating a loop where cultural values are revised to align with broader ethical insights. This perspective allows for a deeper appreciation of how societal values are dynamic, shaped by recursive processes of collective reflection.

Legal Systems and the Recursion of Justice: Legal systems formalise ethical principles into laws that govern behaviour, balancing individual rights with collective well-being. Recursive Realism suggests that legal frameworks involve self-referential processes, where laws are evaluated and amended based on societal changes and ethical debates. For example, the evolution of human rights law has involved a loop of revisiting legal precedents, addressing injustices, and expanding rights to previously marginalised groups. This perspective highlights how justice evolves through recursive cycles of interpretation, where laws adapt to emerging ethical awareness.

Collective Responsibility and the Loop of Moral Progress: Societies are often faced with ethical challenges that require collective action, such as addressing climate change or combating inequality. Recursive Realism suggests that collective moral responsibility involves a feedback loop, where societal actions and policies are informed by ethical reflections and adapted to meet new challenges. For example, the global shift towards sustainable practices involves re-evaluating economic models and consumer behaviours, creating a cycle where ethical considerations guide practical actions. This perspective allows for a deeper understanding of how societal progress is shaped by recursively engaging with ethical principles and collective goals.

Ethical thinking is deeply self-referential, involving processes of reflection, empathy, and collective engagement that shape moral decisions and values. Recursive Realism provides a framework for understanding how moral reasoning functions as a dialogue between intuition and reflection, revealing how ethical principles evolve through recursively structured thought. By recognising the recursive nature of moral reasoning, Recursive Realism offers a vision of ethics as a dynamic process, where individuals and societies continually refine their understanding of what it means to do good.

Recursive Realism and the Interplay Between Ethics and Consciousness

Ethical decisions are deeply tied to conscious awareness, as moral choices require self-reflection, consideration of consequences, and deliberation over values. Recursive Realism suggests that consciousness plays a crucial role in ethical reasoning, providing a framework where awareness engages in recursive loops to evaluate actions and navigate moral challenges. This section explores how self-referential thought shapes ethical consciousness, examining the nature of moral dilemmas, the interplay between free will and recursive thinking, and the role of reflection in moral autonomy.

Moral Dilemmas and the Role of Conscious Awareness

Moral dilemmas, situations where ethical principles conflict, require deep conscious reflection to resolve. Recursive Realism suggests that the experience of facing a dilemma involves a self-referential process, where conscious awareness revisits values, weighs options, and considers implications through recursive loops.

The Recursion of Ethical Self-Interrogation: Conscious awareness allows individuals to interrogate their own motives and evaluate the consistency of their ethical beliefs. Recursive Realism suggests that ethical self-interrogation involves a loop, where thoughts about one's actions are re-examined from different angles to determine the right course. For example, someone considering whether to reveal a painful truth might reflect on the potential hurt, then reconsider the importance of honesty, creating a cycle where each reflection deepens understanding. This perspective highlights how conscious reflection transforms ethical reasoning into a recursive process, where each thought builds upon the previous ones.

Conscious Conflict and the Recursion of Inner Debate: Moral dilemmas often create internal conflicts, where desires, duties, and values clash. Recursive Realism suggests that conscious conflict involves a loop where different voices within the mind engage in a dialogue, creating a dynamic debate. For example, deciding whether to act out of compassion or follow a rule might involve an inner argument, where emotional impulses and principles are weighed against each other. This perspective allows for a richer understanding of how consciousness plays a critical role in resolving moral conflicts, where awareness mediates a recursive negotiation between competing impulses.

Reflective Equilibrium and the Recursion of Ethical Consistency: Reflective equilibrium is the process of adjusting one's beliefs to achieve consistency between moral principles and particular judgments. Recursive Realism suggests that achieving reflective equilibrium involves a self-referential process, where beliefs are revisited and aligned through a loop of self-reflection. For example, someone might adjust their stance on punishment after reflecting on a specific case that challenges their principles, creating a cycle where values and decisions are harmonised. This perspective highlights how conscious reasoning enables ethical consistency, where recursively re-evaluating beliefs allows for a coherent moral framework.

Free Will, Autonomy, and the Recursion of Choice

Free will, the ability to choose actions based on one's own volition, is central to ethical autonomy. Recursive Realism suggests that free will is inherently self-referential, involving feedback loops where choices are considered and reconsidered based on internal reflection and awareness of consequences.

The Recursion of Deliberate Choice: Deliberate decision-making involves pausing to consider the impact of actions before committing to a course. Recursive Realism suggests that deliberation involves a feedback loop, where the mind evaluates possible outcomes, then revisits those evaluations to refine the final decision. For example, choosing whether to take a risky opportunity might involve weighing benefits and risks, then revisiting the decision based on new insights, creating a loop where reflection guides choice. This perspective allows for a deeper appreciation of how free will operates through recursively structured thought, where each reconsideration refines the decision-making process.

Moral Responsibility and the Recursion of Accountability: Taking responsibility for one's actions requires the ability to reflect on how decisions align with personal values. Recursive Realism suggests that moral responsibility involves self-referential thought, where awareness of potential consequences and self-evaluation of motivations create a loop of accountability. For example, apologising for a mistake might involve reflecting on the harm caused, then evaluating how to make amends, creating a cycle where self-awareness fosters ethical accountability. This perspective highlights how autonomy is strengthened through recursive reflection, where the ability to look back on choices enhances moral integrity.

The Illusion of Determinism and the Loop of Creative Agency: Debates about free will often contrast deterministic views of behaviour with the experience of choice. Recursive Realism suggests that while external factors shape influences on behaviour, conscious awareness allows for a recursive loop that introduces creative agency. For example, even if biological predispositions and social context influence decision-making, self-reflection allows individuals to challenge instincts, reconsider values, and choose alternative paths, creating a space for autonomy within deterministic frameworks. This perspective allows for a nuanced understanding of free will, where recursively engaging with one's own thoughts enables the experience of choice.

Ethical Reflection and the Integration of Experience

Ethical reflection is not just about individual decisions but involves integrating past experiences and projecting future consequences into a cohesive understanding of right and wrong. Recursive Realism suggests that moral reflection is inherently self-referential, where awareness of past actions informs present decisions and shapes future choices through a loop of ongoing integration.

Memory, Regret, and the Recursion of Moral Learning: Memory plays a crucial role in ethical reflection, allowing individuals to learn from past mistakes and adapt their behaviour. Recursive Realism suggests that moral learning involves a feedback loop, where memories of past actions create a basis for future adjustments. For example, regret over a missed opportunity to help others might lead to greater compassion in future situations, creating a cycle where memory shapes moral growth. This perspective highlights how conscious awareness of the past guides ethical development, where recursively reflecting on experiences refines moral judgment.

Anticipation, Consequences, and the Loop of Ethical Foresight: Anticipating the consequences of actions is crucial for ethical decision-making, allowing individuals to consider the potential impact of their choices. Recursive Realism suggests that ethical foresight involves a self-referential loop, where the mind envisions future scenarios, then adjusts actions based on expected outcomes. For example, a politician considering a policy change might simulate different outcomes for various groups, creating a loop where ethical foresight guides policy decisions. This perspective allows for a deeper appreciation of how consciousness enables a future-oriented approach to morality, where awareness of potential consequences informs ethical choices.

Holistic Ethics and the Recursion of Self-Integration: Holistic ethical thinking involves integrating one's values, past experiences, and aspirations into a coherent moral framework. Recursive Realism suggests that self-integration is a self-referential process, where awareness of different aspects of the self creates a loop that aligns actions with values. For example, someone might integrate their love for nature with concerns about justice, leading to environmental activism, creating a loop where self-awareness shapes ethical commitments. This perspective highlights how self-reflection enables the synthesis of diverse values, where the recursive mind brings cohesion to moral life.

Ethical consciousness is characterised by self-referential processes, where reflection and awareness enable the navigation of moral dilemmas, the exercise of free will, and the integration of experiences into a cohesive understanding of right and wrong. Recursive Realism provides a framework for understanding how self-reflective thought allows for ethical autonomy, revealing how consciousness is deeply tied to the ability to engage with moral challenges. By recognising the recursive nature of ethical reflection, Recursive Realism offers a vision of moral reasoning as a dynamic interplay between awareness, values, and choices, where the mind's recursive loops shape a richer understanding of what it means to be moral.

In Chapter 32, we have explored how Recursive Realism offers a framework for understanding the ethical dimensions of self-reflection, revealing how moral reasoning, conscious awareness, and free will are shaped by self-referential processes. By examining the nature of moral dilemmas, the role of empathy, and the integration of ethical experiences, Recursive Realism provides a nuanced perspective on how ethics evolves through recursively structured thought. This perspective allows us to appreciate morality as a dynamic process, where consciousness and self-awareness enable the refinement of values and the pursuit of moral integrity.

References

Arendt, H. (1958) *The Human Condition*. Chicago: University of Chicago Press.

Bandura, A. (1977) 'Self-Efficacy: Toward a Unifying Theory of Behavioral Change', *Psychological Review*, 84(2), pp. 191–215.

Chalmers, D.J. (1996) *The Conscious Mind: In Search of a Fundamental Theory*. Oxford: Oxford University Press.

Darwall, S. (2006) *The Second-Person Standpoint: Morality, Respect, and Accountability*. Cambridge, MA: Harvard University Press.

Dennett, D.C. (1984) *Elbow Room: The Varieties of Free Will Worth Wanting*. Cambridge, MA: MIT Press.

Gilligan, C. (1982) *In a Different Voice: Psychological Theory and Women's Development*. Cambridge, MA: Harvard University Press.

Habermas, J. (1990) *Moral Consciousness and Communicative Action*. Cambridge, MA: MIT Press.

Haidt, J. (2001) 'The Emotional Dog and Its Rational Tail: A Social Intuitionist Approach to Moral Judgment', *Psychological Review*, 108(4), pp. 814–834.

Kohlberg, L. (1981) *Essays on Moral Development, Volume 1: The Philosophy of Moral Development*. San Francisco: Harper & Row.

MacIntyre, A. (1981) *After Virtue: A Study in Moral Theory*. Notre Dame: University of Notre Dame Press.

Nagel, T. (1979) *Mortal Questions*. Cambridge: Cambridge University Press.

Nussbaum, M.C. (2001) *Upheavals of Thought: The Intelligence of Emotions*. Cambridge: Cambridge University Press.

Rawls, J. (1971) *A Theory of Justice*. Cambridge, MA: Harvard University Press.

Sandel, M.J. (1982) *Liberalism and the Limits of Justice*. Cambridge: Cambridge University Press.

Scanlon, T.M. (1998) *What We Owe to Each Other*. Cambridge, MA: Harvard University Press.

Singer, P. (1979) *Practical Ethics*. Cambridge: Cambridge University Press.

Taylor, C. (1989) *Sources of the Self: The Making of the Modern Identity*. Cambridge, MA: Harvard University Press.

Velleman, J.D. (2000) *The Possibility of Practical Reason*. Oxford: Oxford University Press.

Waldron, J. (1993) *Liberal Rights: Collected Papers 1981–1991*. Cambridge: Cambridge University Press.

Williams, B. (1985) *Ethics and the Limits of Philosophy*. Cambridge, MA: Harvard University Press.

Wolf, S. (1982) 'Moral Saints', *The Journal of Philosophy*, 79(8), pp. 419–439.

Chapter 33: Recursive Realism and the Structure of Knowledge

The Recursive Nature of Learning and Understanding

Knowledge is not a static repository but a dynamic process that involves building upon previous insights, challenging assumptions, and refining theories. Recursive Realism suggests that learning is inherently self-referential, where the mind engages in loops of assimilation and accommodation, allowing for the integration of new information into existing frameworks. This section explores how Recursive Realism offers a perspective on learning, understanding, and the construction of knowledge, examining the processes through which ideas are formed, tested, and transformed.

The Process of Learning as Recursive Integration

Learning involves the acquisition of new information and its integration into existing knowledge structures. Recursive Realism suggests that learning is a self-referential process, where each new insight is evaluated against prior understanding, creating a loop that allows for continuous refinement.

Assimilation, Accommodation, and the Feedback Loop of Cognitive Growth: Jean Piaget's theory of assimilation and accommodation describes how children integrate new experiences into their understanding of the world. Recursive Realism extends this concept, suggesting that learning involves feedback loops where new information is assimilated into existing schemas, and schemas are accommodated to incorporate novel insights. For example, a student learning a scientific concept like gravity might initially assimilate it into everyday experiences of falling objects, but then adjust their understanding as they encounter more complex phenomena, creating a cycle of cognitive adaptation. This perspective highlights how learning is a recursive process, where each new piece of information refines the overall structure of knowledge.

Metacognition and the Recursion of Self-Directed Learning: Metacognition, the ability to think about one's own thinking, plays a key role in effective learning. Recursive Realism suggests that metacognitive processes are inherently self-referential, where learners reflect on their understanding and adjust their strategies through a loop of self-evaluation. For example, a student preparing for an exam might evaluate which areas they understand well and which require further study, creating a cycle where self-reflection guides the focus of learning. This perspective allows for a deeper appreciation of how metacognition enhances knowledge acquisition, where awareness of one's own learning process fosters cognitive growth.

The Spiral of Learning: Recursion in Skill Development: Skill development often follows a spiral model, where learners revisit concepts at increasing levels of complexity. Recursive Realism suggests that this spiral of learning involves self-referential loops, where each iteration builds on previous mastery while introducing new challenges. For example, learning a musical instrument might involve repeatedly practising scales, but each iteration introduces nuances like dynamics and interpretation, creating a loop where technical proficiency and musical expression evolve together. This perspective highlights how learning is a recursive journey, where progress is achieved through revisiting and expanding upon foundational skills.

Scientific Inquiry and the Recursion of Hypothesis Testing

Science is often described as a methodical pursuit of truth, involving the generation of hypotheses, experimentation, and revision of theories. Recursive Realism suggests that scientific inquiry is inherently self-referential, where theories are continuously tested, challenged, and refined through feedback loops between data and interpretation.

The Scientific Method as a Recursive Process: The scientific method involves a cycle of observation, hypothesis formation, experimentation, and revision. Recursive Realism suggests that this process is self-referential, as each hypothesis is tested against empirical evidence, creating a loop where theories evolve based on new data. For example, the transition from Newtonian physics to Einstein's theory of relativity involved revisiting fundamental assumptions about space and time, creating a cycle where new observations reshaped scientific understanding. This perspective highlights how science progresses through recursive refinement, where each iteration deepens our understanding of natural laws.

Paradigm Shifts and the Feedback Loop of Scientific Revolutions: Thomas Kuhn's concept of paradigm shifts describes how scientific revolutions occur when existing frameworks are replaced by new models that better explain anomalies. Recursive Realism suggests that paradigm shifts involve a self-referential loop, where the limitations of current theories prompt a cycle of re-evaluation and conceptual transformation. For example, the shift from geocentric to heliocentric models of the solar system involved a recursive process where observations of planetary motion challenged prevailing beliefs, leading to a new framework that redefined our place in the universe. This perspective allows for a deeper understanding of how scientific progress is shaped by recursively questioning assumptions, where each shift opens new avenues for exploration.

The Interplay Between Theory and Data: A Recursive Relationship: Scientific theories and empirical data exist in a recursive relationship, where theories guide observations, and observations inform the refinement of theories. Recursive Realism suggests that this interplay involves feedback loops, where hypotheses are tested, and the resulting data leads to the revision of the theoretical model. For example, in cosmology, theories about the nature of dark matter are continually adjusted as new astronomical data becomes available, creating a cycle where theoretical models and empirical evidence evolve together. This perspective highlights how scientific knowledge is not static but evolves through a recursive dialogue between ideas and reality.

The Integration of Knowledge into Coherent Worldviews

Understanding the world involves integrating diverse pieces of knowledge into a coherent framework that makes sense of reality. Recursive Realism suggests that the construction of worldviews is a self-referential process, where new insights are integrated through a loop of reflection and reconciliation with existing beliefs.

Worldview Formation and the Recursion of Belief Systems: Worldviews provide a lens through which individuals interpret reality, integrating scientific understanding, philosophical ideas, and personal experiences. Recursive Realism suggests that worldview formation involves a feedback loop, where new knowledge prompts re-evaluation of core beliefs, creating a cycle of intellectual synthesis. For example, the advent of quantum mechanics prompted philosophical reflections on the nature of reality, creating a loop where scientific discoveries influenced metaphysical ideas. This perspective highlights how knowledge is not compartmentalised but weaves together through recursive reflection, allowing for the construction of comprehensive worldviews.

Interdisciplinary Thinking and the Loop of Conceptual Integration: Interdisciplinary thinking, the ability to connect insights from different fields, is essential for tackling complex problems. Recursive Realism suggests that interdisciplinary integration involves self-referential loops, where concepts from one domain are reframed in the context of another, creating a cycle of conceptual expansion. For example, applying principles of ecology to economics has led to sustainable development models, creating a loop where ideas from natural systems reshape economic theories. This perspective allows for a richer understanding of how knowledge expands, where recursive thinking bridges the gaps between disciplines.

Knowledge and Uncertainty: Recursion in the Exploration of the Unknown: The pursuit of knowledge involves engaging with uncertainty, where new discoveries often reveal further mysteries. Recursive Realism suggests that exploring the unknown involves a self-referential process, where each answer leads to new questions, creating a loop of intellectual curiosity. For example, understanding the human genome has led to new insights into genetics, but also revealed complex regulatory networks that remain unexplained, creating a cycle of discovery and inquiry. This perspective highlights how the quest for knowledge is never-ending, where recursion drives the continuous expansion of understanding.

Learning and knowledge creation are deeply self-referential, involving processes of reflection, synthesis, and re-evaluation that allow for the integration of new insights into a coherent understanding of the world. Recursive Realism provides a framework for understanding how cognitive growth, scientific progress, and worldview formation are shaped by recursive loops, revealing how knowledge evolves through cycles of engagement with ideas and evidence. By recognising the recursive nature of learning, Recursive Realism offers a vision of intellectual growth as an iterative process, where each new discovery adds depth to our understanding of reality.

Recursive Realism and the Dynamics of Scientific and Philosophical Inquiry

Scientific theories and philosophical ideas are not just collections of facts or logical constructs but represent frameworks for understanding reality. Recursive Realism suggests that the development of theories and philosophical perspectives involves self-referential loops, where ideas are refined, challenged, and expanded through cycles of reflection and empirical testing. This section explores how Recursive Realism offers a perspective on the evolution of scientific thought, the relationship between abstract reasoning and practical discovery, and the role of recursion in advancing human understanding.

The Evolution of Scientific Theories Through Recursive Inquiry

Scientific theories serve as models of reality, providing explanations for natural phenomena and guiding experimentation. Recursive Realism suggests that the evolution of theories is a self-referential process, where theories are tested, revised, and reconceptualised through a loop of empirical engagement and theoretical refinement.

Theories as Models and the Recursion of Conceptual Revisions: Theories can be understood as models that represent aspects of reality, making predictions and offering explanations. Recursive Realism suggests that theoretical models evolve through a feedback loop, where new data prompts revisions to the model, which in turn shapes future observations. For example, Darwin's theory of evolution by natural selection has been continually refined by genetic discoveries, creating a loop where the model of evolution is expanded by new insights into molecular biology. This perspective highlights how scientific theories are not static truths but dynamic constructs that evolve through recursive reflection and empirical testing.

Hypothesis Testing and the Recursion of Falsification: Karl Popper's concept of falsification emphasises the importance of testing theories through attempts to disprove them. Recursive Realism suggests that falsification is inherently self-referential, involving a loop where theories are subjected to critical examination, leading to refinement or replacement. For example, the theory of general relativity was tested through observations of gravitational lensing and time dilation, creating a cycle where theoretical predictions were verified or challenged. This perspective allows for a richer understanding of how scientific progress is driven by recursively testing ideas, where the process of inquiry refines the boundaries of knowledge.

Emergent Theories and the Feedback Loop of Cross-Disciplinary Insights: Emergent theories often arise when ideas from different disciplines converge to form new models of understanding. Recursive Realism suggests that cross-disciplinary integration involves a self-referential loop, where concepts from one field are reframed in the context of another, creating a cycle that expands theoretical horizons. For example, the integration of information theory with neuroscience has led to new models of brain function, creating a loop where concepts of entropy and computation shape our understanding of cognition. This perspective highlights how scientific inquiry is enriched by recursively blending ideas from different fields, where the synthesis of concepts leads to the emergence of new paradigms.

Philosophy, Science, and the Recursive Exploration of Abstract Concepts

Philosophy and science often interact in the pursuit of knowledge, with philosophy offering conceptual frameworks and science providing empirical grounding. Recursive Realism suggests that the relationship between philosophy and science is self-referential, where philosophical questions inspire scientific investigations, and scientific discoveries reshape philosophical perspectives.

Philosophical Inquiry and the Loop of Conceptual Clarification: Philosophy often seeks to clarify concepts and explore foundational questions about existence, knowledge, and values. Recursive Realism suggests that philosophical inquiry involves a loop of reflective questioning, where each answer leads to new questions, creating a cycle of conceptual refinement. For example, the philosophical problem of free will has evolved through engagement with scientific insights into neuroscience and determinism, creating a dialogue where ideas about autonomy and causality are continually refined. This perspective highlights how philosophy functions as a recursive dialogue, where concepts are explored through successive layers of reflection.

Science as a Response to Philosophical Questions: Many scientific investigations have their origins in philosophical curiosity about the nature of reality. Recursive Realism suggests that science and philosophy engage in a feedback loop, where philosophical questions about time, space, and existence guide scientific exploration, and scientific findings prompt new philosophical inquiries. For example, questions about the nature of time have led to scientific theories in relativity and quantum mechanics, while discoveries about quantum entanglement have, in turn, inspired philosophical discussions about non-locality and causality. This perspective allows for a richer appreciation of how philosophy and science co-evolve through recursive engagement, where each discipline expands the scope of the other.

The Philosophy of Science and the Recursion of Methodological Reflection: The philosophy of science examines the methods and principles underlying scientific practice, providing a framework for understanding the nature of inquiry. Recursive Realism suggests that methodological reflection is a self-referential process, where scientists and philosophers critique the assumptions of scientific methods, creating a loop that guides the evolution of scientific practice. For example, debates about the role of objectivity and the limits of empiricism have led to new approaches in quantitative research and interpretive methodologies, creating a cycle where scientific methods are refined through philosophical reflection. This perspective highlights how methodological progress is shaped by recursively examining the principles of inquiry, where philosophical engagement enriches scientific practice.

The Dynamics of Paradigm Shifts and Intellectual Transformation

Paradigm shifts, transformations in dominant models of understanding, represent key moments in intellectual history. Recursive Realism suggests that paradigm shifts involve self-referential processes, where existing frameworks are challenged and replaced through a loop of critical reflection and conceptual innovation.

Intellectual Revolutions and the Feedback Loop of Conceptual Breakthroughs: Intellectual revolutions often occur when new ideas challenge deeply held assumptions, leading to the emergence of new paradigms. Recursive Realism suggests that conceptual breakthroughs involve a loop where criticism of existing theories prompts the development of new models that redefine the field. For example, the Copernican Revolution shifted the cosmological model from geocentrism to heliocentrism, creating a recursive dialogue between observation and philosophical reflection that reshaped our understanding of the universe. This perspective allows for a deeper appreciation of how paradigm shifts emerge through recursive questioning, where new frameworks are born from cycles of challenge and refinement.

Dialectics and the Recursion of Thesis and Antithesis: Dialectical reasoning, the process of reconciling opposing ideas, has played a key role in philosophical and scientific progress. Recursive Realism suggests that dialectics involves a self-referential process, where thesis and antithesis engage in a loop, leading to the synthesis of new perspectives. For example, Hegel's dialectical model proposed that ideas evolve through a cycle of contradiction and resolution, influencing thinkers like Marx and Einstein. This perspective highlights how dialectical thinking drives intellectual transformation, where recursively engaging with contradictions leads to the evolution of understanding.

The Role of Recursion in Intellectual Adaptability: Intellectual adaptability, the ability to adjust theories and beliefs in response to new evidence, is crucial for the advancement of knowledge.

Recursive Realism suggests that adaptability involves a feedback loop, where theories are revisited and reformulated as understanding deepens. For example, the adjustments to evolutionary theory in light of genetic discoveries demonstrate how scientific models evolve through recursively integrating new data. This perspective highlights how intellectual flexibility is not merely a response to change but is driven by recursive processes that allow the mind to reorganise its understanding of the world.

Scientific theories and philosophical perspectives are shaped by self-referential processes, where theories evolve through cycles of testing, reflection, and conceptual synthesis. Recursive Realism provides a framework for understanding how the pursuit of knowledge is driven by recursively questioning assumptions, revealing how scientific inquiry and philosophical thought interact to create a deeper understanding of reality. By recognising the recursive nature of theoretical innovation, Recursive Realism offers a vision of intellectual growth as an ongoing dialogue, where each new insight is part of a broader process of intellectual refinement.

In Chapter 33, we have explored how Recursive Realism offers a framework for understanding the structure of knowledge, revealing how learning, scientific inquiry, and philosophical reflection are shaped by self-referential processes. By examining the nature of theoretical refinement, the interplay between science and philosophy, and the dynamics of paradigm shifts, Recursive Realism provides a nuanced perspective on how knowledge evolves through recursively structured thought. This perspective allows us to appreciate knowledge as a dynamic process, where ideas and theories grow through cycles of reflection, empirical engagement, and conceptual expansion.

References

Bachelard, G. (1984) *The New Scientific Spirit*. Boston: Beacon Press.

Barrow, J.D. (1991) *Theories of Everything: The Quest for Ultimate Explanation*. Oxford: Oxford University Press.

Bateson, G. (1972) *Steps to an Ecology of Mind*. Chicago: University of Chicago Press.

Chalmers, A.F. (1999) *What Is This Thing Called Science?* 3rd edn. Indianapolis: Hackett Publishing.

Collins, R. (1994) *The Sociology of Philosophies: A Global Theory of Intellectual Change*. Cambridge, MA: Harvard University Press.

Dennett, D.C. (1995) *Darwin's Dangerous Idea: Evolution and the Meanings of Life*. New York: Simon & Schuster.

Feyerabend, P. (1975) *Against Method: Outline of an Anarchistic Theory of Knowledge*. London: Verso.

Gardner, H. (1983) *Frames of Mind: The Theory of Multiple Intelligences*. New York: Basic Books.

Gopnik, A. (1996) 'The Scientist as Child', *Philosophy of Science*, 63(4), pp. 485–514.

Kuhn, T.S. (1962) *The Structure of Scientific Revolutions*. Chicago: University of Chicago Press.

Lakatos, I. (1976) *Proofs and Refutations: The Logic of Mathematical Discovery*. Cambridge: Cambridge University Press.

Lorenz, K. (1977) *Behind the Mirror: A Search for a Natural History of Human Knowledge*. London: Methuen.

Maturana, H.R. and Varela, F.J. (1987) *The Tree of Knowledge: The Biological Roots of Human Understanding*. Boston: Shambhala.

Popper, K.R. (1959) *The Logic of Scientific Discovery*. London: Hutchinson.

Prigogine, I. and Stengers, I. (1984) *Order Out of Chaos: Man's New Dialogue with Nature*. New York: Bantam Books.

Rescher, N. (1996) *Process Metaphysics: An Introduction to Process Philosophy*. Albany: State University of New York Press.

Sacks, O. (2010) *The Mind's Eye*. London: Picador.

Schön, D.A. (1983) *The Reflective Practitioner: How Professionals Think in Action*. New York: Basic Books.

Senge, P.M. (1990) *The Fifth Discipline: The Art and Practice of the Learning Organization*. New York: Doubleday.

Snow, C.P. (1959) *The Two Cultures and the Scientific Revolution*. Cambridge: Cambridge University Press.

Wigner, E.P. (1960) 'The Unreasonable Effectiveness of Mathematics in the Natural Sciences', *Communications on Pure and Applied Mathematics*, 13(1), pp. 1–14.

Wittgenstein, L. (1953) *Philosophical Investigations*. Oxford: Blackwell.

Chapter 34: Recursive Realism and the Limits of Human Understanding

The Concept of the Unknowable

Human understanding has always encountered boundaries, regions where knowledge fades into mystery and questions remain unanswered. Recursive Realism suggests that the concept of the unknowable is deeply tied to self-referential processes, where the mind confronts the edges of its own capacity to comprehend. This section explores how Recursive Realism offers a perspective on the unknowable, examining the nature of paradox, the limits of explanation, and the role of mystery in intellectual life.

The Paradox of Self-Reference and the Limits of Explanation

Paradoxes, logical statements that challenge coherence, often highlight the boundaries of our understanding. Recursive Realism suggests that self-reference is at the heart of many philosophical paradoxes, revealing the limits of explanation and the edges of logic.

Gödel's Incompleteness Theorems and the Recursion of Mathematical Limits: Kurt Gödel's incompleteness theorems demonstrate that within any formal mathematical system, there are true statements that cannot be proven within the system itself. Recursive Realism suggests that Gödel's theorems reveal a self-referential limit, where mathematics confronts its own boundaries through a loop of self-reference. For example, Gödel's proof relies on constructing statements that reference their own unprovability, creating a paradox that demonstrates the incompleteness of mathematical logic. This perspective highlights how self-referential processes reveal the limits of formal systems, where the attempt to describe reality ultimately encounters the boundaries of what can be articulated.

The Liar Paradox and the Loop of Self-Contradiction: Philosophical paradoxes like the Liar Paradox, which states, "This statement is false", illustrate the problems of self-reference. Recursive Realism suggests that such paradoxes expose the limits of language and logic, revealing how self-referential statements can defy coherence. For example, if the statement is true, then it must be false, but if it is false, then it must be true, creating an infinite loop of contradiction. This perspective allows for a deeper appreciation of how the attempt to understand reality through self-reference leads to paradoxical boundaries, where logical systems reach their limits.

The Problem of Consciousness and the Limits of Self-Knowledge: Consciousness, the experience of being aware, presents a profound philosophical challenge, as awareness tries to understand itself. Recursive Realism suggests that the problem of consciousness involves a self-referential loop, where the mind attempts to grasp the nature of its own awareness, encountering limits in the process. For example, the question of how subjective experience arises from physical processes remains elusive, creating a boundary where self-reflective inquiry reaches a point of mystery. This perspective highlights how self-referential processes can reveal the inherent limits of understanding, where the mind's attempt to know itself encounters profound mysteries.

The Boundaries of Human Perception

Perception is the gateway through which we interact with the world, yet it is also limited by the constraints of our sensory systems and cognitive frameworks. Recursive Realism suggests that the boundaries of perception are shaped by self-referential processes, where the mind continually interprets and reinterprets sensory data, creating a loop that both enhances and restricts our view of reality.

The Recursion of Sensory Filtering: Human perception relies on filtering out vast amounts of information, focusing on what is most relevant. Recursive Realism suggests that perception involves self-referential feedback loops, where sensory inputs are interpreted based on expectations and previous experiences, creating a cycle that shapes awareness. For example, visual perception relies on cognitive shortcuts to recognise familiar patterns, but this filtering also limits our ability to perceive subtle variations. This perspective allows for a richer understanding of how the mind's recursive processes create a bounded view of the world, where perception is shaped by both its capabilities and its constraints.

Illusions and the Recursion of Cognitive Bias: Cognitive biases and perceptual illusions reveal the limitations of our mental models, showing how the mind's expectations can distort reality. Recursive Realism suggests that illusions are a result of self-referential loops, where the mind's interpretations of sensory data create a feedback loop that overrides raw perception. For example, the Müller-Lyer illusion, where lines of equal length appear unequal due to contextual cues, demonstrates how the mind constructs a reality that does not always align with objective measurement. This perspective highlights how perception is shaped by recursively structured thought, where our understanding of the world is influenced by patterns that both reveal and obscure reality.

The Spectrum of the Unseen: The Limits of Human Sensory Range: Human senses are limited to certain ranges of light, sound, and touch, leaving vast aspects of the universe beyond direct experience. Recursive Realism suggests that the limitations of our senses create a feedback loop, where scientific instruments extend our perception and then reframe how we interpret reality. For example, telescopes and microscopes have revealed phenomena far beyond naked-eye observation, such as distant galaxies or microscopic life, but these tools also reveal the boundaries of what we can directly perceive. This perspective allows for a deeper understanding of how the expansion of perception through technology still involves self-referential limits, where our view of the universe remains conditioned by the constraints of our tools.

Mystery, Humility, and the Role of Intellectual Awe

Encountering the unknown often evokes a sense of mystery, prompting humility in the face of the universe's complexity. Recursive Realism suggests that the recognition of limits is inherently self-referential, where awareness of our own ignorance creates a loop that deepens intellectual curiosity while acknowledging the boundaries of what can be known.

Intellectual Humility and the Recursion of Questioning: Humility in intellectual pursuits involves recognising the limitations of one's own understanding and remaining open to new possibilities. Recursive Realism suggests that intellectual humility involves a self-referential process, where awareness of ignorance leads to a deeper engagement with mystery, creating a loop where each question opens new horizons. For example, the philosophical attitude of Socratic ignorance, the recognition that wisdom comes from knowing how little one knows, reflects a recursive awareness that deepens the pursuit of knowledge. This perspective highlights how self-reflective thought can foster a sense of wonder, where intellectual humility becomes a gateway to deeper inquiry.

Mystery as a Source of Intellectual Awe: Mystery, the sense that there are truths beyond human comprehension, has often inspired scientific and philosophical exploration. Recursive Realism suggests that mystery involves a feedback loop, where encountering the unknowable prompts a reflection on the nature of understanding itself. For example, the mystery of dark energy and dark matter, which constitute most of the universe's mass-energy content yet remain poorly understood, creates a loop where scientific inquiry deepens our sense of cosmic mystery. This perspective allows for a richer appreciation of how the unknown inspires intellectual awe, where the recognition of limits is not a barrier but a source of motivation for exploration.

The Balance Between Certainty and Openness: Certainty provides a sense of stability in our understanding, while openness to mystery allows for growth and exploration. Recursive Realism suggests that the balance between certainty and openness is a self-referential process, where intellectual inquiry moves between established knowledge and the willingness to embrace the unknown. For example, while scientific theories provide a framework for understanding natural laws, frontier areas like the nature of consciousness remind us that mystery remains. This perspective highlights how the quest for knowledge is shaped by a recursive tension between what is known and what remains to be discovered, where the mind's recursive loops enable a deeper engagement with the mysteries of existence.

The limits of human understanding are deeply tied to self-referential processes, where consciousness confronts its own boundaries and recognises the mysteries that lie beyond its grasp. Recursive Realism provides a framework for understanding how paradoxes, the limits of perception, and the sense of mystery shape our intellectual journey, revealing how self-reflective thought can both expand and limit our understanding of reality. By recognising the recursive nature of engaging with the unknown, Recursive Realism offers a vision of intellectual life as a dialogue between curiosity and humility, where each new insight deepens the mystery of existence.

Recursive Realism and the Boundaries of Conscious Exploration

Conscious exploration involves seeking answers to fundamental questions about existence, purpose, and reality. However, the depth of such inquiries often leads to the recognition of limits, where certainty gives way to ambiguity. Recursive Realism suggests that engaging with these questions involves self-referential processes, where the mind turns back on itself, examining its own assumptions and limitations. This section explores how Recursive Realism frames the pursuit of meaning, the nature of metaphysical boundaries, and the role of recursion in embracing uncertainty.

The Pursuit of Meaning and the Recursion of Self-Reflection

The search for meaning has been central to philosophical inquiry, involving questions about purpose, identity, and the nature of reality. Recursive Realism suggests that the pursuit of meaning is inherently self-referential, where consciousness reflects on its own existence, creating a loop that deepens our engagement with existential questions.

Self-Reflection and the Loop of Existential Inquiry: Questions of meaning often prompt individuals to reflect on their place in the cosmos, leading to self-examination. Recursive Realism suggests that existential reflection involves a feedback loop, where the mind considers its own thoughts, revisits past experiences, and explores potential purposes. For example, someone pondering the meaning of life might reflect on their experiences, consider the nature of human consciousness, and question the nature of reality, creating a recursive dialogue that shapes their philosophical outlook. This perspective highlights how self-reflection is central to the search for meaning, where recursive engagement allows for the exploration of existential depths.

Identity and the Recursion of Self-Definition: The concept of identity, what it means to be oneself, is a key aspect of self-reflective inquiry. Recursive Realism suggests that identity is not static but involves a self-referential process, where individuals continually reinterpret their sense of self through reflection and experience. For example, as a person encounters new experiences, they might reconsider their values, reframe their life story, and redefine their sense of purpose, creating a loop where identity evolves through self-awareness. This perspective allows for a deeper appreciation of how self-definition is shaped by recursive thought, where the mind's reflection on its own nature leads to a dynamic understanding of who we are.

Meaning-Making and the Recursive Interpretation of Reality: Making sense of the world involves interpreting experiences and finding connections between disparate events. Recursive Realism suggests that meaning-making is a self-referential process, where the mind integrates new experiences into a cohesive narrative, creating a loop of interpretation. For example, during a moment of crisis, an individual might re-evaluate their past choices, reconsider their beliefs, and reframe their future goals, creating a cycle where meaning is reconstructed. This perspective highlights how meaning emerges from recursively engaging with life's challenges, where each reflection contributes to a broader understanding of one's place in the world.

Metaphysical Inquiry and the Boundaries of Knowledge

Metaphysical questions, those concerning the nature of reality, time, and existence, often extend beyond the reach of empirical verification, dwelling in the realm of speculation and abstract thought. Recursive Realism suggests that engaging with metaphysical questions involves a loop of self-referential reasoning, where the mind explores its own capacity to conceive of realities beyond immediate experience.

The Infinite Regress Problem and the Recursion of Causality: Metaphysical questions often encounter the problem of infinite regress, where each explanation leads to a further question about its origins. Recursive Realism suggests that infinite regress reveals the self-referential nature of causal reasoning, where the mind attempts to trace the origins of reality through a loop that never fully resolves. For example, questions like "What caused the universe?" lead to further inquiries about the cause of that cause, creating a cycle where answers remain elusive. This perspective allows for a deeper appreciation of how the search for first causes is shaped by recursive reasoning, where the pursuit of ultimate explanations inevitably encounters the boundaries of understanding.

Ontology and the Recursion of Being: Ontology, the study of being, involves questions about what exists and the nature of existence itself. Recursive Realism suggests that ontological reflection is a self-referential process, where the mind considers the nature of being through a loop of self-inquiry. For example, pondering the nature of consciousness involves reflecting on what it means to exist as a conscious entity, creating a dialogue where the mind examines its own existence. This perspective highlights how ontological questions challenge the boundaries of human thought, where self-reflective reasoning reveals the depths and limits of our understanding of being.

The Concept of the Infinite and the Recursion of Abstract Thought: The infinite is a concept that eludes full comprehension, yet it is central to mathematics, cosmology, and metaphysics. Recursive Realism suggests that thinking about the infinite involves a self-referential loop, where the mind tries to grasp an unbounded concept through finite thought. For example, contemplating the nature of infinite space or infinite time involves a loop where each attempt to conceptualise infinity reveals the limitations of our finite perspective. This perspective highlights how the exploration of infinity is both a challenge and an inspiration, where recursively engaging with the infinite leads to a deeper awareness of the mysteries of existence.

Navigating the Unknown: Recursion, Humility, and the Open Mind

Embracing the unknown is essential for intellectual growth, requiring an open mind that is willing to explore beyond the familiar. Recursive Realism suggests that engaging with uncertainty involves a self-referential process, where the mind continuously re-evaluates its beliefs, challenges assumptions, and remains open to new possibilities.

Intellectual Openness and the Recursion of Inquiry: Intellectual openness involves the willingness to question established beliefs and consider alternative perspectives. Recursive Realism suggests that openness is a self-referential stance, where the mind engages in a loop of continual questioning and exploration. For example, a scientist exploring a new theory might challenge their own assumptions, re-evaluate data, and remain open to unexpected findings, creating a cycle of intellectual flexibility. This perspective highlights how intellectual progress is driven by recursive engagement with the unknown, where each inquiry deepens our openness to discovery.

Exploring the Edge of Understanding and the Loop of Conceptual Expansion: The edge of understanding, where the known meets the unknown, is often where the most profound discoveries are made. Recursive Realism suggests that exploring these edges involves a feedback loop, where each new insight reveals further questions and challenges. For example, discovering the structure of DNA opened new realms of genetic research and evolutionary biology, creating a cycle where each answer led to new lines of inquiry. This perspective allows for a deeper understanding of how intellectual exploration is an iterative process, where recursion drives the expansion of knowledge into the unknown.

Accepting Uncertainty and the Recursion of Intellectual Humility: Accepting uncertainty is often necessary for engaging with complex questions, recognising that not all mysteries will be resolved. Recursive Realism suggests that intellectual humility involves a self-referential process, where awareness of the limits of understanding creates a loop that embraces mystery.

For example, cosmologists confronting the mysteries of the origins of the universe might acknowledge the limits of current models, creating a stance where curiosity is balanced with humility. This perspective highlights how the acceptance of uncertainty can be a source of strength, where recursively reflecting on the unknown fosters a deeper connection to the vastness of existence.

Conscious exploration is shaped by self-referential processes, where the mind grapples with the unknown through cycles of reflection, questioning, and engagement with mystery. Recursive Realism provides a framework for understanding how existential inquiry, metaphysical reflection, and intellectual openness are shaped by recursively structured thought, revealing how the search for meaning and the encounter with limits are deeply intertwined. By recognising the recursive nature of engaging with the unknown, Recursive Realism offers a vision of intellectual life as a journey through uncertainty, where each question deepens our appreciation of the mysteries that lie beyond understanding.

In Chapter 34, we have explored how Recursive Realism offers a framework for understanding the limits of human understanding, revealing how self-referential processes shape our engagement with the unknown. By examining the nature of paradox, the boundaries of perception, and the role of humility, Recursive Realism provides a nuanced perspective on how consciousness navigates the depths of reality, revealing how recursively structured thought allows for both the expansion and recognition of limits. This perspective allows us to appreciate intellectual life as an exploration of both what we can know and the mysteries that elude us, where curiosity and humility are intertwined in the pursuit of deeper understanding.

References

Baars, B.J. (1997) *In the Theater of Consciousness: The Workspace of the Mind*. Oxford: Oxford University Press.

Bateson, G. (1972) *Steps to an Ecology of Mind*. Chicago: University of Chicago Press.

Chalmers, D.J. (1995) 'Facing Up to the Problem of Consciousness', *Journal of Consciousness Studies*, 2(3), pp. 200–219.

Descartes, R. (1641/1996) *Meditations on First Philosophy*. Cambridge: Cambridge University Press.

Gödel, K. (1931) 'Über formal unentscheidbare Sätze der Principia Mathematica und verwandter Systeme I', *Monatshefte für Mathematik und Physik*, 38(1), pp. 173–198.

Heidegger, M. (1927/1962) *Being and Time*. New York: Harper & Row.

Hofstadter, D.R. (1979) *Gödel, Escher, Bach: An Eternal Golden Braid*. New York: Basic Books.

James, W. (1902/1985) *The Varieties of Religious Experience: A Study in Human Nature*. Cambridge, MA: Harvard University Press.

Kuhn, T.S. (1962) *The Structure of Scientific Revolutions*. Chicago: University of Chicago Press.

Nagel, T. (1974) 'What Is It Like to Be a Bat?', *The Philosophical Review*, 83(4), pp. 435–450.

Penrose, R. (1989) *The Emperor's New Mind: Concerning Computers, Minds and the Laws of Physics*. Oxford: Oxford University Press.

Prigogine, I. (1997) *The End of Certainty: Time, Chaos, and the New Laws of Nature*. New York: Free Press.

Popper, K.R. (1945) *The Open Society and Its Enemies*. London: Routledge.

Russell, B. (1948) *Human Knowledge: Its Scope and Limits*. New York: Simon and Schuster.

Schrödinger, E. (1944) *What Is Life? The Physical Aspect of the Living Cell*. Cambridge: Cambridge University Press.

Socrates (Plato) (2003) *The Apology of Socrates*. Translated by B. Jowett. New York: Dover Publications.

Sosa, E. (1991) *Knowledge in Perspective: Selected Essays in Epistemology*. Cambridge: Cambridge University Press.

Stewart, I. (1990) *Does God Play Dice? The New Mathematics of Chaos*. Oxford: Blackwell.

Tegmark, M. (2014) *Our Mathematical Universe: My Quest for the Ultimate Nature of Reality*. New York: Alfred A. Knopf.

Varela, F.J., Thompson, E. and Rosch, E. (1991) *The Embodied Mind: Cognitive Science and Human Experience*. Cambridge, MA: MIT Press.

Wittgenstein, L. (1953) *Philosophical Investigations*. Oxford: Blackwell.

Chapter 35: Recursive Realism and the Interconnectedness of All Things

The Interconnectedness of Natural Phenomena

The universe is a web of interdependent phenomena, where complex systems interact and mutually shape each other. Recursive Realism suggests that understanding interconnectedness involves recognising the self-referential loops that create patterns of unity across nature, life, and cosmic processes. This section explores how Recursive Realism provides insights into the connections between living systems, the dynamics of ecosystems, and the recursive structures that link the individual to the universe.

Ecosystems as Self-Referential Systems

Ecosystems, networks of organisms and their environments, exemplify the interconnected nature of life. Recursive Realism suggests that ecosystems are self-referential systems, where each component influences and is influenced by the others, creating a loop of mutual adaptation and evolution.

The Recursion of Feedback Loops in Ecological Systems: Ecosystems operate through feedback loops, where changes in one species or environmental factor affect others, leading to reciprocal adjustments. Recursive Realism suggests that these feedback loops are inherently self-referential, where predator-prey dynamics, resource cycles, and climate interactions create patterns that maintain ecological balance. For example, in a forest ecosystem, the population of herbivores is regulated by predators, while vegetation influences soil quality, creating a cycle of mutual dependence. This perspective highlights how ecosystems are dynamic systems shaped by recursively structured interactions, where each organism is interconnected with the whole.

Symbiosis and the Loop of Co-Evolution: Symbiotic relationships, mutually beneficial interactions between species, demonstrate how life forms adapt to each other's presence. Recursive Realism suggests that symbiosis involves a self-referential process, where each organism adapts to the needs and responses of its partner, creating a cycle of co-evolution. For example, bees and flowering plants have evolved together, with flowers adapting their shapes and colours to attract pollinators, while bees develop specialised structures for collecting nectar, creating a loop that sustains both populations. This perspective allows for a deeper appreciation of how co-evolutionary relationships reflect the recursive nature of biological adaptation, where life forms evolve in harmony with each other.

Biodiversity and the Recursion of Resilience: Biodiversity, the variety of life in ecosystems, contributes to resilience, allowing systems to adapt to changes and disturbances. Recursive Realism suggests that biodiversity creates a self-referential network of interdependent species, where each organism plays a role in maintaining ecological stability. For example, in coral reef ecosystems, the diversity of fish, invertebrates, and plants creates a complex web that buffers the system against environmental changes, such as temperature fluctuations or nutrient shifts, creating a loop where diversity sustains balance. This perspective highlights how the interconnectedness of life is amplified by diverse interactions, where recursion contributes to the adaptability and stability of natural systems.

The Mind-Matter Connection and the Recursion of Consciousness

The relationship between mind and matter, how consciousness interacts with the physical world, has long been a subject of philosophical inquiry. Recursive Realism suggests that the mind-matter connection involves self-referential processes, where consciousness reflects on its own experiences and interacts with the material universe, creating a loop of perception and reality.

Perception as a Recursive Interaction with Reality: Perception involves interpreting sensory information to create a coherent understanding of the external world. Recursive Realism suggests that perception is inherently self-referential, where the mind engages in a loop of interpreting stimuli, testing hypotheses, and revising its understanding. For example, the brain's ability to recognise familiar faces involves processing visual information, comparing it with memory templates, and adjusting interpretations based on contextual cues, creating a cycle where perception continuously revises itself. This perspective highlights how the mind and the world are interconnected through recursive perception, where each experience deepens our interaction with reality.

Consciousness and the Feedback Loop of Self-Awareness: Consciousness is often described as the awareness of one's own existence and thoughts. Recursive Realism suggests that self-awareness is a self-referential process, where the mind reflects on its own state, creating a loop of inner observation. For example, meditative practices that involve observing one's thoughts and emotions illustrate how consciousness can turn inward, creating a cycle of awareness that deepens self-knowledge. This perspective allows for a deeper understanding of how consciousness is connected to the physical world through recursive processes, where the act of observing oneself shapes the experience of being.

Mind and Matter as a Unified System: The Recursion of Embodiment: Embodied cognition suggests that thoughts and emotions are influenced by the physical body and its interaction with the environment. Recursive Realism proposes that mind and body form a self-referential loop, where physical experiences influence mental states, and mental states shape physical actions. For example, the experience of anxiety involves a loop where physical sensations like a racing heart influence thought patterns, which in turn amplify bodily reactions, creating a cycle that affects the whole being. This perspective highlights how the mind and the body are not separate, but interconnected through recursively structured interactions that shape conscious experience.

Cosmic Patterns: Linking the Microcosm to the Macrocosm

The universe exhibits patterns that connect the smallest scales of matter to vast cosmic structures. Recursive Realism suggests that the patterns of the microcosm and macrocosm are self-referential, where the same principles are reflected across scales, revealing the unity of cosmic processes.

Fractals and the Recursion of Nature's Patterns: Fractals are geometric patterns that repeat at different scales, appearing in natural phenomena such as coastlines, cloud formations, and vascular systems. Recursive Realism suggests that fractal structures reflect self-referential processes, where patterns are replicated across levels of scale, creating a loop that links the micro to the macro. For example, the branching pattern of trees resembles the structure of river systems, where each part reflects the shape of the whole. This perspective highlights how nature's patterns are unified through recursively structured forms, where the same principles appear from the smallest cells to cosmic formations.

The Principle of Symmetry and the Recursion of Cosmic Order: Symmetry is a fundamental principle in physics, underlying the laws of nature and the structure of matter. Recursive Realism suggests that symmetry involves self-referential properties, where patterns of balance and proportion are reflected throughout the universe. For example, atomic structures exhibit symmetry in electron orbits, while galaxies and stellar systems follow symmetrical arrangements in their rotations and formations, creating a loop where the same principles govern both atoms and stars. This perspective allows for a richer appreciation of how the universe is interconnected through symmetrical patterns, where recursively structured order extends across scales.

Cosmic Evolution and the Recursion of Emergent Complexity: Cosmic evolution, the process by which simple elements formed stars, planets, and life, illustrates how complexity emerges from interconnected processes. Recursive Realism suggests that the evolution of the cosmos is shaped by self-referential loops, where each stage builds upon the conditions created by previous stages. For example, the formation of heavy elements within stars creates the building blocks for planetary systems, which in turn provide the conditions for life's emergence, creating a cycle where matter evolves into complex forms. This perspective highlights how the history of the universe is a story of recursive emergence, where simple interactions give rise to complex structures through cycles of transformation.

The universe is a web of interconnected systems, where life, consciousness, and cosmic structures are united through self-referential processes. Recursive Realism provides a framework for understanding how ecosystems, the mind-matter connection, and cosmic patterns are shaped by recursively structured interactions, revealing how the microcosm and macrocosm are linked through the same underlying principles. By recognising the recursive nature of interconnectedness, Recursive Realism offers a vision of the universe as a unified whole, where each part reflects the patterns of the greater cosmos.

Recursive Realism and the Unity of Human Experience

Human experience is often characterised by a sense of connection to others, to nature, and to larger realities beyond the self. Recursive Realism suggests that this sense of unity arises from self-referential processes, where individual awareness engages in a loop of reflection and connection, creating a deeper understanding of interdependence. This section explores how Recursive Realism frames the unity of experience, examining empathy, transcendence, and the role of recursion in spiritual and collective awareness.

Empathy and the Loop of Shared Awareness

Empathy, the ability to understand and share the feelings of others, is a fundamental aspect of human connection. Recursive Realism suggests that empathy is inherently self-referential, involving a loop where one's own experiences serve as a reference for understanding others, creating a bridge between individual and collective consciousness.

Empathy as a Recursive Reflection of Self and Other: Empathy involves the ability to project oneself into another's perspective, creating a recursive loop between self-awareness and awareness of others. Recursive Realism suggests that empathic understanding is shaped by self-referential processes, where the mind uses its own experiences to simulate the inner world of others. For example, when witnessing someone's grief, an individual might recall their own experiences of loss, creating a loop that allows for a deeper connection to the other's emotions. This perspective highlights how empathy is not just an automatic reaction but a recursive process, where self-reflection enables the bridging of inner worlds.

Mirror Neurons and the Biological Recursion of Empathy: Neuroscientific research on mirror neurons suggests that observing another's actions activates similar neural pathways as performing those actions oneself. Recursive Realism proposes that this mirror mechanism is a biological foundation for self-referential empathy, where the brain's ability to reflect the experiences of others creates a feedback loop of understanding. For example, seeing someone smile might trigger neural responses that simulate the feeling of smiling, creating a loop where emotional states are shared. This perspective allows for a richer understanding of how the mind is wired for empathy, where biological recursion supports the connection between individuals.

Collective Empathy and the Loop of Social Resonance: Empathy extends beyond individual interactions, shaping the dynamics of communities and societies. Recursive Realism suggests that collective empathy involves a self-referential loop, where shared emotions create a resonance that amplifies connections within groups. For example, during social movements or tragedies, the shared experience of emotion can create a sense of unity and purpose among large groups, creating a loop where individual feelings merge into collective awareness. This perspective highlights how empathy contributes to social cohesion, where recursively shared emotions foster a sense of belonging and mutual support.

Transcendence and the Experience of Unity

Transcendence, the experience of moving beyond the individual self, is often described as a connection to something greater, whether it be nature, cosmos, or spiritual realms. Recursive Realism suggests that transcendence involves self-referential processes, where consciousness reflects on its own boundaries, creating a loop that allows for a sense of connection to the larger whole.

Meditation and the Loop of Self-Transcendence: Meditative practices often involve focusing inward to transcend individual thoughts, creating a state of awareness that feels connected to the broader reality. Recursive Realism suggests that meditation involves a self-referential loop, where the mind observes its own activity and lets go of identification with specific thoughts, creating a sense of expanded awareness. For example, mindfulness meditation focuses on observing thoughts without attachment, creating a cycle where self-reflection leads to a feeling of being part of a larger whole. This perspective allows for a deeper appreciation of how spiritual experiences can emerge from recursively engaging with the nature of consciousness.

Mystical Experiences and the Recursion of Unity: Mystical experiences often involve a feeling of oneness with the universe, where the boundaries between self and world dissolve. Recursive Realism suggests that such experiences are self-referential, where consciousness reflects on its own nature to recognise its connection to the larger cosmos. For example, during peak experiences described by mystics, individuals might feel deeply connected to all life, creating a loop where self-awareness merges into universal awareness. This perspective highlights how transcendence is shaped by recursively structured reflection, where the mind's ability to observe itself allows for a sense of cosmic unity.

Nature, Awe, and the Recursion of Cosmic Belonging: Experiences of awe, such as witnessing a vast landscape or gazing at the stars, often evoke a feeling of connection to the cosmos. Recursive Realism suggests that awe involves a self-referential loop, where awareness of the vastness of the universe prompts reflection on one's place within it, creating a sense of belonging to a larger reality. For example, the feeling of insignificance under a star-filled sky might lead to a deeper awareness of the interconnectedness of all things, creating a loop where personal identity expands into cosmic identity. This perspective allows for a richer appreciation of how experiences of awe foster a sense of unity, where the mind's recursive reflection connects the individual to the cosmic whole.

Collective Consciousness and the Recursion of Cultural Identity

Collective consciousness, the shared beliefs and values that shape societies, emerges from the interaction of individual minds. Recursive Realism suggests that collective awareness is shaped by self-referential processes, where cultural norms and shared experiences create a loop that binds individuals into a larger social reality.

Cultural Identity and the Recursion of Shared Narratives: Cultural identity involves shared stories, myths, and rituals that define a group's sense of itself. Recursive Realism suggests that these narratives are self-referential, where societies reflect on their own history and values through a loop of cultural expression. For example, national myths and legends often retell foundational stories, creating a cycle where each generation reinterprets the narrative to align with contemporary values. This perspective highlights how cultural identity is not fixed but evolves through recursively re-engaging with shared narratives, where the past continually informs the present.

Rituals and the Loop of Collective Belonging: Rituals, from religious ceremonies to secular celebrations, create a sense of unity by reinforcing shared values. Recursive Realism suggests that rituals are self-referential, where collective practices create a feedback loop that reaffirms group identity. For example, annual festivals and communal gatherings allow individuals to participate in shared traditions, creating a loop where social bonds are strengthened through recurrent acts. This perspective allows for a deeper understanding of how rituals contribute to social cohesion, where recursively shared practices foster a sense of belonging to a community.

Global Consciousness and the Recursion of Interconnected Ideals: Global consciousness, the recognition of shared human experiences across nations and cultures, has become increasingly important in a connected world. Recursive Realism suggests that global awareness is shaped by a self-referential process, where local identities are influenced by global narratives, creating a loop of mutual influence. For example, awareness of global challenges like climate change creates a shared narrative that shapes local actions, creating a cycle where collective responsibility emerges. This perspective highlights how the emergence of global consciousness is a recursively structured process, where each part of the world contributes to a larger sense of unity.

The experience of unity, whether through empathy, transcendence, or collective awareness, is shaped by self-referential processes, where individual consciousness engages in loops of reflection that connect the self to the larger reality. Recursive Realism provides a framework for understanding how human connection, spiritual experiences, and cultural identity are shaped by recursively structured thought, revealing how the mind's ability to reflect on its own nature creates a sense of interdependence with the world. By recognising the recursive nature of human experience, Recursive Realism offers a vision of unity that is both individual and universal, where each act of self-reflection deepens our connection to the greater whole.

In Chapter 35, we have explored how Recursive Realism offers a framework for understanding the interconnectedness of natural systems, human consciousness, and cosmic patterns, revealing how self-referential processes create unity across different scales of reality. By examining the dynamics of ecosystems, the mind-matter connection, and the experience of transcendence, Recursive Realism provides a nuanced perspective on how all things are connected, revealing how the mind's recursive nature allows for an appreciation of the unity that underlies diverse phenomena. This perspective allows us to see the universe as a tapestry woven from self-referential patterns, where the smallest details reflect the structure of the cosmos itself.

References

Abrams, E. and Primack, J.R. (2011) *The New Universe and the Human Future: How a Shared Cosmology Could Transform the World*. New Haven: Yale University Press.

Bateson, G. (1972) *Steps to an Ecology of Mind*. Chicago: University of Chicago Press.

Capra, F. (1996) *The Web of Life: A New Scientific Understanding of Living Systems*. New York: Anchor Books.

Chalmers, D.J. (1996) *The Conscious Mind: In Search of a Fundamental Theory*. Oxford: Oxford University Press.

Darwin, C. (1859) *On the Origin of Species by Means of Natural Selection*. London: John Murray.

Dawkins, R. (1976) *The Selfish Gene*. Oxford: Oxford University Press.

Deacon, T.W. (2011) *Incomplete Nature: How Mind Emerged from Matter*. New York: W.W. Norton & Company.

Hawking, S. (1988) *A Brief History of Time: From the Big Bang to Black Holes*. New York: Bantam Books.

Heisenberg, W. (1958) *Physics and Philosophy: The Revolution in Modern Science*. New York: Harper & Row.

Kauffman, S. (1993) *The Origins of Order: Self-Organization and Selection in Evolution*. Oxford: Oxford University Press.

Levine, J.M. et al. (2001) 'Biodiversity and Ecosystem Stability', *Nature*, 405, pp. 234–242.

Margulis, L. and Sagan, D. (2002) *Acquiring Genomes: A Theory of the Origins of Species.* New York: Basic Books.

Prigogine, I. and Stengers, I. (1984) *Order Out of Chaos: Man's New Dialogue with Nature.* New York: Bantam Books.

Rosen, R. (1991) *Life Itself: A Comprehensive Inquiry into the Nature, Origin, and Fabrication of Life.* New York: Columbia University Press.

Sagan, C. (1994) *Pale Blue Dot: A Vision of the Human Future in Space.* New York: Random House.

Stewart, I. and Cohen, J. (1999) *Figments of Reality: The Evolution of the Curious Mind.* Cambridge: Cambridge University Press.

Thompson, W.I. (1996) *Gaia, A Way of Knowing: Political Implications of the New Biology.* Hudson, NY: Lindisfarne Press.

Varela, F.J., Thompson, E. and Rosch, E. (1991) *The Embodied Mind: Cognitive Science and Human Experience.* Cambridge, MA: MIT Press.

Wilson, E.O. (1984) *Biophilia.* Cambridge, MA: Harvard University Press.

Zohar, D. and Marshall, I. (1994) *The Quantum Society: Mind, Physics and a New Social Vision.* London: Bloomsbury.

Chapter 36: Recursive Realism and the Future of Knowledge

The Future of Science Through Recursive Realism

Science has been a powerful tool for unveiling the mysteries of the natural world, yet it has also revealed the limits of empirical understanding. Recursive Realism suggests that the future of scientific inquiry will involve embracing the self-referential nature of knowledge creation, where the process of discovery is seen as an evolving loop that continually refines understanding. This section explores how Recursive Realism can shape the next phase of scientific exploration, focusing on new models of inquiry, the role of complexity, and the integration of consciousness into the scientific worldview.

Integrating Complexity into Scientific Models

Understanding complex systems, from ecosystems to the human brain, requires a shift from linear models to those that capture the recursive interactions that shape dynamic phenomena. Recursive Realism suggests that future scientific models will embrace self-referential complexity, recognising how feedback loops create patterns of order and emergence.

Nonlinear Dynamics and the Recursion of Predictive Models: Classical science often relied on linear models to describe natural laws, but many phenomena involve nonlinear interactions that create unpredictable outcomes. Recursive Realism suggests that the future of science will focus on nonlinear dynamics and complexity theory, using self-referential models that can capture the emergent properties of systems. For example, in climate science, recognising how ocean currents, atmospheric patterns, and biosphere interactions create feedback loops can provide more accurate models of climate change. This perspective highlights how embracing recursion allows for a deeper understanding of dynamic systems, where each interaction shapes the whole.

The Shift Toward Holistic Science and Recursive Synthesis: Holistic approaches to science focus on understanding systems as integrated wholes, rather than isolated parts. Recursive Realism suggests that holistic science will involve self-referential synthesis, where different disciplines are integrated into a unified framework. For example, the study of consciousness might combine neuroscience, quantum mechanics, and philosophy of mind, creating a loop where each field informs the others. This perspective allows for a richer appreciation of how future scientific models will transcend traditional boundaries, using recursive thinking to integrate insights across varied domains.

Emergent Technologies and the Recursion of Innovation: Emergent technologies, such as artificial intelligence, synthetic biology, and quantum computing, are reshaping the boundaries of what is possible. Recursive Realism suggests that technological innovation involves a feedback loop, where each breakthrough opens new possibilities for exploration and redefines the limits of knowledge. For example, advancements in AI create new tools for modelling complex phenomena, which in turn refine the algorithms that drive AI research, creating a recursive cycle of discovery. This perspective highlights how future science will be shaped by recursively expanding technologies, where each step forward leads to new horizons of understanding.

Consciousness in the Scientific Worldview

Consciousness has historically been considered a challenge for empirical science, often relegated to philosophy or the humanities. Recursive Realism suggests that the next phase of scientific inquiry will involve embracing the self-referential nature of consciousness, recognising how awareness itself is a dynamic system that can be studied using recursive principles.

Consciousness as a Self-Referential System: Consciousness involves the ability to reflect on itself, creating a loop of awareness that is both subject and object. Recursive Realism suggests that future scientific models of consciousness will explore the self-referential nature of awareness, using theories that recognise its recursive structure. For example, integrated information theory (IIT) proposes that consciousness arises from the integration of information in a system, creating a feedback loop that reflects the mind's capacity for self-awareness. This perspective allows for a deeper appreciation of how consciousness can be understood scientifically, where self-referential models reveal the structure of awareness.

Bridging Subjective Experience and Objective Science: One of the key challenges for science is bridging the gap between subjective experience and objective measurement. Recursive Realism suggests that this bridge involves a self-referential approach, where science recognises the role of consciousness in constructing reality. For example, the study of perceptual phenomena might involve examining how the brain constructs reality through feedback loops between sensory inputs and interpretation, creating a recursive relationship between mind and world. This perspective highlights how future science can integrate subjective and objective realms, using recursive models to understand the interplay between consciousness and reality.

Expanding the Scope of Inquiry: From Neural Networks to Quantum Consciousness: Emerging theories suggest that consciousness might be linked not only to neural networks but also to fundamental aspects of the universe, such as quantum processes. Recursive Realism suggests that exploring these frontiers will involve a self-referential approach, where the nature of awareness is considered in relation to the deepest levels of reality. For example, theories of quantum consciousness propose that quantum entanglement might be a basis for the unity of experience, creating a loop where the mind is connected to the structure of space-time. This perspective allows for a richer understanding of how future science can expand its horizons, using recursive thinking to explore the connections between mind and cosmos.

Integrating the Human Experience into Scientific Paradigms

Science has often been characterised as objective, focusing on external phenomena while excluding human subjectivity. Recursive Realism suggests that the future of scientific paradigms will involve recognising the self-referential nature of human experience, where the observer and the observed are interconnected through a loop of awareness.

The Observer Effect and the Recursion of Measurement: The observer effect in quantum mechanics, where the act of observation influences the state of a particle, illustrates how subjectivity and objectivity are intertwined. Recursive Realism suggests that future scientific paradigms will recognise the role of the observer as a self-referential process, where measurement and observation are part of the system being studied. For example, understanding consciousness as a participant in physical processes might involve recognising how the mind shapes its perception of quantum phenomena, creating a cycle where the observer and the observed influence each other. This perspective highlights how future science can embrace the role of the observer, using recursive principles to understand the interaction between mind and matter.

Human Subjectivity and the Recursion of Meaning: Meaning and purpose are often considered subjective dimensions that lie outside the scope of scientific inquiry. Recursive Realism suggests that meaning can be understood as a self-referential process, where individuals and communities create a feedback loop between experience and interpretation. For example, the study of human well-being might explore how psychological states are influenced by belief systems, creating a cycle where worldviews shape individual well-being and vice versa. This perspective allows for a richer understanding of how science can integrate human meaning, recognising how subjective experience is part of the fabric of reality.

Transdisciplinary Science and the Loop of Integrated Knowledge: Transdisciplinary approaches seek to integrate knowledge across fields, recognising that complex problems require multiple perspectives. Recursive Realism suggests that transdisciplinary science involves a self-referential loop, where the insights of one field are reframed in the context of another, creating a cycle of knowledge integration. For example, combining ecological models with cognitive science can lead to new insights into how human societies interact with natural systems, creating a loop where understanding is enriched through cross-pollination of ideas. This perspective highlights how future science can be more inclusive, using recursive thinking to bridge the gaps between the natural and social sciences.

The future of scientific inquiry involves embracing complexity, integrating consciousness, and recognising the self-referential nature of knowledge creation. Recursive Realism provides a framework for understanding how science can advance by acknowledging the dynamic feedback loops that shape natural phenomena, human awareness, and cosmic processes. By recognising the recursive structure of scientific thought, Recursive Realism offers a vision of scientific progress as an evolving dialogue between mind, matter, and the mysteries of the universe.

Recursive Realism and the Future of Philosophy and Spirituality

Philosophy and spirituality have long sought to grapple with fundamental questions about existence, purpose, and the nature of reality. Recursive Realism suggests that the evolution of these disciplines will involve embracing the self-referential nature of conscious reflection, where inquiry turns back on itself to explore the deeper structures of being and meaning. This section explores how Recursive Realism provides a framework for rethinking philosophical approaches, deepening ethical considerations, and revisiting spiritual perspectives in a way that bridges ancient wisdom with modern understanding.

Philosophy as a Self-Referential Discipline

Philosophy is often characterised as the love of wisdom, a discipline that seeks to ask fundamental questions and challenge assumptions. Recursive Realism suggests that the future of philosophy will involve a focus on the self-referential nature of thought, where reflection and self-examination guide the evolution of ideas and frameworks for understanding reality.

Metaphilosophy and the Recursion of Reflective Inquiry: Metaphilosophy, the study of the nature of philosophy itself, involves examining the methods, assumptions, and goals of philosophical thought. Recursive Realism suggests that metaphilosophy will play a key role in the future of philosophical inquiry, where philosophers reflect on their own methods and the structure of their thinking, creating a loop that deepens philosophical clarity. For example, questions about the limits of reason or the nature of human understanding might involve a recursive analysis of how different philosophical traditions approach these issues, creating a cycle where ideas evolve through self-reflection. This perspective highlights how philosophy can remain vibrant and relevant by continually revisiting its foundations and redefining its approach to timeless questions.

Existentialism and the Recursion of Self-Definition: Existential philosophy focuses on questions of individual freedom, meaning, and the nature of human existence. Recursive Realism suggests that existential inquiry is inherently self-referential, where individuals confront the nature of their own being through a loop of reflection and self-definition. For example, existential thinkers like Jean-Paul Sartre and Simone de Beauvoir explored how consciousness shapes its own identity through acts of choice, creating a cycle where freedom and self-awareness are intertwined. This perspective allows for a deeper appreciation of how existential thought can continue to evolve, where philosophical reflection becomes a dialogue between the self and the world.

Process Philosophy and the Recursion of Becoming: Process philosophy, as developed by thinkers like Alfred North Whitehead, emphasises the dynamic nature of reality and the idea that existence is characterised by change and evolution. Recursive Realism suggests that process philosophy involves self-referential thinking, where each moment is understood as part of a larger process of transformation and becoming. For example, Whitehead's idea that reality consists of events rather than static substances can be seen as a recursive vision, where each event is connected to the past and the potential future, creating a continuous loop of becoming. This perspective highlights how philosophy can evolve through embracing change, where recursive reflection reveals the interconnectedness of all processes of existence.

Ethics and the Recursion of Moral Reflection

Ethics involves considering what is right and good, often requiring self-reflection and engagement with complex dilemmas. Recursive Realism suggests that moral reasoning is shaped by self-referential processes, where awareness of consequences, empathy, and personal values create a loop that guides ethical decisions.

Moral Reflection and the Loop of Self-Examination: Making ethical decisions often involves reflecting on one's values and considering the impact of actions. Recursive Realism suggests that moral reflection is inherently self-referential, where individuals engage in a loop of weighing values, anticipating consequences, and revisiting principles. For example, ethical deliberations about social justice might involve considering how one's actions align with principles of fairness, creating a cycle where reflection shapes choices. This perspective highlights how ethics is not a fixed set of rules but a dynamic process, where recursively engaging with moral questions allows for personal growth and adaptation.

Virtue Ethics and the Recursion of Character Development: Virtue ethics, as developed by Aristotle and modern philosophers, focuses on cultivating moral character through habits and self-reflection. Recursive Realism suggests that character development involves a feedback loop, where each ethical decision shapes one's character, and character, in turn, guides future decisions. For example, the practice of compassion might involve reflecting on moments of kindness and seeking opportunities to act empathetically, creating a loop where self-awareness leads to the deepening of virtues. This perspective allows for a richer understanding of how ethics can be transformative, where self-referential reflection cultivates a sense of purpose and moral clarity.

Global Ethics and the Loop of Interconnected Responsibility: In a globalised world, ethical challenges often require considering the impact of actions on distant others. Recursive Realism suggests that global ethics involves a self-referential process, where awareness of interconnectedness prompts a loop of responsibility that extends beyond the self. For example, ethical considerations about climate change involve recognising the global consequences of local actions, creating a cycle where individual responsibility is connected to planetary well-being. This perspective highlights how ethics can evolve to address global challenges, using recursive thinking to expand the scope of moral concern.

Spirituality and the Recursion of Transcendent Experience

Spirituality often involves seeking a sense of connection to the divine, the universe, or the deeper dimensions of life. Recursive Realism suggests that spiritual exploration is inherently self-referential, where the journey inward leads to a greater awareness of the unity of all things. This section explores how Recursive Realism provides a framework for reimagining spirituality, offering a perspective that integrates ancient practices with modern insights.

Mysticism and the Loop of Inner Exploration: Mystical traditions across cultures focus on the inner journey, seeking direct experiences of the divine or the unity of existence. Recursive Realism suggests that mystical experiences involve a self-referential loop, where inner reflection creates a sense of connection to something greater. For example, in Sufism, the practice of zikr (remembrance) involves repeating sacred phrases, creating a rhythmic cycle that focuses awareness inward, leading to a sense of unity with the divine. This perspective allows for a deeper appreciation of how spiritual practices use recursive techniques to deepen awareness, where the mind's reflection becomes a path to transcendence.

Integrating Science and Spirituality: The Recursion of Awe: Modern spirituality often seeks to bridge insights from science with spiritual perspectives, creating a sense of wonder about the cosmos. Recursive Realism suggests that awe involves a self-referential loop, where awareness of the universe's vastness deepens the sense of spiritual connection. For example, contemplating the scale of the universe through cosmology can evoke a sense of mystery that aligns with spiritual reflections on the infinite. This perspective highlights how spirituality and science can be mutually enriching, using recursive thinking to integrate insights about existence and the human spirit.

Reimagining Spiritual Purpose: The Loop of Self and Cosmos: Spirituality often involves seeking a purpose that transcends individual desires, focusing on a larger reality. Recursive Realism suggests that spiritual purpose is a self-referential process, where awareness of the self and awareness of the universe are interconnected. For example, Eastern philosophies like Buddhism emphasise the dissolution of the self into the flow of existence, creating a cycle where self-awareness leads to a deeper sense of interconnection. This perspective allows for a richer understanding of how spiritual traditions can be reinterpreted in light of recursive principles, where the exploration of inner depth aligns with a cosmic sense of belonging.

Philosophy and spirituality are shaped by self-referential processes, where the mind reflects on its own nature and seeks deeper connections to the cosmos. Recursive Realism provides a framework for understanding how philosophical inquiry, ethical reflection, and spiritual exploration can evolve through embracing the recursive nature of thought and awareness. By recognising the loops of reflection that shape existential questions and spiritual experiences, Recursive Realism offers a vision of human inquiry as an ongoing dialogue between self and the greater reality.

In Chapter 36, we have explored how Recursive Realism offers a framework for understanding the evolution of knowledge, revealing how self-referential processes shape the future of science, philosophy, and spirituality. By examining the role of complexity, the integration of consciousness, and the quest for meaning, Recursive Realism provides a perspective on how human understanding can advance through embracing recursion, where each new insight deepens our connection to the mysteries of existence. This perspective allows us to envision a future where knowledge is not just about facts, but about engaging with the deeper questions that connect the self to the universe.

References

Capra, F. (1996) *The Web of Life: A New Scientific Understanding of Living Systems*. New York: Anchor Books.

Chalmers, D.J. (1996) *The Conscious Mind: In Search of a Fundamental Theory*. Oxford: Oxford University Press.

Deacon, T.W. (2011) *Incomplete Nature: How Mind Emerged from Matter*. New York: W.W. Norton & Company.

Gleick, J. (1987) *Chaos: Making a New Science*. New York: Viking.

Gödel, K. (1931) 'On Formally Undecidable Propositions of Principia Mathematica and Related Systems I', *Monatshefte für Mathematik und Physik*, 38, pp. 173–198.

Hawking, S. (1988) *A Brief History of Time: From the Big Bang to Black Holes*. New York: Bantam Books.

Kuhn, T.S. (1962) *The Structure of Scientific Revolutions*. Chicago: University of Chicago Press.

Margulis, L. and Sagan, D. (2002) *Acquiring Genomes: A Theory of the Origins of Species*. New York: Basic Books.

Prigogine, I. and Stengers, I. (1984) *Order Out of Chaos: Man's New Dialogue with Nature*. New York: Bantam Books.

Sartre, J.-P. (1943) *Being and Nothingness*. New York: Washington Square Press.

Schrödinger, E. (1944) *What is Life? The Physical Aspect of the Living Cell*. Cambridge: Cambridge University Press.

Searle, J.R. (1980) 'Minds, Brains, and Programs', *Behavioral and Brain Sciences*, 3(3), pp. 417–457.

Varela, F.J., Thompson, E., and Rosch, E. (1991) *The Embodied Mind: Cognitive Science and Human Experience*. Cambridge, MA: MIT Press.

Whitehead, A.N. (1929) *Process and Reality: An Essay in Cosmology*. New York: Free Press.

Zohar, D. and Marshall, I. (1994) *The Quantum Society: Mind, Physics and a New Social Vision*. London: Bloomsbury.

Chapter 37: Recursive Realism and the Ethics of Knowledge Creation

The Balance Between Curiosity and Humility

Curiosity drives the pursuit of knowledge, inspiring exploration and discovery, but it must be balanced with humility, acknowledging the limitations of human understanding. Recursive Realism suggests that ethical knowledge creation involves a self-referential process, where the desire to know is tempered by an awareness of the broader consequences and the boundaries of inquiry. This section explores how Recursive Realism frames the relationship between curiosity and humility, focusing on intellectual responsibility, the recognition of limits, and the ethical challenges of engaging with the unknown.

Intellectual Responsibility and the Loop of Reflective Inquiry

The pursuit of knowledge carries with it a responsibility to consider the impacts of new discoveries on society, the environment, and future generations. Recursive Realism suggests that intellectual responsibility involves self-referential reflection, where scientists, philosophers, and thinkers continually examine the consequences of their inquiries and innovations.

Curiosity as a Double-Edged Sword: Curiosity can lead to groundbreaking discoveries, but it can also create unintended consequences when ethical considerations are overlooked. Recursive Realism suggests that ethical inquiry involves a feedback loop, where the drive to explore is balanced by reflection on the broader impacts of knowledge. For example, the development of nuclear technology involved a quest to understand atomic energy, but its application as a weapon created ethical dilemmas about the responsible use of scientific knowledge. This perspective highlights how curiosity must be tempered by reflective awareness, where the potential risks of discovery are considered alongside its benefits.

The Role of Ethical Reflection in Scientific Exploration: Scientific progress often involves pushing boundaries and challenging assumptions, but ethical reflection is crucial to ensure that advances serve the common good. Recursive Realism suggests that scientific inquiry should involve a self-referential process, where scientists consider the long-term impacts of their research through cycles of reflection. For example, biomedical research into gene editing technologies like CRISPR raises questions about how far humans should intervene in genetic processes, creating a loop where scientific possibilities are weighed against ethical concerns. This perspective highlights how ethical reflection can guide scientific inquiry, ensuring that curiosity does not outpace our responsibility to society.

Acknowledging Ignorance: The Loop of Intellectual Humility: Intellectual humility involves recognising the limits of what we can know and acknowledging the mysteries that remain beyond our grasp. Recursive Realism suggests that humility is a self-referential process, where awareness of our ignorance creates a loop that encourages caution and restraint. For example, the recognition that human cognition has limitations when it comes to understanding the vast complexity of ecological systems might lead to more cautious approaches to environmental interventions. This perspective allows for a deeper understanding of how intellectual humility can serve as a check on reckless exploration, where awareness of our limits fosters a sense of responsibility in the pursuit of knowledge.

The Ethical Dimensions of Engaging with the Unknown

Engaging with the unknown is a core aspect of human curiosity, yet it also involves ethical challenges when the consequences of new discoveries are uncertain. Recursive Realism suggests that exploring the unknown involves a self-referential process, where the thrill of discovery is balanced by an awareness of potential risks and unknown outcomes.

The Ethics of Risk and Uncertainty: Scientific exploration often involves entering uncharted territory, where the outcomes of experiments and technological applications are not fully predictable. Recursive Realism suggests that ethical inquiry into uncertainty involves a loop of assessing risks and weighing potential benefits, where each new step in exploration is revisited with a critical perspective. For example, the development of artificial intelligence raises questions about unintended consequences related to automation, data privacy, and the nature of human agency, creating a feedback loop where each advancement is evaluated in terms of its broader societal impacts. This perspective highlights how responsible exploration requires engaging recursively with the ethical dimensions of the unknown.

Balancing Innovation with Precaution: The Precautionary Principle: The precautionary principle suggests that when potential risks are significant, caution should guide decision-making even if full certainty is lacking. Recursive Realism suggests that this principle involves a self-referential approach, where awareness of possible dangers leads to a loop of revisiting decisions and adjusting strategies. For example, in the context of environmental preservation, the precautionary principle might suggest limiting certain types of development until their impacts on biodiversity are better understood, creating a cycle where scientific innovation is balanced by considerations of ecological health. This perspective allows for a richer understanding of how ethical considerations can guide technological progress, using recursive thinking to ensure that curiosity is aligned with the values of sustainability.

The Ethics of Discovery: Reverence for Mystery: Engaging with the mysteries of existence, whether through cosmology, quantum physics, or biological research, often evokes a sense of wonder and reverence. Recursive Realism suggests that ethical exploration involves recognising the sacredness of the unknown, creating a loop where curiosity is tempered by respect for the mysteries of nature. For example, studying the origins of life or the potential for extraterrestrial life might involve considering the philosophical implications of discovering new forms of existence, creating a cycle where scientific inquiry is shaped by a sense of humility. This perspective highlights how reverence for mystery can be an ethical guide, where the thrill of discovery is balanced by a deep respect for the unknown.

Curiosity and humility are intertwined in the pursuit of knowledge, where the desire to explore is balanced by an awareness of the broader consequences and the limitations of human understanding. Recursive Realism provides a framework for understanding how ethical inquiry involves self-referential processes, where reflecting on the impacts of new discoveries and acknowledging the unknown are central to intellectual responsibility. By recognising the recursive nature of engaging with the unknown, Recursive Realism offers a vision of knowledge creation that respects the complexities of reality and the mysteries that remain beyond comprehension.

Recursive Realism and the Ethics of Knowledge in a Changing World

The world today is characterised by rapid technological advancement, shifting social norms, and complex global challenges that require innovative solutions. Recursive Realism suggests that the ethics of knowledge in this evolving context involves self-referential processes, where the impact of new ideas and technologies is continually reassessed. This section explores how Recursive Realism provides a framework for ethical knowledge creation, focusing on technological impacts, intergenerational responsibility, and the role of knowledge in addressing global challenges.

Technological Advancement and the Feedback Loop of Innovation

Technology has transformed every aspect of life, from communication to healthcare and education. Yet, with innovation comes ethical dilemmas about the consequences of technological change. Recursive Realism suggests that managing the impact of technology involves a self-referential process, where each new development is evaluated through a loop of reflection, adaptation, and reconsideration.

Artificial Intelligence and the Ethics of Autonomy: Artificial intelligence (AI) represents a paradigm shift in automation and decision-making, raising questions about the ethical boundaries of machine autonomy. Recursive Realism suggests that the development of AI should involve a feedback loop where the capabilities of machines are balanced with ethical considerations about their role in human society. For example, the use of AI in law enforcement or autonomous vehicles involves ethical questions about bias, accountability, and the potential for harm, creating a loop where each advancement prompts a reevaluation of ethical frameworks. This perspective highlights how technological innovation can be guided by recursive reflection, ensuring that new technologies serve human well-being rather than disrupting social values.

Genetic Engineering and the Loop of Biological Responsibility: Genetic engineering has opened up possibilities for curing diseases, enhancing crops, and even altering human traits, but it also raises ethical questions about the limits of intervention. Recursive Realism suggests that biotechnological advances should involve a loop of evaluating potential benefits and considering the long-term consequences for individuals and ecosystems. For example, CRISPR technology allows for precise genetic modifications, but its use in editing human embryos creates a feedback loop of ethical considerations about the implications for human evolution, genetic diversity, and social inequality. This perspective allows for a deeper understanding of how ethical reflection can guide biotechnology, using recursive thinking to balance innovation with respect for the integrity of life.

Digital Technologies and the Recursion of Human Connection: Digital technologies, from social media to virtual reality, have transformed how humans connect, but they also introduce challenges related to privacy, mental health, and social cohesion. Recursive Realism suggests that the ethical use of digital platforms involves a feedback loop, where the impact of these technologies on human relationships is continually reassessed. For example, social media platforms have the potential to foster global connections, but they also contribute to polarisation and misinformation, creating a cycle where each new tool requires reflection on its effects on public discourse. This perspective highlights how digital innovation can be ethically managed through recursive evaluation, ensuring that technological progress enhances human flourishing rather than undermining social bonds.

Intergenerational Responsibility and the Ethics of Legacy

Future generations will inherit the consequences of the choices made today, making intergenerational responsibility a crucial aspect of ethical knowledge creation. Recursive Realism suggests that considering the needs of future generations involves a self-referential loop, where the present is continuously reevaluated in light of its impacts on the future.

Sustainability and the Loop of Ecological Stewardship: Environmental sustainability involves protecting natural resources and maintaining ecological balance for the benefit of future generations. Recursive Realism suggests that sustainable practices involve a feedback loop, where the impacts of human activity on the environment are regularly reassessed to ensure long-term balance. For example, policies on deforestation, water usage, and carbon emissions create a cycle where present actions are evaluated for their long-term effects on biodiversity and climate stability. This perspective allows for a richer understanding of how sustainability is a dynamic process, where self-reflective decision-making ensures that the needs of future generations are considered alongside immediate benefits.

Ethical Innovation and the Legacy of Progress: Technological and scientific advances have the potential to transform societies, but they also create responsibilities to ensure that progress benefits future generations. Recursive Realism suggests that the legacy of innovation involves a self-referential process, where new developments are evaluated for their long-term social and environmental impacts. For example, developing renewable energy technologies involves considering how these innovations will shape economies and reduce reliance on fossil fuels, creating a feedback loop where today's advances contribute to a more sustainable future. This perspective highlights how intergenerational responsibility can guide innovation, using recursive thinking to ensure that progress serves both the present and the future.

Education and the Recursion of Knowledge Transmission: Passing on knowledge to future generations is a core aspect of human culture, shaping how societies evolve and adapt over time. Recursive Realism suggests that education involves a feedback loop, where each generation revisits the knowledge of the past, adapts it to new contexts, and transmits it forward. For example, teaching ecological principles in schools creates a cycle where students learn about environmental responsibility, apply it in local contexts, and carry these values into adulthood, shaping the next generation's approach to stewardship. This perspective allows for a deeper understanding of how education can be a tool for intergenerational ethics, using recursive learning to ensure that future generations inherit both knowledge and values that support long-term well-being.

Knowledge, Power, and the Ethics of Global Challenges

Knowledge has the power to transform societies and shape the world, but it also involves ethical responsibilities when addressing global challenges like climate change, social inequality, and technological disruption. Recursive Realism suggests that using knowledge responsibly involves a self-referential approach, where the implications of new ideas are continually revisited in light of their impact on humanity.

Climate Change and the Recursion of Collective Action: Addressing climate change requires global cooperation and a commitment to change at individual, national, and international levels. Recursive Realism suggests that climate action involves a feedback loop, where each generation's efforts to reduce carbon emissions and preserve ecosystems are evaluated and adapted over time. For example, international agreements like the Paris Climate Accord create a cycle of revisiting commitments, tracking progress, and adjusting policies to meet global goals, creating a loop where collective knowledge guides sustainable action. This perspective highlights how tackling global challenges requires a recursive approach, where knowledge and action are continuously aligned to meet the needs of the planet.

Social Justice and the Loop of Reflective Progress: Social justice movements seek to address inequalities and promote human rights, often challenging entrenched systems of power and privilege. Recursive Realism suggests that the pursuit of justice involves a self-referential process, where societal values and policies are revisited in light of new understandings of equity and human dignity. For example, movements for racial equality and gender rights create a cycle of challenging discrimination, revisiting legal frameworks, and adjusting cultural norms, creating a loop where each generation builds on the progress of the past. This perspective highlights how social change is not linear but involves recursively engaging with the challenges of the present, ensuring that knowledge serves the goal of a more just society.

The Ethics of Knowledge in a Global Society: In a connected world, the impact of knowledge extends across borders, making the ethical use of knowledge a matter of global concern. Recursive Realism suggests that global knowledge sharing involves a feedback loop, where local innovations are shared globally, adapted to different contexts, and reintegrated into global strategies. For example, the global response to pandemics like COVID-19 involves sharing scientific knowledge about vaccines, treatments, and public health strategies, creating a loop where collective learning guides response efforts worldwide. This perspective allows for a richer understanding of how global challenges can be addressed through recursive thinking, where knowledge is harnessed responsibly to benefit the entire human community.

Knowledge in a rapidly evolving world comes with ethical responsibilities, requiring self-reflective thinking and a commitment to the well-being of future generations. Recursive Realism provides a framework for understanding how technological advances, intergenerational stewardship, and global challenges can be managed ethically, using self-referential processes to ensure that curiosity aligns with the values of sustainability, justice, and collective progress. By recognising the recursive nature of addressing global issues, Recursive Realism offers a vision of knowledge creation as a dynamic process that respects the complexities of life and the interconnectedness of all things.

In Chapter 37, we have explored how Recursive Realism provides a framework for understanding the ethics of knowledge, revealing how self-referential processes shape the balance between curiosity and humility, the responsibility to future generations, and the role of knowledge in addressing global challenges. By examining the impact of technology, the dynamics of social change, and the pursuit of sustainable futures, Recursive Realism offers insights into how intellectual progress can align with ethical principles, ensuring that the quest for understanding serves the greater good. This perspective allows us to envision a world where knowledge is a force for positive change, guided by reflective awareness and a commitment to the well-being of all life.

References

Beck, U. (1992) *Risk Society: Towards a New Modernity*. London: Sage Publications.

Capra, F. (2002) *The Hidden Connections: A Science for Sustainable Living*. New York: Anchor Books.

Dewey, J. (1938) *Experience and Education*. New York: Macmillan.

Jonas, H. (1984) *The Imperative of Responsibility: In Search of an Ethics for the Technological Age*. Chicago: University of Chicago Press.

Kahneman, D. (2011) *Thinking, Fast and Slow*. New York: Farrar, Straus and Giroux.

Kant, I. (1785) *Groundwork for the Metaphysics of Morals*. Cambridge: Cambridge University Press (translated edition, 1997).

Latour, B. (2004) *Politics of Nature: How to Bring the Sciences into Democracy*. Cambridge, MA: Harvard University Press.

MacIntyre, A. (1981) *After Virtue: A Study in Moral Theory*. Notre Dame, IN: University of Notre Dame Press.

Rawls, J. (1971) *A Theory of Justice*. Cambridge, MA: Harvard University Press.

Sandel, M.J. (2009) *Justice: What's the Right Thing to Do?*. New York: Farrar, Straus and Giroux.

Sarewitz, D. (2016) *Frontiers of Illusion: Science, Technology, and the Politics of Progress*. Philadelphia: Temple University Press.

Senge, P.M. (1990) *The Fifth Discipline: The Art and Practice of the Learning Organization*. New York: Doubleday.

UNESCO (2017) *Ethics of Artificial Intelligence*. Paris: United Nations Educational, Scientific and Cultural Organization.

Von Bertalanffy, L. (1968) *General System Theory: Foundations, Development, Applications*. New York: George Braziller.

World Commission on Environment and Development (1987) *Our Common Future*. Oxford: Oxford University Press.

Chapter 38: Recursive Realism and the Role of Imagination in Knowledge Creation

The Nature of Imagination and Recursive Thinking

Imagination is central to the human capacity for abstract thinking, problem-solving, and envisioning new possibilities. It involves the ability to reconfigure reality and project new scenarios, creating a mental space where ideas can take shape before they are manifested in the world. Recursive Realism suggests that imagination is inherently self-referential, where the mind generates internal models that are refined and revisited, creating a feedback loop between what is known and what could be. This section explores the mechanisms of imagination, its role in cognitive processes, and how recursive thinking allows us to transcend immediate experience.

The Mechanics of Imagination as a Recursive Process

Imagination involves constructing mental images, conceptual models, and narratives that extend beyond direct perception. Recursive Realism suggests that imagination is a self-referential process, where the mind revisits and refines its own constructs, creating a loop that deepens the potential for innovation and discovery.

Imagination as a Feedback Loop of Conceptualisation: Imaginative thought involves creating new scenarios by combining elements of memory, perception, and creative associations. Recursive Realism suggests that this process is self-referential, where the mind cycles through conceptual possibilities, refining ideas with each iteration. For example, when designing a new scientific theory, a researcher might envision different models, compare them with existing data, and refine their concepts based on feedback, creating a loop where imagination and empirical evidence shape the evolving idea. This perspective highlights how imagination is not just a flight of fancy but a recursive tool for understanding reality.

Mental Simulation and the Recursion of Hypothetical Scenarios: Mental simulation allows individuals to imagine possible outcomes, test ideas in their minds, and anticipate future events. Recursive Realism suggests that mental simulation is inherently self-referential, where the mind generates hypothetical scenarios, evaluates them, and revises assumptions based on the imagined feedback. For example, when planning a complex project, an engineer might visualise different stages, anticipate potential challenges, and adjust plans accordingly, creating a cycle where imagination guides practical action. This perspective allows for a deeper appreciation of how imagination enables strategic thinking, using recursive loops to explore multiple possibilities before acting.

Imagination and the Recursion of Abstract Concepts: Imaginative thinking allows for the exploration of abstract concepts that extend beyond direct experience, such as higher dimensions in physics or metaphysical ideas in philosophy. Recursive Realism suggests that abstract thinking involves a self-referential process, where the mind loops through different layers of abstraction, creating a conceptual space where new ideas can emerge. For example, in mathematics, visualising multi-dimensional spaces requires imagining forms that are not directly observable, creating a recursive dialogue between visualisation and mathematical reasoning. This perspective highlights how imagination can be a powerful tool for understanding the unobservable, where self-referential reflection allows for conceptual breakthroughs.

Imagination as a Catalyst for Innovation

Innovation in science, technology, and the arts often begins with a vision that transcends current limitations, allowing for new possibilities to be explored. Recursive Realism suggests that imagination plays a key role in innovation, where self-referential thinking allows for the reimagining of what is possible.

Scientific Imagination and the Recursion of Hypothesis Formation: Formulating scientific hypotheses involves imagining potential explanations for natural phenomena, where the mind envisions how things could be based on current knowledge. Recursive Realism suggests that scientific imagination involves a feedback loop, where initial ideas are tested, refined, and revisited through experimentation and theoretical adjustment. For example, Einstein's thought experiments on relativity involved imagining scenarios like riding alongside a beam of light, creating a loop where imagination and mathematical reasoning deepened his understanding of space-time. This perspective highlights how imagination is integral to scientific progress, where recursively exploring possibilities allows for the emergence of new theories.

Artistic Creativity and the Loop of Self-Expression: Artistic creation involves exploring new forms, expressing emotions, and envisioning alternative realities. Recursive Realism suggests that artistic imagination involves self-referential loops, where the artist revisits their own work, explores different interpretations, and redefines their vision. For example, a painter might experiment with different colours and compositions, creating a cycle where each iteration informs the next, leading to a refined expression of a deeper concept. This perspective allows for a richer appreciation of how imagination drives creativity, where recursive reflection allows for the development of new artistic languages.

Technological Innovation and the Recursion of Design Thinking: Design thinking involves imagining new solutions to practical problems, focusing on user experience, functionality, and aesthetics. Recursive Realism suggests that design innovation involves a self-referential process, where concepts are tested, iterated, and refined through a cycle of feedback and improvement. For example, designing a new software interface might involve imagining different user interactions, testing prototypes, and revisiting design choices based on user feedback, creating a loop where imagination guides practical refinement. This perspective highlights how imagination can be a driving force in technological progress, using recursive thinking to balance creativity with functionality.

The Role of Imagination in Expanding Intellectual Horizons

Imagination allows for the exploration of ideas that challenge the status quo, pushing the boundaries of what is known and exploring the unknown. Recursive Realism suggests that the expansion of intellectual horizons involves a feedback loop, where imagination opens new pathways for thought that are then refined through reflection and dialogue.

The Power of Metaphor and the Recursion of Conceptual Bridges: Metaphors and analogies are often used in philosophy, science, and art to make complex ideas more accessible. Recursive Realism suggests that metaphorical thinking involves a self-referential process, where one concept is mapped onto another, creating a loop that deepens understanding through comparison. For example, using the metaphor of a "web" to describe the interconnectedness of ecosystems allows for a deeper understanding of ecological relationships, creating a cycle where metaphor serves as a bridge to new insights. This perspective highlights how imagination uses metaphorical thinking to expand intellectual horizons, where recursively comparing concepts leads to innovative perspectives.

Speculative Philosophy and the Recursion of Thought Experiments: Philosophical thought experiments, such as the brain in a vat or Schrödinger's cat, allow for the exploration of hypothetical scenarios that challenge existing frameworks. Recursive Realism suggests that thought experiments involve a self-referential process, where imagination revisits the boundaries of understanding, creating a loop that questions what is possible. For example, exploring the nature of consciousness through imagining artificial intelligence or alien minds allows for a deeper reflection on the nature of awareness, creating a cycle where imagination and philosophical reasoning shape the debate. This perspective allows for a richer appreciation of how speculative thinking can drive philosophical inquiry, using recursive reflection to explore the edges of reality.

Dreaming of the Future: Imagination as a Visionary Tool: Imagination plays a key role in envisioning future possibilities, from science fiction to futuristic architecture. Recursive Realism suggests that dreaming of the future involves a self-referential process, where the mind projects scenarios that are then revisited to align with emerging realities. For example, science fiction has often predicted technological advancements that later become reality, creating a loop where imaginative visions inspire real-world innovations. This perspective highlights how visionary thinking can shape the trajectory of knowledge, using recursive imagination to explore new frontiers before they materialise.

Imagination is a core aspect of human cognition, allowing for the exploration of possibilities, the expansion of knowledge, and the reimagining of reality. Recursive Realism provides a framework for understanding how imagination is inherently self-referential, where the mind engages in loops of conceptualisation, reflection, and innovation. By recognising the recursive nature of imaginative thought, Recursive Realism offers a vision of creativity as a dynamic process that shapes intellectual life, revealing how imagination allows us to envision realities beyond the immediately observable and create new worlds of understanding.

Recursive Realism and the Creative Power of Imagination

Creativity plays a central role in driving human progress, whether through artistic innovation, cultural evolution, or scientific breakthroughs. Recursive Realism suggests that the creative process is shaped by self-referential loops, where each new idea is revisited, refined, and reinterpreted, leading to the emergence of new forms and concepts. This section explores how Recursive Realism frames the power of imagination, focusing on art, cultural change, and the integration of creative thinking into the pursuit of knowledge.

Artistic Expression and the Recursion of Creative Vision

Art involves exploring new ways of expressing reality, using forms, colours, sounds, and symbols to capture the essence of human experience. Recursive Realism suggests that artistic creation is inherently self-referential, where the artist engages in a loop of creating, revisiting, and reimagining their work, leading to a deeper exploration of the self and the world.

The Creative Process as a Loop of Experimentation: Creating art often involves a process of trial and error, where initial ideas are explored and then revisited to find new forms of expression. Recursive Realism suggests that artistic experimentation is self-referential, where each iteration builds upon previous attempts, creating a cycle of refinement. For example, a sculptor might experiment with different materials, reshape forms, and adjust their vision based on the feedback of the medium, creating a loop where the process itself shapes the final outcome. This perspective highlights how creativity thrives through recursive engagement, where each phase of creation is an opportunity for new insights.

The Recursion of Meaning in Abstract Art: Abstract art often challenges viewers to interpret forms and find meaning that is not immediately obvious. Recursive Realism suggests that abstract art involves a self-referential process, where meaning is constructed through multiple layers of interpretation. For example, a painter working in abstract styles might create a series of overlapping shapes, inviting viewers to engage in a cycle of seeing and reseeing, where each new interpretation reveals a different aspect of the work's depth. This perspective allows for a deeper appreciation of how art can create a dialogue between the artist and the audience, using recursive imagery to explore complex ideas.

Performance Art and the Loop of Audience Interaction: Performance art involves engaging directly with audiences, creating an experience that is shaped by real-time feedback. Recursive Realism suggests that performance is a self-referential loop, where the artist's actions and the audience's reactions create a cycle that evolves the meaning of the performance. For example, a theatre production might involve actors responding to audience reactions, adapting their delivery in real-time, creating a dynamic loop where the performance is co-created with the audience. This perspective highlights how creativity in performance involves recursively engaging with the energy of the moment, creating a unique experience each time.

Cultural Evolution and the Recursion of Shared Narratives

Cultures evolve through the exchange of stories, symbols, and traditions, shaping the collective identity of communities. Recursive Realism suggests that cultural change involves self-referential loops, where new ideas and cultural practices are adopted, reinterpreted, and transformed through cycles of collective reflection.

Myths and Legends: The Recursion of Cultural Memory: Myths and legends form the bedrock of many cultural identities, providing stories that convey values, morals, and shared history. Recursive Realism suggests that mythology involves a self-referential process, where stories are retold and adapted to new contexts, creating a loop of cultural continuity. For example, ancient myths like the story of Prometheus have been reinterpreted in modern literature and philosophy, creating a dialogue between past and present. This perspective highlights how myths serve as recursive structures in cultural memory, where each retelling deepens the collective understanding of universal themes.

Cultural Renaissance and the Loop of Artistic Revitalisation: Cultural renaissances occur when new movements breathe fresh life into traditional forms, leading to a flourishing of creativity.

Recursive Realism suggests that renaissance periods involve a feedback loop, where past artistic styles are revisited and reimagined to create new expressions. For example, the Italian Renaissance revisited classical Greek and Roman ideals, creating a cycle where the revival of ancient knowledge led to new innovations in art, science, and philosophy. This perspective allows for a richer appreciation of how cultural evolution involves recursively reengaging with heritage, where each new era brings new interpretations of timeless ideas.

Globalisation and the Recursion of Cultural Exchange: Globalisation has created a world where cultures interact on an unprecedented scale, leading to new hybrids of art, music, and beliefs. Recursive Realism suggests that cultural exchange involves a self-referential process, where local traditions adapt to global influences, creating a loop of mutual transformation. For example, the fusion of musical genres like hip-hop with traditional sounds from around the world creates a dynamic dialogue where cultural identities are expanded through shared creativity. This perspective highlights how cultural change is not a one-way process but a recursive cycle of adaptation and reinvention, where new forms emerge from the interaction of diverse traditions.

Creativity in Science and Philosophy: The Recursion of Paradigm Shifts

Scientific revolutions and philosophical breakthroughs often involve a shift in perspective, where old frameworks are challenged and new paradigms emerge. Recursive Realism suggests that paradigm shifts involve a self-referential process, where existing ideas are reexamined through a loop of critical thinking, imagination, and reinterpretation.

Paradigm Shifts in Science: The Recursion of Conceptual Revolutions: Scientific revolutions, such as the shift from Newtonian mechanics to Einstein's relativity, often involve revisiting fundamental assumptions and imagining new ways to conceptualise reality. Recursive Realism suggests that paradigm shifts involve a feedback loop, where new models are tested against empirical evidence, refined, and reinterpreted until they reshape the foundations of knowledge. For example, quantum mechanics required physicists to imagine a world where particles exist in superposition, creating a cycle of theoretical exploration that continues to challenge our understanding of the universe. This perspective highlights how scientific progress is driven by imagination, where recursively refining hypotheses allows for the discovery of new truths.

Philosophical Innovation and the Recursion of Critical Reflection: Philosophical inquiry often involves challenging existing paradigms, using critical reflection to reimagine the nature of reality, ethics, and consciousness. Recursive Realism suggests that philosophical breakthroughs involve a self-referential loop, where ideas are critically examined, reinterpreted, and reconstructed through cycles of debate and reflection. For example, the shift from Cartesian dualism to phenomenology involved revisiting the nature of consciousness, creating a dialogue between mind and body that continues to shape contemporary thought. This perspective allows for a deeper understanding of how philosophy evolves, using recursive thinking to challenge assumptions and expand the boundaries of thought.

Creativity as a Unifying Force in Knowledge Creation: Creativity serves as a bridge between different disciplines, fostering interdisciplinary approaches that integrate insights from science, philosophy, and the arts. Recursive Realism suggests that creativity involves a feedback loop, where the insights of one field inspire new approaches in others, creating a cycle of cross-pollination. For example, neuroscientific research into the brain's creative processes might inspire artists to explore the nature of perception, creating a dialogue that enriches both art and science. This perspective highlights how creativity is a unifying force, where recursive reflection allows for new connections to be made across the entire spectrum of human knowledge.

Imagination is a transformative force that drives creativity, cultural evolution, and the expansion of knowledge. Recursive Realism provides a framework for understanding how the creative process is inherently self-referential, where ideas are refined and reimagined through cycles of reflection. By recognising the recursive nature of artistic expression, cultural change, and philosophical inquiry, Recursive Realism offers a vision of creativity as a dynamic process that shapes the evolution of human thought, revealing how imaginative thinking can unlock new realms of understanding and inspire visions of what is possible.

In Chapter 38, we have explored how Recursive Realism offers a framework for understanding the role of imagination in expanding knowledge, revealing how self-referential processes shape art, culture, and intellectual breakthroughs. By examining the nature of imagination, the power of creativity, and the dynamics of cultural evolution, Recursive Realism provides insights into how imaginative thinking is a driving force in the pursuit of understanding, allowing for new possibilities to be envisioned and realised. This perspective allows us to see imagination as a cornerstone of human progress, where the mind's recursive nature enables the creation of new worlds of thought and meaning.

References

Bachelard, G. (1994) *The Poetics of Space*. Boston: Beacon Press.

Barad, K. (2007) *Meeting the Universe Halfway: Quantum Physics and the Entanglement of Matter and Meaning*. Durham: Duke University Press.

Boden, M. A. (2004) *The Creative Mind: Myths and Mechanisms*. 2nd edn. London: Routledge.

Damasio, A. (1999) *The Feeling of What Happens: Body and Emotion in the Making of Consciousness*. London: Harcourt Brace.

Dennett, D. C. (1991) *Consciousness Explained*. Boston: Little, Brown and Company.

Einstein, A. and Infeld, L. (1938) *The Evolution of Physics: From Early Concepts to Relativity and Quanta*. New York: Simon & Schuster.

Fauconnier, G. and Turner, M. (2002) *The Way We Think: Conceptual Blending and the Mind's Hidden Complexities*. New York: Basic Books.

Koestler, A. (1964) *The Act of Creation*. London: Hutchinson & Co.

Lakoff, G. and Johnson, M. (1980) *Metaphors We Live By*. Chicago: University of Chicago Press.

Popper, K. (1963) *Conjectures and Refutations: The Growth of Scientific Knowledge*. London: Routledge & Kegan Paul.

Said, E. W. (1994) *Culture and Imperialism*. New York: Vintage Books.

Sartre, J.-P. (1943) *Being and Nothingness*. Paris: Gallimard (translated edition, 1956, New York: Philosophical Library).

Whitehead, A. N. (1929) *Process and Reality: An Essay in Cosmology*. Cambridge: Cambridge University Press.

Zeki, S. (1999) *Inner Vision: An Exploration of Art and the Brain*. Oxford: Oxford University Press.

Zohar, D. and Marshall, I. (2001) *Spiritual Intelligence: The Ultimate Intelligence*. London: Bloomsbury Publishing.

Chapter 39: Recursive Realism and the Quest for Meaning

The Self-Referential Nature of Meaning

Meaning is central to human existence, shaping how we interpret our lives, understand our place in the world, and connect with others. Recursive Realism suggests that the construction of meaning is inherently self-referential, where individuals engage in a loop of reflection that continually reinterprets experiences in light of new insights. This section explores how meaning is created through the recursive processes of personal reflection, social narratives, and cosmic contemplation.

Personal Reflection and the Recursion of Self-Understanding

The search for meaning often begins with personal reflection, where individuals consider their own experiences, values, and purpose. Recursive Realism suggests that self-reflection involves a feedback loop, where thoughts and emotions are revisited and reinterpreted, leading to a deeper understanding of one's identity and life's significance.

Self-Reflection as a Loop of Reinterpretation: Personal reflection allows individuals to revisit past experiences, evaluate their significance, and integrate them into a coherent sense of self. Recursive Realism suggests that self-reflection is a self-referential process, where each new experience prompts a reassessment of one's story. For example, a person might reflect on a life-changing event, such as a loss or a new opportunity, and reinterpret its meaning over time, creating a loop where each layer of reflection adds new dimensions to their understanding. This perspective highlights how meaning is not fixed but evolves through recursively engaging with the narrative of one's life.

Identity and the Recursion of Self-Construction: Identity is often seen as the core of personal meaning, encompassing the roles, values, and beliefs that define who we are. Recursive Realism suggests that identity formation involves a feedback loop, where self-perception and external feedback create a cycle of self-definition. For example, an individual's sense of cultural identity might be shaped by interactions with community values, personal introspection, and historical narratives, creating a loop where self-understanding is continuously shaped by internal and external reflections. This perspective allows for a richer appreciation of how identity is a dynamic construct, where recursively reflecting on one's place in the world contributes to a deeper sense of meaning.

Existential Questions and the Recursion of Purpose: Existential reflection involves confronting questions about the nature of life, death, and the purpose of existence. Recursive Realism suggests that these questions are inherently self-referential, where each inquiry prompts a loop of contemplation that deepens one's understanding of what it means to exist. For example, asking why life matters might lead to reflecting on personal values, which in turn prompts consideration of the larger context of the universe and humanity's place within it, creating a cycle of reexamining purpose. This perspective highlights how the quest for meaning involves recursively revisiting fundamental questions, where each cycle of reflection adds new layers to our understanding of existence.

Social Narratives and the Recursion of Collective Meaning

Meaning is not only a personal construct but is also shaped by the stories and values that we share with others. Recursive Realism suggests that collective meaning involves a self-referential process, where social narratives are created, revisited, and adapted to new contexts, creating a loop that binds individuals into a shared sense of purpose.

The Role of Stories in Shaping Collective Meaning: Stories, from mythologies to national histories, serve as repositories of meaning, providing frameworks for understanding the world and our place within it. Recursive Realism suggests that storytelling is self-referential, where each generation retells and reinterprets stories in light of new experiences, creating a loop that transmits values while adapting them to the needs of the present. For example, heroic myths might be reinterpreted in contemporary contexts, creating a dialogue between the timeless themes of courage and sacrifice and the specific challenges of modern life. This perspective highlights how stories serve as recursively structured vehicles for meaning-making, where each retelling deepens the connection between individuals and their cultural heritage.

Tradition and the Recursion of Cultural Values: Traditions, rituals, celebrations, and cultural practices, serve as markers of meaning, providing continuity across generations. Recursive Realism suggests that traditions are self-referential, where practices are revisited and adapted in a loop that allows for both preservation and renewal. For example, the celebration of holidays like Diwali or Christmas involves repeating rituals, but each iteration might take on new meanings based on contemporary values or personal experiences, creating a cycle where tradition evolves while maintaining its core significance. This perspective allows for a deeper understanding of how traditions serve as living sources of meaning, where recursively engaging with cultural values allows for the renewal of community bonds.

Social Movements and the Loop of Collective Identity: Social movements, from civil rights struggles to climate activism, often involve redefining collective meaning, challenging established norms and advocating for new values. Recursive Realism suggests that social change involves a feedback loop, where collective identity is reexamined and redefined in response to new ideas. For example, the feminist movement has evolved through waves of reinterpretation, where each generation revisits the meaning of gender equality, creating a cycle of social reflection that deepens the understanding of justice. This perspective highlights how collective meaning is not static but dynamic, where recursively engaging with social narratives allows for the reimagining of what is possible.

Cosmic Perspectives and the Recursion of Universal Meaning

The search for meaning often extends beyond the self and society, seeking a connection to the cosmos and the deeper structure of reality. Recursive Realism suggests that cosmic contemplation involves a self-referential process, where awareness of the universe's vastness prompts a loop of reflection on our place within it, creating a sense of belonging to something greater.

The Experience of Awe and the Recursion of Cosmic Wonder: Awe is often evoked by encounters with the vastness of nature, such as gazing at the stars or witnessing natural phenomena. Recursive Realism suggests that awe involves a feedback loop, where the recognition of the universe's scale deepens self-reflection, creating a sense of connection to the greater cosmos. For example, the experience of standing before a towering mountain range might lead to a deeper contemplation of life's transience and the interconnectedness of existence, creating a cycle where awe deepens understanding. This perspective allows for a richer appreciation of how cosmic perspectives can contribute to a sense of meaning, where self-referential contemplation allows us to feel connected to the universe's unfolding story.

Astrobiology and the Recursion of Existential Questions: The search for life beyond Earth raises profound questions about the uniqueness of life and the nature of consciousness in the universe. Recursive Realism suggests that exploring the potential for extraterrestrial life involves a self-referential process, where questions about life on other planets prompt a reexamination of life's meaning on Earth. For example, the discovery of microbial life on Mars would lead to a cycle of reflection on what it means to be alive, our place in the cosmos, and the potential for universal life processes, creating a loop where scientific discoveries deepen existential reflections. This perspective highlights how cosmic inquiry can transform our understanding of meaning, using recursive thinking to expand the scope of existential questions.

Spirituality and the Loop of Universal Unity: Spirituality often involves seeking a sense of unity with the cosmos, focusing on the interconnectedness of all things. Recursive Realism suggests that spiritual reflections involve a self-referential process, where awareness of cosmic patterns deepens the sense of connection to a larger reality. For example, contemplating the nature of time and space might lead to a sense of spiritual wonder, creating a cycle where awareness of the universe's order fosters a deeper connection to the rhythms of existence. This perspective allows for a deeper understanding of how spirituality contributes to the quest for meaning, where recursively reflecting on cosmic truths deepens the sense of unity with the universe.

The search for meaning is shaped by self-referential processes, where personal reflection, social narratives, and cosmic perspectives create loops of understanding that deepen our connection to the self, the world, and the universe. Recursive Realism provides a framework for understanding how meaning emerges from the interplay between experience and reflection, revealing how each cycle of contemplation allows for a deeper appreciation of existence. By recognising the recursive nature of the quest for meaning, Recursive Realism offers a vision of purpose that is dynamic, expansive, and interconnected.

Recursive Realism and the Meaning of Existence

Existence has long been a central focus of philosophy and spiritual inquiry, prompting questions about the nature of reality, the purpose of life, and the underlying principles that govern the universe. Recursive Realism suggests that exploring existence involves self-referential thinking, where awareness loops back on itself to consider the nature of being and reality. This section explores how Recursive Realism provides a framework for understanding existence, focusing on the interplay between mind and matter, the nature of being, and the metaphysical dimensions of life's purpose.

The Mind-Matter Relationship and Recursive Perception

The relationship between mind and matter, between consciousness and the physical universe, is a central question in philosophy and science. Recursive Realism suggests that the mind and the material world are connected through self-referential loops, where awareness engages in a dialogue with reality, shaping perception and understanding.

The Mind as a Self-Referential Mirror: Consciousness involves the ability to reflect on experiences, creating a feedback loop between perception and awareness. Recursive Realism suggests that the mind functions as a mirror that reflects reality back onto itself, creating a loop where awareness shapes the understanding of the world. For example, observing a natural landscape might involve not only perceiving the physical features but also reflecting on the feelings and thoughts evoked by the scene, creating a cycle where inner experience and external reality interact. This perspective highlights how the mind is not separate from the world but is part of a recursive process that allows for a deeper understanding of existence.

The Material World as a Feedback System: The universe is often described in scientific terms as a complex system governed by physical laws, yet Recursive Realism suggests that matter itself can be understood as part of a recursive structure, where the interactions of particles, forces, and energies create self-sustaining loops. For example, biological systems, from the molecular dynamics of DNA to the self-organising patterns of ecosystems, demonstrate how physical processes create feedback loops that sustain life. This perspective allows for a richer understanding of how the material world is not static, but dynamic, where recursive interactions shape the emergence of complex structures.

Mind and Matter as Interdependent Realities: Recursive Realism suggests that the mind and matter are interdependent, where awareness and physical reality exist in a loop of mutual influence. For example, quantum mechanics suggests that the act of observation influences the behaviour of particles, creating a feedback loop between consciousness and physical phenomena. This perspective highlights how existence involves a recursive dialogue between the observer and the observed, revealing how the nature of reality is shaped by the mind's ability to reflect upon itself and engage with the physical world.

Being and the Recursive Nature of Existence

The concept of being, what it means to exist, is central to philosophical inquiry, involving questions about the nature of self, the flow of time, and the essence of life. Recursive Realism suggests that the experience of being is shaped by self-referential processes, where awareness loops through different layers of existence, revealing the depth of what it means to be.

Being as a Flow of Recursive Awareness: Existence can be understood as a continuous flow, where each moment of awareness is connected to past experiences and future possibilities. Recursive Realism suggests that being involves a feedback loop, where awareness engages in a continuous cycle of reflecting on the present, revisiting the past, and anticipating the future. For example, meditative practices that focus on mindfulness involve returning to the present moment, creating a loop where awareness deepens through a focus on the now. This perspective allows for a deeper appreciation of how existence is not static but dynamic, where the experience of being is shaped by recursively engaging with the unfolding of time.

The Self as a Recursive Entity: The concept of selfhood is often seen as a core aspect of being, yet Recursive Realism suggests that the self is not a fixed entity but a dynamic process that involves continually reexamining and redefining its nature. For example, the philosophical idea of "becoming", as explored by existentialists like Heidegger, suggests that the self is constantly in flux, shaped by choices and reflections on what it means to exist. This perspective highlights how being involves a recursive loop, where each moment of self-awareness contributes to the ongoing construction of one's identity and purpose.

Time, Change, and the Recursion of Existence: The flow of time is often associated with change, where existence unfolds through a sequence of events. Recursive Realism suggests that time can be understood as a self-referential loop, where each moment is connected to the past and the potential future, creating a cycle of becoming. For example, the growth of a tree can be seen as a process where each ring reflects its history while shaping its future form, creating a recursive relationship between past growth and future potential. This perspective allows for a richer understanding of how existence involves a dynamic interplay between stability and change, where each moment of being is part of a larger recursive process.

Metaphysical Dimensions and the Recursion of Life's Purpose

Metaphysical inquiry involves exploring the nature of reality beyond the physical realm, seeking a deeper understanding of the principles that underlie existence. Recursive Realism suggests that metaphysical exploration involves self-referential thinking, where awareness turns inward to consider the nature of reality, creating a dialogue between the known and the unknowable.

Metaphysical Reflection as a Loop of Inquiry: Philosophical questions about the nature of reality, such as the existence of free will, the nature of consciousness, or the origins of the universe, involve recursive reflection, where each question leads to new layers of inquiry. Recursive Realism suggests that metaphysical reflection involves a feedback loop, where ideas are revisited and refined in light of new insights. For example, the question of whether the universe has a purpose might involve considering different philosophical perspectives, creating a cycle where each interpretation deepens the mystery rather than resolving it. This perspective highlights how metaphysical questions are not necessarily solvable but serve as a space for endless reflection.

Life's Purpose and the Recursion of Value: Questions about the purpose of life often involve considering what gives life value, whether through individual achievements, relationships, or a sense of connection to the larger universe. Recursive Realism suggests that the search for purpose involves a self-referential process, where each individual revisits their sense of value through a cycle of action, reflection, and recommitment. For example, finding meaning in work or creative expression might involve a feedback loop where each new accomplishment leads to a reevaluation of one's goals and a deeper sense of purpose. This perspective allows for a richer understanding of how purpose is not a destination but an ongoing process, where life's meaning is continuously shaped by recursively reflecting on what matters most.

The Universe as a Recursive Whole: Metaphysical perspectives often explore the idea that the universe itself is a unified whole, where all phenomena are interconnected through underlying principles. Recursive Realism suggests that the universe can be understood as a self-referential system, where each part reflects the whole, creating a loop of interconnection. For example, the holographic principle in theoretical physics suggests that information about a system is encoded in every part, creating a recursive structure where each element reflects the entire cosmos. This perspective highlights how metaphysical inquiry can reveal the deeper connections between individual existence and the structure of reality, using recursive thinking to explore the mysteries of the universe.

The exploration of existence involves self-referential processes, where awareness loops through layers of being, perception, and reality to uncover deeper meanings. Recursive Realism provides a framework for understanding how the nature of mind, the flow of time, and the metaphysical dimensions of the universe are interconnected through recursive thinking, revealing how life's purpose is shaped by the interplay between the known and the mysterious. By embracing the self-referential nature of existence, Recursive Realism offers a vision of being that is dynamic, reflective, and deeply connected to the unfolding story of the universe.

In Chapter 39, we have explored how Recursive Realism provides a framework for understanding the quest for meaning, revealing how self-referential processes shape personal reflection, collective narratives, and cosmic perspectives. By examining the interplay between mind and matter, the dynamic nature of being, and the metaphysical dimensions of existence, Recursive Realism offers insights into how meaning is constructed through cycles of reflection, allowing us to see life's purpose as an evolving dialogue between self, society, and the cosmos. This perspective allows us to envision a world where the search for meaning is not a quest for certainty, but a journey of continual discovery, guided by the mind's recursive nature and its capacity to explore the depths of existence.

References

Arendt, H. (1958) *The Human Condition*. Chicago: University of Chicago Press.

Baumeister, R. F. (1991) *Meanings of Life*. New York: Guilford Press.

Camus, A. (1942) *The Myth of Sisyphus*. Paris: Gallimard (translated edition, 1955, New York: Vintage International).

Cassirer, E. (1946) *The Myth of the State*. New Haven: Yale University Press.

Dennett, D. C. (1995) *Darwin's Dangerous Idea: Evolution and the Meanings of Life*. New York: Simon & Schuster.

Frankl, V. E. (1946) *Man's Search for Meaning*. Vienna: Deuticke (translated edition, 1959, Boston: Beacon Press).

Heidegger, M. (1927) *Being and Time*. Tubingen: Max Niemeyer Verlag (translated edition, 1962, New York: Harper & Row).

Jaspers, K. (1953) *The Origin and Goal of History*. London: Routledge.

Lovelock, J. (1979) *Gaia: A New Look at Life on Earth*. Oxford: Oxford University Press.

Nagel, T. (1987) *What Does It All Mean? A Very Short Introduction to Philosophy*. Oxford: Oxford University Press.

Nietzsche, F. (1883-1885) *Thus Spoke Zarathustra*. Leipzig: E. W. Fritzsch (translated edition, 1961, New York: Penguin Books).

Sagan, C. (1994) *Pale Blue Dot: A Vision of the Human Future in Space*. New York: Random House.

Sartre, J.-P. (1943) *Being and Nothingness*. Paris: Gallimard (translated edition, 1956, New York: Philosophical Library).

Teilhard de Chardin, P. (1955) *The Phenomenon of Man*. Paris: Éditions du Seuil (translated edition, 1959, New York: Harper & Row).

Tillich, P. (1952) *The Courage to Be*. New Haven: Yale University Press.

Whitehead, A. N. (1929) *Process and Reality: An Essay in Cosmology*. Cambridge: Cambridge University Press.

Chapter 40: Recursive Realism and the Evolution of Consciousness

The Emergence of Self-Referential Awareness

Consciousness is often seen as a hallmark of human evolution, but its roots extend deep into the history of life on Earth, shaped by gradual developments in perception, self-awareness, and cognitive complexity. Recursive Realism suggests that the evolution of consciousness involves self-referential processes, where organisms develop the ability to reflect on their own states, creating a feedback loop that enhances awareness and adaptive behaviour. This section explores the origins of consciousness, focusing on the emergence of self-reference in early life forms and the role of recursion in the evolution of awareness.

Early Life and the Recursion of Perceptual Awareness

The foundations of consciousness lie in the capacity of organisms to perceive and respond to their environment, creating the basis for more complex forms of awareness. Recursive Realism suggests that even simple life forms exhibit basic recursive patterns of interaction, where sensory input and adaptive responses create a feedback loop that shapes behaviour.

Perception as a Primitive Feedback Loop: Perception allows organisms to sense changes in their environment and respond accordingly, forming a basic loop of stimulus and response. Recursive Realism suggests that perception is inherently self-referential, where each sensory input leads to a response that in turn modifies future perceptions. For example, single-celled organisms like amoebae can sense changes in chemical gradients and adjust their movement, creating a loop where perception guides adaptive action. This perspective highlights how the roots of awareness lie in simple feedback systems, where recursively engaging with environmental stimuli leads to increasingly refined responses.

The Evolution of Sensory Systems and Recursion: As life evolved, sensory systems became more complex, allowing for a richer interaction with the environment. Recursive Realism suggests that the development of specialised senses, such as vision, hearing, and touch, involved recursive adaptations, where each improvement in sensory ability led to new behaviours that further enhanced perception. For example, the evolution of eyes in early aquatic animals allowed for detection of light patterns, creating a feedback loop where the ability to see shaped the evolution of neural structures dedicated to processing visual information. This perspective allows for a deeper understanding of how sensory evolution is not linear but involves recursively refining the ability to perceive and interpret the environment.

Early Nervous Systems and the Recursion of Reflexes: The development of nervous systems marked a significant step in the evolution of consciousness, allowing for more coordinated responses to environmental stimuli. Recursive Realism suggests that early nervous systems functioned as self-referential networks, where reflex actions created a feedback loop between sensory input and motor output. For example, in primitive animals like jellyfish, simple neural circuits enable reflexive responses to touch, creating a loop where each sensory signal directly influences movement. This perspective highlights how the emergence of nervous systems set the stage for more complex forms of awareness, where recursively processing sensory information allows for the coordination of behaviour.

The Rise of Self-Awareness and Cognitive Complexity

As nervous systems became more sophisticated, some organisms developed the ability to reflect on their own states, marking the emergence of self-awareness. Recursive Realism suggests that self-awareness is inherently self-referential, where organisms develop a sense of their own experience, creating a loop that enhances cognitive abilities and social interactions.

The Mirror Test and the Recursion of Self-Recognition: Self-recognition is often considered a marker of self-awareness, with the mirror test serving as a classic experiment to determine whether animals can recognise themselves in a reflection. Recursive Realism suggests that self-recognition involves a feedback loop, where an individual recognises their own reflection as an image of themselves, creating a recursive understanding of the self. For example, great apes, dolphins, and elephants have been shown to pass the mirror test, indicating a level of self-awareness that allows them to distinguish themselves from their surroundings. This perspective highlights how self-awareness involves recursively reflecting on one's own presence, leading to a deeper understanding of identity and agency.

Theory of Mind and the Recursion of Social Awareness: Theory of mind, the ability to understand that others have thoughts, intentions, and perspectives, represents a more advanced form of self-awareness that is crucial for social interaction. Recursive Realism suggests that theory of mind involves a self-referential loop, where individuals reflect on their own mental states to understand the minds of others. For example, children develop the ability to understand false beliefs around age four, indicating a shift where they can conceptualise how others perceive the world differently from their own perspective, creating a cycle of empathy and understanding. This perspective allows for a deeper appreciation of how social cognition is shaped by recursive reflection, where self-awareness becomes a tool for navigating social dynamics.

The Evolution of Memory and the Recursion of Personal Narrative: Memory allows organisms to store information about past experiences, creating a continuity of self over time. Recursive Realism suggests that memory involves a feedback loop, where each recollection revisits the past and reshapes the narrative of selfhood. For example, episodic memory, the ability to recall specific events, creates a loop where individuals can reflect on their life story, revisit memories, and integrate them into a cohesive identity. This perspective highlights how memory contributes to the evolution of consciousness, allowing for a recursive sense of past, present, and future that deepens self-awareness.

The Evolutionary Role of Recursion in Problem-Solving and Creativity

The capacity for complex thought and problem-solving is a hallmark of higher consciousness, allowing organisms to adapt to new challenges and explore innovative solutions. Recursive Realism suggests that advanced cognitive abilities are shaped by self-referential processes, where the mind engages in a loop of hypothetical thinking, simulation, and creative exploration.

Tool Use and the Recursion of Problem-Solving: Tool use in animals, from chimpanzees using sticks to crows fashioning hooks, indicates a level of problem-solving ability that involves imagining solutions to practical challenges. Recursive Realism suggests that problem-solving involves a feedback loop, where an organism envisions the use of an object, tests its effectiveness, and adjusts the technique based on outcomes. For example, a chimpanzee using a stick to extract termites must visualise how the stick functions and adjust their technique as they encounter different obstacles, creating a cycle of adaptive thinking. This perspective allows for a richer understanding of how cognitive evolution involves recursively refining strategies for interacting with the environment.

Imagination and the Recursion of Mental Simulation: Imaginative thinking, the ability to mentally simulate scenarios that have not yet occurred, allows for advanced planning and creative problem-solving. Recursive Realism suggests that imagination involves a self-referential loop, where the mind generates hypothetical scenarios, tests them mentally, and revises these simulations based on anticipated outcomes. For example, a raven planning how to access food from a complex setup might visualise different approaches, create mental models of how the environment might react, and adjust its strategy based on these mental trials. This recursive loop allows for an iterative refinement of thought before action, highlighting how imagination provides a cognitive advantage in solving new problems. This perspective deepens our understanding of how recursively engaging in mental simulations has been instrumental in the evolution of higher-order cognition.

Creativity as a Recursive Exploration: Creativity involves the ability to combine elements in new ways, leading to novel solutions, artistic expressions, and innovative thinking. Recursive Realism suggests that creativity is inherently self-referential, where ideas are generated, revisited, and recombined in a cycle that deepens the potential for innovation. For example, human language itself is a product of creative recursion, where simple sounds were recombined into words, words into sentences, and sentences into complex narratives, creating a feedback loop where language became a tool for expressing abstract ideas. This perspective highlights how creativity has been central to the evolution of consciousness, where recursive processes enable the exploration of new possibilities, leading to advances in culture, technology, and understanding.

The evolution of consciousness has been shaped by self-referential processes, where the ability to perceive, reflect, and imagine has progressively deepened the complexity of awareness. Recursive Realism provides a framework for understanding how early perceptual systems, the emergence of self-awareness, and the development of cognitive complexity are connected through feedback loops that enhance adaptive behaviour and mental capacities. By recognising the recursive nature of evolutionary processes, Recursive Realism offers a vision of consciousness as a dynamic, evolving system, where each cycle of reflection adds new dimensions to the experience of being.

Recursive Realism and the Evolutionary Pathways to Human Consciousness

Human consciousness represents a culmination of millions of years of cognitive evolution, leading to abilities that include complex language, self-reflection, and the capacity for conceptual thinking. Recursive Realism suggests that these abilities are deeply rooted in self-referential processes, where the mind's capacity to reflect on itself has led to progressively richer forms of awareness. This section examines how key aspects of human thought, including language development, metacognition, and abstract reasoning, have been shaped by recursive patterns, revealing how evolutionary feedback loops enabled the emergence of uniquely human ways of understanding the world.

The Evolution of Language and the Recursive Structure of Communication

Language is one of the most distinctive features of human cognition, allowing for complex communication, abstract thinking, and the transmission of culture. Recursive Realism suggests that language evolved through self-referential processes, where early forms of communication gradually developed recursive structures that enabled the expression of increasingly complex ideas.

Syntax as a Recursive Framework: Human language is characterised by syntax, the ability to organise words into sentences that convey specific meanings. Recursive Realism suggests that syntax is inherently self-referential, allowing phrases to be nested within phrases, creating a structure where language can express relationships between ideas. For example, in a sentence like "The child who is playing in the garden is my niece," the phrase structure allows for embedding information about who the child is and what they are doing, creating a loop that layers meaning. This perspective highlights how the evolution of syntax enabled humans to communicate complex concepts, using recursively structured sentences to articulate thoughts in ways that go beyond simple expressions.

The Development of Symbolic Thought and Recursive Metaphor: Language is not just a tool for communication, but also a vehicle for symbolic thought, allowing humans to use metaphors, analogies, and symbols to convey abstract ideas. Recursive Realism suggests that symbolic thinking involves self-referential loops, where one concept is mapped onto another, creating new ways to interpret the world. For example, the use of metaphors like "time is a river" allows for conceptualising time in terms of the flow of water, creating a feedback loop where the abstract concept of time is understood through the concrete experience of flow. This perspective allows for a deeper understanding of how language and symbolic thought enabled humans to create mental models of the world, using recursively structured ideas to explore concepts that extend beyond direct experience.

Narrative as a Recursive Tool for Understanding: Storytelling is a central aspect of human culture, allowing for the sharing of experiences, the preservation of knowledge, and the construction of identity. Recursive Realism suggests that narrative thinking involves a self-referential process, where stories revisit events, reframe experiences, and connect themes into a cohesive whole. For example, myths and legends often contain recursive elements, where stories within stories convey layered meanings about the nature of life, morality, and the cosmos. This perspective highlights how narrative thinking allowed humans to reflect on their own existence, creating a loop where the act of storytelling deepens understanding and fosters a sense of continuity within communities.

Metacognition and the Recursion of Self-Reflection

Metacognition, the ability to think about one's own thinking, is a key feature of human consciousness, allowing for introspection, critical self-evaluation, and the ability to adapt thoughts based on self-awareness. Recursive Realism suggests that metacognition is inherently self-referential, involving a loop where awareness turns inward to reflect on cognitive processes, creating a space for higher-order reasoning.

Introspection and the Loop of Self-Knowledge: Introspection allows individuals to reflect on their own thoughts, emotions, and motives, creating a deeper understanding of the self. Recursive Realism suggests that introspection involves a feedback loop, where awareness revisits mental states, assesses them, and reframes understanding based on these reflections. For example, meditative practices that focus on observing thoughts without judgment involve a recursive process where the mind continually returns to its own patterns, creating a cycle of self-discovery. This perspective allows for a richer appreciation of how self-reflection contributes to personal growth, where recursively engaging with one's own mind allows for greater emotional resilience and mental clarity.

Self-Evaluation and the Recursion of Learning: Metacognition plays a crucial role in learning, allowing individuals to assess their own understanding, identify gaps in knowledge, and adjust strategies for problem-solving. Recursive Realism suggests that self-evaluation involves a feedback loop, where each learning experience is revisited and reinterpreted, creating a cycle of continuous improvement. For example, a student might reflect on their performance after a challenging exam, identify areas of weakness, and adjust their study methods based on this reflection, creating a loop where self-awareness enhances learning outcomes. This perspective highlights how metacognition allows for adaptive thinking, where recursively engaging with cognitive processes leads to more effective learning and problem-solving.

Moral Reflection and the Recursion of Ethical Awareness: Metacognition also plays a role in moral reasoning, allowing individuals to reflect on their values, consider the consequences of actions, and revisit ethical principles. Recursive Realism suggests that moral reflection involves a self-referential loop, where each decision is evaluated in light of one's values and the impact on others, creating a cycle of moral development. For example, ethical dilemmas often require individuals to rethink their positions, weigh conflicting principles, and adjust their perspective, creating a loop where moral awareness deepens through each encounter with complexity. This perspective highlights how metacognition enables moral growth, using recursive thinking to navigate the intricacies of human values and responsibilities.

Abstract Reasoning and the Recursion of Conceptual Thought

Abstract reasoning allows humans to conceptualise ideas that go beyond direct experience, such as mathematical principles, philosophical concepts, and scientific theories. Recursive Realism suggests that abstract reasoning involves self-referential processes, where ideas are revisited, refined, and reintegrated into more complex frameworks of understanding.

Mathematics and the Recursion of Logical Structures: Mathematics involves creating models of reality that use abstract symbols and logical rules to describe relationships between quantities and structures. Recursive Realism suggests that mathematical thought is inherently self-referential, where equations and theorems build upon each other, creating a cycle of proof and refinement. For example, the development of calculus required revisiting the concept of limits, creating a recursive understanding of change and infinity that transformed the study of motion and natural phenomena. This perspective allows for a richer understanding of how mathematics is a product of recursive thinking, where abstract concepts are layered and revisited to uncover deeper truths about the universe.

Philosophical Reasoning and the Loop of Conceptual Exploration: Philosophy often involves exploring concepts that challenge our understanding of reality, such as the nature of time, consciousness, and the foundations of morality. Recursive Realism suggests that philosophical reasoning is self-referential, where each argument is examined, countered, and reexamined to deepen understanding. For example, dialectical reasoning, as seen in the works of Hegel, involves a process of thesis-antithesis-synthesis, creating a loop where each stage of thought builds upon the previous one, leading to a more comprehensive perspective. This perspective highlights how philosophical inquiry is shaped by recursive thinking, allowing for the expansion of conceptual boundaries through cycles of reflection.

Scientific Theories and the Recursion of Hypothesis Testing: Science is grounded in the process of forming hypotheses, testing them, and revisiting theories in light of new evidence. Recursive Realism suggests that scientific inquiry involves a feedback loop, where theories are continuously refined through experimentation and revision. For example, Darwin's theory of evolution has been revisited and refined through modern genetics, creating a cycle where new discoveries in DNA and genomics reshape our understanding of natural selection. This perspective highlights how science uses recursive processes to deepen knowledge, where each new finding adds to the cumulative understanding of the natural world.

The evolution of human consciousness has been shaped by self-referential processes, where the mind's ability to reflect on itself enabled the emergence of language, metacognition, and abstract reasoning. Recursive Realism provides a framework for understanding how these cognitive abilities evolved through feedback loops, revealing how recursively structured thought allowed humans to explore new concepts, create complex social systems, and expand the boundaries of knowledge. By recognising the recursive nature of human cognitive evolution, Recursive Realism offers a vision of consciousness that is dynamic, expansive, and deeply intertwined with the history of life on Earth.

In Chapter 40, we have explored how Recursive Realism provides insights into the evolution of consciousness, revealing how self-referential processes shaped the development of awareness from early life forms to complex human thought. By examining the emergence of perceptual systems, the rise of self-awareness, and the evolution of abstract reasoning, Recursive Realism offers a framework for understanding how consciousness evolved as a recursive phenomenon, enabling the mind to explore the depths of existence. This perspective allows us to see consciousness as a product of evolutionary loops, where each cycle of reflection deepens our capacity for understanding and engagement with the mysteries of life.

References

Bateson, G. (1972) *Steps to an Ecology of Mind: Collected Essays in Anthropology, Psychiatry, Evolution, and Epistemology*. San Francisco: Chandler.

Buckner, C. (2015) 'The origins of self-recognition: A review of Povinelli's theory and recent developments,' *Animal Cognition*, 18(3), pp. 443–455.

Carruthers, P. (2011) *The Opacity of Mind: An Integrative Theory of Self-Knowledge*. Oxford: Oxford University Press.

Deacon, T. (1997) *The Symbolic Species: The Co-Evolution of Language and the Brain*. New York: W. W. Norton & Company.

Dennett, D. C. (1991) *Consciousness Explained*. Boston: Little, Brown and Company.

Gazzaniga, M. S. (2008) *Human: The Science Behind What Makes Us Unique*. New York: HarperCollins.

Haidt, J. (2012) *The Righteous Mind: Why Good People Are Divided by Politics and Religion*. New York: Pantheon.

Lakoff, G. and Johnson, M. (1980) *Metaphors We Live By*. Chicago: University of Chicago Press.

Pinker, S. (1994) *The Language Instinct: How the Mind Creates Language*. New York: William Morrow.

Premack, D. and Woodruff, G. (1978) 'Does the chimpanzee have a theory of mind?' *Behavioral and Brain Sciences*, 1(4), pp. 515–526.

Roth, G. and Dicke, U. (2005) 'Evolution of the brain and intelligence,' *Trends in Cognitive Sciences*, 9(5), pp. 250–257.

Tomasello, M. (2019) *Becoming Human: A Theory of Ontogeny*. Cambridge, MA: Harvard University Press.

Varela, F. J., Thompson, E., and Rosch, E. (1991) *The Embodied Mind: Cognitive Science and Human Experience*. Cambridge, MA: MIT Press.

Wheeler, M. (2011) 'Martin Heidegger,' in Zalta, E. N. (ed.) *The Stanford Encyclopedia of Philosophy* (Fall 2011 Edition).

Wittgenstein, L. (1953) *Philosophical Investigations*. Oxford: Blackwell.

Zeki, S. (1999) *Inner Vision: An Exploration of Art and the Brain*. Oxford: Oxford University Press.

Chapter 41: Recursive Realism and the Dynamics of Human Culture

The Evolution of Cultural Practices Through Recursive Interactions

Culture is a product of human interaction, shaped by the exchange of ideas, rituals, stories, and shared values. Recursive Realism suggests that the development of culture involves self-referential processes, where each cultural practice is revisited, adapted, and reinterpreted across generations, creating a loop that shapes the evolution of societies. This section explores how cultural practices are formed, transmitted, and transformed through recursive interactions, focusing on communication, rituals, and the continuity of tradition.

Communication and the Recursion of Language in Culture

Language serves as the foundation of cultural exchange, allowing humans to share knowledge, express emotions, and preserve traditions. Recursive Realism suggests that communication involves self-referential loops, where language shapes cultural norms while being shaped by them in return.

Language as a Recursive Cultural System: Language is not only a means of communication but also a carrier of cultural values, stories, and ways of thinking. Recursive Realism suggests that language evolves through a feedback loop, where words and expressions reflect cultural values, and cultural values are transmitted and reshaped through the use of language. For example, phrases that carry moral connotations, such as proverbs, encapsulate cultural wisdom, but their meanings can shift as social contexts change, creating a loop where language adapts to the evolving needs of society. This perspective highlights how language serves as a dynamic cultural system, where recursively using and reinterpreting words allows for the transmission and renewal of cultural identity.

Oral Traditions and the Recursion of Storytelling: Oral traditions, from folktales to epic poetry, have long been a vehicle for preserving cultural heritage, creating a shared narrative that connects generations. Recursive Realism suggests that storytelling involves self-referential processes, where stories are retold and adapted over time, creating a feedback loop between past wisdom and present interpretations. For example, ancient myths often have multiple versions, where each retelling adds a new layer of cultural meaning, reflecting the concerns and values of the contemporary audience. This perspective allows for a richer understanding of how oral traditions function as recursively evolving narratives, where the act of storytelling deepens the connection between individuals and their cultural roots.

Symbolic Language and the Recursion of Cultural Identity: Symbols, from national flags to religious icons, play a key role in shaping cultural identity, providing visual representations of shared values. Recursive Realism suggests that symbolic language involves a feedback loop, where symbols are imbued with meaning through cultural practices and rituals, and these meanings are reinterpreted as society evolves. For example, the meaning of national symbols like flags can shift over time, reflecting changes in political identity, social movements, or historical events, creating a cycle where symbols are continually redefined. This perspective highlights how symbols serve as recursively structured expressions of cultural values, where each generation reengages with symbolic meanings to reinforce or challenge identity.

Rituals and the Recursion of Collective Practices

Rituals, from religious ceremonies to civic traditions, play a central role in defining cultural life, providing structure and meaning to collective experiences. Recursive Realism suggests that rituals are shaped by self-referential processes, where repetition and adaptation create a loop that reinforces cultural norms while allowing for innovation.

Repetition and the Loop of Cultural Continuity: Rituals often involve repetition, where specific actions are performed in a consistent manner, creating a sense of continuity across time. Recursive Realism suggests that the repetition of rituals creates a feedback loop, where the act of repeating deepens the connection to shared values. For example, annual religious festivals like Ramadan or Easter involve repeating specific practices that reinforce beliefs, but each observance can take on new meanings based on personal experiences or current events, creating a cycle of cultural renewal. This perspective highlights how rituals serve as recursively structured activities, where the repetition of practices allows for the preservation of cultural identity while adapting to changing contexts.

Ritual Innovation and the Recursion of Cultural Adaptation: While rituals are often associated with tradition, they can also be adapted and reimagined to address new needs and circumstances. Recursive Realism suggests that ritual innovation involves a self-referential process, where communities revisit their practices to align them with contemporary values, creating a loop that allows for the evolution of cultural expression. For example, marriage ceremonies have evolved in many cultures to reflect changing attitudes towards gender roles and family structures, creating a cycle where traditional elements are blended with new customs. This perspective allows for a deeper appreciation of how rituals can adapt while maintaining continuity, using recursively structured practices to negotiate the balance between preservation and change.

Rituals of Remembrance and the Recursion of Memory: Rituals of remembrance, such as memorial services or anniversaries of historical events, play a key role in maintaining collective memory, allowing societies to reflect on their past and honour their heritage. Recursive Realism suggests that these rituals involve a feedback loop, where each act of remembrance revisits past events, creating a cycle that shapes how history is understood. For example, memorials for events like World War II or civil rights movements often involve retelling the stories of those who lived through them, creating a loop where each generation engages with the legacy of the past in new ways. This perspective highlights how rituals of remembrance contribute to the continuity of cultural identity, using recursively engaging practices to ensure that collective memory remains a living part of cultural life.

The Continuity of Tradition and the Recursion of Cultural Values

Traditions, from family customs to national celebrations, serve as cultural anchors, providing a sense of stability and continuity in a changing world. Recursive Realism suggests that tradition is a self-referential process, where practices are revisited and reinterpreted, creating a loop that allows for both preservation and renewal.

The Recursion of Tradition Through Generational Transmission: Traditions are often passed down from one generation to the next, creating a sense of continuity that connects past and present. Recursive Realism suggests that the transmission of tradition involves a feedback loop, where each generation reengages with customs and redefines their meaning. For example, cultural festivals like the Lunar New Year involve rituals and symbols that have been passed down for centuries, but each new celebration adds a unique touch, creating a loop where tradition evolves while maintaining its core essence. This perspective highlights how tradition serves as a recursive structure, where the act of passing on practices allows for the continuity of culture while adapting to new times.

The Recursion of Values in Cultural Reinvention: Cultural values, such as hospitality, honour, and community, are often seen as core aspects of cultural identity, guiding behaviour and social norms. Recursive Realism suggests that values are not static but are continually revisited through a cycle of cultural reflection. For example, the value of democracy has been reinterpreted in different historical periods, adapting to new challenges while maintaining its foundational principles, creating a loop where each generation refines the concept of what democracy means. This perspective allows for a richer understanding of how values serve as recursively structured ideals, where each cultural moment contributes to the ongoing conversation about what matters most.

The Rituals of Reintegration and the Cycle of Cultural Renewal: Periods of cultural change, such as social movements or revolutions, often involve the reinvention of traditions to reflect new ideals. Recursive Realism suggests that cultural renewal involves a feedback loop, where old practices are revisited and reinterpreted to align with emerging values. For example, the reclamation of indigenous traditions in many parts of the world involves revisiting cultural practices that were suppressed or marginalised, creating a cycle where traditions are reimagined as symbols of resilience and cultural pride. This perspective highlights how cultural renewal is not a break from tradition but a recursive engagement with the past, where revisiting heritage allows for the creation of new cultural identities.

Culture evolves through self-referential processes, where language, rituals, and traditions create a feedback loop that shapes the continuity and adaptation of cultural life. Recursive Realism provides a framework for understanding how cultural practices are both preserved and transformed, revealing how recursively engaging with shared values and customs allows for the creation of cohesive societies. By recognising the recursive nature of cultural evolution, Recursive Realism offers a vision of culture that is dynamic, resilient, and deeply connected to the collective memory of humanity.

Recursive Realism and the Evolution of Social Norms and Collective Identity

Social norms and collective identities form the backbone of human societies, guiding behaviour, regulating interactions, and providing a sense of belonging. Recursive Realism suggests that the creation and adaptation of social norms involve self-referential processes, where rules and expectations are revisited and reinterpreted through cycles of reflection and social interaction. This section explores how self-referential dynamics shape the evolution of norms, the integration of individuals into communities, and the role of recursion in driving social change.

The Formation of Social Norms and the Recursion of Behavioural Expectations

Social norms define acceptable behaviour within a group or society, creating a framework that regulates how individuals interact. Recursive Realism suggests that the establishment of norms involves a feedback loop, where individual actions and collective expectations influence each other, creating a cycle that shapes the social fabric.

The Recursion of Norm Enforcement: Social norms are maintained through mechanisms of enforcement, where deviations from expected behaviour prompt corrective actions, reinforcing the boundaries of acceptable conduct. Recursive Realism suggests that norm enforcement involves a self-referential loop, where each instance of enforcement reaffirms the standard and shapes future behaviour. For example, in traditional communities, rituals of punishment or social shaming for rule violations create a cycle where individuals learn the consequences of nonconformity, reinforcing the collective agreement on what is acceptable. This perspective highlights how social norms are not fixed rules but are constantly reinforced through recursive social processes, where the interaction between individuals and group expectations shapes normative boundaries.

Norm Innovation and the Recursion of Social Adaptation: While social norms often provide stability, they are also subject to change as society evolves. Recursive Realism suggests that norm innovation involves a feedback loop, where individuals challenge existing norms and these challenges prompt a reassessment of cultural standards. For example, movements for social justice, such as civil rights or gender equality, often involve individuals questioning the fairness of existing norms, creating a cycle where the collective response leads to the adaptation of social values. This perspective allows for a richer understanding of how social norms are both maintained and transformed through recursively revisiting and reinterpreting societal expectations.

The Recursion of Social Learning and Cultural Transmission: Social norms are often passed down through generations via social learning, where younger members observe and imitate the behaviour of elders. Recursive Realism suggests that social learning involves a feedback loop, where cultural practices are revisited and reinforced through each generation's interpretation. For example, manners and etiquette, such as how to greet others or show respect, are taught through repeated interactions, creating a cycle where each generation contributes to the continuity of social norms. This perspective highlights how social learning serves as a mechanism for maintaining cultural cohesion, using recursively structured interactions to preserve and adapt norms over time.

Collective Identity and the Recursion of Group Belonging

Collective identity is central to how individuals understand their place within society, providing a sense of belonging and shared purpose. Recursive Realism suggests that the formation of group identity involves self-referential processes, where individual experiences and group narratives create a feedback loop that shapes social cohesion and collective values.

Identity Formation as a Recursively Reinforced Process: Group identity is often reinforced through shared stories, symbols, and rituals that connect individuals to a larger narrative. Recursive Realism suggests that identity formation involves a feedback loop, where individuals internalise group values, and their participation in cultural practices reaffirms the collective identity. For example, national identity is often reinforced through national holidays, symbols, and historical narratives that create a sense of unity, creating a cycle where each person's engagement with national symbols deepens their sense of belonging. This perspective highlights how collective identity is not static but is constantly redefined through recursively engaging with the stories and symbols that define group membership.

The Role of Exclusion and Inclusion in Identity Recursion: Group identity is often defined not only by who belongs but also by who is excluded, creating boundaries between in-group and out-group members. Recursive Realism suggests that the processes of inclusion and exclusion involve a self-referential loop, where the act of defining boundaries shapes group cohesion and reinforces identity. For example, religious communities may define themselves through shared beliefs and rituals, but also through distinctions from other faiths, creating a loop where the reinforcement of boundaries solidifies a shared sense of purpose. This perspective allows for a deeper appreciation of how collective identities are shaped by the dynamics of belonging, where recursively drawing and redrawing boundaries helps to define group identity.

Cultural Resilience and the Recursion of Identity Through Challenges: Collective identities often become most visible during periods of challenge, such as times of war, social upheaval, or economic crisis. Recursive Realism suggests that cultural resilience involves a feedback loop, where the collective response to external challenges reaffirms the core values of the group. For example, resistance movements often draw on cultural symbols and shared narratives to inspire solidarity and resilience, creating a cycle where each act of resistance strengthens the collective identity. This perspective highlights how collective identity is not only preserved through shared rituals, but is also deepened through recursively responding to external pressures.

The Dynamics of Social Change and the Recursion of Cultural Evolution

Societies are not static entities but are constantly evolving, adapting to new ideas, technologies, and social challenges. Recursive Realism suggests that social change involves self-referential processes, where cultural norms and collective identities are revisited and reinterpreted through cycles of reflection, leading to new social structures and values.

The Role of Social Movements in Recursively Reframing Norms: Social movements, from abolitionism to environmentalism, play a key role in shaping social change, challenging existing norms and advocating for new values. Recursive Realism suggests that social movements involve a feedback loop, where each act of protest or advocacy prompts a collective re-examination of social values. For example, the civil rights movement in the United States involved recursively challenging racial segregation, leading to a cycle where each victory redefined the meaning of equality and justice. This perspective allows for a deeper understanding of how social change is not linear but involves recursively engaging with cultural ideals to reshape society.

Cultural Evolution Through Technological Recursion: Technological advances, from the printing press to the internet, have often transformed social structures, creating new ways for people to connect and share ideas. Recursive Realism suggests that technological change involves a feedback loop, where new technologies change cultural practices, which in turn influence the development of further technologies. For example, the rise of social media has created a loop where online platforms reshape how people communicate, and these new communication patterns influence the evolution of digital technologies. This perspective highlights how technology serves as a recursive driver of cultural evolution, where each new tool redefines the possibilities for social interaction.

Recursive Reflections on Tradition and Modernity: Modern societies often grapple with the tension between maintaining traditions and embracing new ideas. Recursive Realism suggests that navigating this tension involves a feedback loop, where cultural practices are revisited and recontextualised in light of contemporary values. For example, the preservation of indigenous languages in the face of globalisation involves recursively engaging with traditional practices while integrating modern tools, such as digital archives or language apps, creating a cycle where heritage is maintained through new methods. This perspective allows for a deeper appreciation of how societies evolve by recursively balancing the past and the present, using reflection to integrate tradition with innovation.

Social norms and collective identities are shaped by self-referential processes, where individual actions and group expectations create a feedback loop that guides the evolution of societies. Recursive Realism provides a framework for understanding how norms are established, identities are formed, and social change occurs, revealing how recursively engaging with cultural ideals allows for both stability and adaptation. By recognising the recursive nature of social processes, Recursive Realism offers a vision of culture that is dynamic, reflective, and capable of transforming itself in response to new challenges.

In Chapter 41, we have explored how Recursive Realism offers insights into the dynamics of culture, revealing how self-referential processes shape the evolution of social norms, collective identities, and cultural practices. By examining how language, rituals, and social movements contribute to cultural continuity and adaptation, Recursive Realism provides a framework for understanding how societies balance tradition and innovation, using recursive reflection to navigate change. This perspective allows us to envision culture as a living system, where each cycle of reflection contributes to the ongoing evolution of human values and social structures.

References

Appadurai, A. (1996) *Modernity at Large: Cultural Dimensions of Globalization*. Minneapolis: University of Minnesota Press.

Barthes, R. (1977) *Image-Music-Text*. Translated by S. Heath. London: Fontana.

Bourdieu, P. (1990) *The Logic of Practice*. Translated by R. Nice. Stanford: Stanford University Press.

Durkheim, E. (1915) *The Elementary Forms of Religious Life*. Translated by J. W. Swain. London: Allen & Unwin.

Geertz, C. (1973) *The Interpretation of Cultures: Selected Essays*. New York: Basic Books.

Giddens, A. (1984) *The Constitution of Society: Outline of the Theory of Structuration*. Berkeley: University of California Press.

Habermas, J. (1984) *The Theory of Communicative Action*. Translated by T. McCarthy. Boston: Beacon Press.

Hall, S. (1997) *Representation: Cultural Representations and Signifying Practices*. London: Sage.

Hobsbawm, E. and Ranger, T. (eds) (1983) *The Invention of Tradition*. Cambridge: Cambridge University Press.

Inglehart, R. (1997) *Modernization and Postmodernization: Cultural, Economic, and Political Change in 43 Societies*. Princeton: Princeton University Press.

Jenkins, R. (2008) *Social Identity*. 3rd edn. London: Routledge.

Malinowski, B. (1922) *Argonauts of the Western Pacific*. London: Routledge & Kegan Paul.

Mead, G. H. (1934) *Mind, Self, and Society: From the Standpoint of a Social Behaviorist*. Edited by C. W. Morris. Chicago: University of Chicago Press.

Sahlins, M. (1976) *Culture and Practical Reason*. Chicago: University of Chicago Press.

Turner, V. (1969) *The Ritual Process: Structure and Anti-Structure*. Chicago: Aldine Publishing.

Weber, M. (1930) *The Protestant Ethic and the Spirit of Capitalism*. Translated by T. Parsons. London: Allen & Unwin.

Williams, R. (1977) *Marxism and Literature*. Oxford: Oxford University Press.

Wittgenstein, L. (1953) *Philosophical Investigations*. Oxford: Blackwell.

Zerubavel, E. (1997) *Social Mindscapes: An Invitation to Cognitive Sociology*. Cambridge, MA: Harvard University Press.

Chapter 42: Recursive Realism and the Nature of Time

The Subjective Experience of Time and Recursive Reflection

Time is a fundamental aspect of human experience, shaping how we perceive change, anticipate the future, and understand our place in the flow of existence. Recursive Realism suggests that our perception of time is inherently self-referential, where awareness loops through experiences of past, present, and future, creating a sense of continuity that allows for a deeper understanding of the temporal nature of reality. This section explores how time is experienced subjectively, focusing on the role of memory, the dynamics of anticipation, and the recursive nature of temporal awareness.

Memory and the Recursion of the Past

Memory allows individuals to revisit past experiences, creating a sense of continuity that connects the present self to its history. Recursive Realism suggests that memory is a self-referential process, where each act of recollection involves revisiting and reinterpreting past events, creating a loop that shapes our understanding of the past and its relationship to the present.

Episodic Memory as a Recursive Reengagement: Episodic memory allows for the recall of specific events, creating a narrative of personal history. Recursive Realism suggests that episodic memory involves a feedback loop, where each recollection reshapes the memory itself, adding new layers of meaning based on current understanding. For example, remembering a childhood experience might involve reinterpreting the emotions and details of the event from an adult perspective, creating a cycle where the memory is enriched through the act of remembering. This perspective highlights how memory is not a static record but a dynamic process, where recursively engaging with the past allows for a deeper integration of experience into the self.

Collective Memory and the Recursion of Cultural History: Memory is not only a personal phenomenon but also a collective one, shaping the identity of communities through shared stories and historical events. Recursive Realism suggests that collective memory involves a feedback loop, where societies revisit and reinterpret historical events, creating a cycle that influences how the past is understood. For example, the commemoration of historical events like Independence Day or World War anniversaries involves retelling the story of the past in ways that resonate with current values, creating a loop where history is continually reshaped. This perspective highlights how collective memory serves as a recursively evolving narrative, where each generation contributes to the reinterpretation of cultural history.

Memory's Role in Shaping Identity Through Time: Memory is crucial for maintaining a sense of identity, allowing individuals to see themselves as the same person over different moments in time. Recursive Realism suggests that identity is shaped by a self-referential loop, where memories are integrated into a cohesive narrative of who we are. For example, reflecting on significant life events like graduation, marriage, or loss allows individuals to revisit these moments and interpret their impact on personal growth, creating a cycle where memory shapes identity and identity shapes how memory is understood. This perspective allows for a deeper appreciation of how the experience of time is woven into the narrative of selfhood, using recursive reflection to create a sense of continuity across the temporal flow.

Anticipation and the Recursion of the Future

Anticipation allows individuals to project forward in time, imagining future possibilities and planning actions based on expectations. Recursive Realism suggests that anticipation is a self-referential process, where the mind engages in a loop of simulating scenarios, revisiting expectations, and adjusting plans, creating a cycle that shapes how the future is perceived.

Future Simulations as Recursive Imagination: Anticipating the future often involves mentally simulating potential outcomes, using past experiences to imagine what might happen. Recursive Realism suggests that future simulations involve a feedback loop, where each imagined scenario is revised based on new information and changing circumstances. For example, planning a major life event like moving to a new city might involve imagining different challenges and adapting plans as more details become clear, creating a loop where expectations are continually refined. This perspective highlights how the ability to anticipate is not static but involves recursively adjusting mental models of the future.

The Role of Uncertainty and the Recursion of Hope and Fear: Anticipating the future often involves navigating uncertainty, where hope and fear shape how possibilities are imagined. Recursive Realism suggests that hope and fear are part of a self-referential loop, where expectations influence emotional responses, which in turn shape how the future is visualised. For example, anticipating the outcome of a medical test might involve alternating between hopeful and fearful thoughts, creating a cycle where each emotional response informs the interpretation of possibilities. This perspective allows for a deeper understanding of how the mind's recursive nature shapes our emotional engagement with the future, using cycles of reflection to navigate the unknown.

Anticipation and Decision-Making Through Temporal Recursion: Decision-making is often influenced by the ability to anticipate outcomes, where plans are evaluated based on potential consequences. Recursive Realism suggests that decision-making involves a feedback loop, where future scenarios are imagined, actions are taken, and results are revisited to inform future decisions. For example, a business leader might project future market trends, make strategic decisions, and then reassess their plans based on the outcomes, creating a cycle where each decision refines the ability to anticipate. This perspective highlights how the mind's recursive engagement with the future allows for more adaptive decision-making, using feedback loops to adjust expectations and navigate change.

The Present Moment and the Recursion of Now

The present moment is often seen as the point of contact between past memories and future anticipation, creating a sense of being grounded in the flow of time. Recursive Realism suggests that our awareness of the present involves self-referential loops, where the mind engages with the unfolding moment while reflecting on its connection to past and future.

Mindfulness and the Loop of Present Awareness: Mindfulness practices, such as meditation or deep focus, emphasise the importance of being present, allowing individuals to fully engage with the moment without distraction. Recursive Realism suggests that mindfulness involves a feedback loop, where the mind returns to the present moment, acknowledges thoughts without clinging to them, and reengages with the now. For example, focusing on the breath involves repeatedly returning to the sensation of breathing, creating a cycle where awareness becomes more attuned to the subtleties of the present. This perspective highlights how mindfulness serves as a recursive practice, deepening the experience of now through cycles of reflection.

The Experience of Flow and the Recursion of Action: Flow states, moments when individuals are fully immersed in an activity, represent a heightened sense of the present where time seems to slow down or disappear. Recursive Realism suggests that flow involves a feedback loop, where focus on the task creates a cycle of action and adjustment, allowing awareness to become fully engaged with the present challenge. For example, a musician might experience flow while playing an instrument, where each note informs the next in a seamless loop, creating a sense of timeless engagement. This perspective allows for a deeper appreciation of how the experience of the present can be intensified through recursive interaction with the activity at hand.

The Recursion of the Present in the Context of Past and Future: The present moment is often experienced in relation to the past and future, where awareness cycles between reflection and anticipation. Recursive Realism suggests that the experience of now is shaped by a loop that connects what has been to what is yet to come, allowing for a sense of time's continuity. For example, reflecting on a recent conversation while planning for a meeting creates a cycle where the present task is shaped by memories and future goals, creating a complex awareness of time's flow. This perspective highlights how the present moment is not isolated but is woven into a recursive fabric, where past, present, and future are interconnected through the loops of awareness.

Time is experienced through self-referential processes, where memory, anticipation, and present awareness create a feedback loop that shapes our perception of the temporal flow. Recursive Realism provides a framework for understanding how the mind engages with time, revealing how cycles of reflection create a sense of continuity and depth in our experience of existence. By recognising the recursive nature of temporal awareness, Recursive Realism offers a vision of time that is dynamic, reflective, and deeply integrated into the fabric of consciousness.

Recursive Realism and the Philosophical and Scientific Dimensions of Time

Time has been a central theme in philosophy and science, prompting questions about its nature, directionality, and relation to reality. Recursive Realism suggests that understanding time involves self-referential processes, where awareness engages in a dialogue between empirical observations and metaphysical interpretations, creating a feedback loop that deepens our comprehension of the temporal structure of the universe. This section explores how Recursive Realism can bridge the subjective experience of time with philosophical inquiry and scientific theories, revealing how self-referential thinking is crucial to grasping the complexities of time's nature.

The Philosophical Exploration of Time and Recursive Reflection

Philosophers have long grappled with the nature of time, debating whether it is a fundamental aspect of reality or a construct of the human mind. Recursive Realism suggests that philosophical inquiry into time involves a self-referential process, where each perspective invites a reflection on its implications, creating a loop that leads to a more nuanced understanding of time's essence.

The Problem of Time and the Recursion of Being: Philosophers like Aristotle, Kant, and Heidegger have explored time as it relates to the nature of being, questioning whether time exists independently of human experience or is inextricably tied to our perception of change. Recursive Realism suggests that the problem of time involves a feedback loop, where the concept of time is re-examined through the lens of existence, leading to a deeper exploration of how time shapes being. For example, Heidegger's notion of being-toward-death involves recursively reflecting on one's mortality as a way to understand the flow of time, creating a loop where the awareness of time's passage becomes central to the experience of being. This perspective highlights how philosophical explorations of time rely on self-referential reflection, using recursive inquiry to probe the relationship between time and existence.

Presentism, Eternalism, and the Recursion of Temporal Perspectives: Presentism, the view that only the present moment exists, and eternalism, which suggests that past, present, and future are equally real, represent two opposing views on the nature of time. Recursive Realism suggests that debating these perspectives involves a self-referential loop, where the mind shifts between different models of temporal reality, creating a cycle that deepens the understanding of how time might be structured. For example, eternalism invites a reflection on how all moments are fixed within a temporal block, while presentism emphasises the dynamic unfolding of the present, leading to a recursive dialogue between static and dynamic views of time. This perspective allows for a richer appreciation of how philosophical debates about time are not resolved but deepened through the process of recursive inquiry, where each viewpoint sheds light on different aspects of temporal experience.

Time as a Construct and the Recursion of Human Perception: Kant famously argued that time is a construct of the mind, a way of organising experiences rather than a feature of reality itself. Recursive Realism suggests that understanding time as a construct involves a feedback loop, where the act of perceiving time shapes our conceptualisation of what time is, creating a cycle where our mental models of time interact with our experience of its passage. For example, the subjective experience of time slowing down during moments of intense focus or speeding up during periods of routine creates a loop where perception influences our understanding of how time flows. This perspective highlights how philosophical reflections on time as a construct reveal the recursive nature of temporal awareness, where the mind's interpretation of time is shaped by the very act of experiencing it.

Scientific Perspectives on Time and Recursive Patterns in Physics

Modern physics has transformed our understanding of time, revealing it to be a fundamental component of the universe that is intertwined with space, energy, and matter. Recursive Realism suggests that scientific theories of time, such as relativity and quantum mechanics, involve self-referential insights, where each new discovery revisits the assumptions about the nature of time, creating a feedback loop that continually refines our understanding.

Relativity and the Recursion of Space-Time: Einstein's theory of relativity fundamentally changed how we understand time, revealing that it is not an absolute constant but is relative to motion and gravitational fields. Recursive Realism suggests that the concept of space-time involves a feedback loop, where our understanding of time is re-examined in light of its relationship with space and matter. For example, time dilation, where time slows down in strong gravitational fields or at high velocities, creates a loop where the perception of time shifts depending on one's reference frame, challenging intuitive notions of a universal clock. This perspective highlights how theories of relativity use recursive insights to rethink time, showing that the experience of time is deeply interwoven with the fabric of the universe.

Quantum Mechanics and the Recursion of Temporal Uncertainty: Quantum mechanics introduces a level of uncertainty to our understanding of time, revealing that particles can exist in superpositions and that events at the quantum level do not always have a definite order. Recursive Realism suggests that quantum theories involve a self-referential process, where each experiment with particles invites a re-evaluation of how time behaves at the smallest scales. For example, entanglement suggests that particles can be correlated across space and time, creating a loop where the state of one particle instantly affects the state of another, challenging classical ideas of time-bound causality. This perspective allows for a deeper appreciation of how quantum mechanics uses recursive thinking to explore the mysterious nature of time, revealing how the fundamental unpredictability of quantum events reshapes our understanding of temporal order.

Cosmology and the Recursion of Time's Origin: Cosmological theories, from the Big Bang to the nature of black holes, explore the origins and ultimate fate of time itself. Recursive Realism suggests that cosmological inquiry involves a feedback loop, where each new discovery about the universe's structure prompts a re-examination of how time began and how it might end. For example, the concept of time before the Big Bang leads to a recursive question of whether time emerged with the universe or if there existed a form of time beyond our current understanding. Similarly, the nature of time near the event horizons of black holes suggests that time might behave differently at extreme gravitational fields, creating a loop of inquiry into the fundamental nature of space-time. This perspective highlights how cosmological theories use recursive reflection to probe the mysteries of time's beginning and its role in the evolution of the cosmos.

Metaphysical Dimensions of Time and Recursive Realism

Beyond the empirical focus of science, metaphysical inquiry into time involves questions about its ultimate nature, its role in consciousness, and its connection to reality itself. Recursive Realism suggests that exploring the metaphysics of time involves self-referential processes, where each reflection on the nature of time creates a loop that deepens our understanding of its connection to existence.

The Arrow of Time and the Recursion of Entropy: The concept of the arrow of time, the idea that time moves in one direction, from past to future, is closely related to the second law of thermodynamics, which states that entropy increases over time. Recursive Realism suggests that the arrow of time involves a feedback loop, where the progression of entropy shapes our experience of time's flow, and our awareness of time's direction influences how we interpret physical processes. For example, the irreversibility of certain processes, like the mixing of gases or the aging of living organisms, reinforces the perception that time moves forward, creating a cycle where the unfolding of natural events aligns with our sense of time's irreversibility. This perspective allows for a deeper understanding of how metaphysical reflections on the nature of time's direction are connected to recursive principles of thermodynamics.

Time as a Metaphysical Connector and the Recursion of Reality: Philosophical perspectives often explore the idea that time serves as a connector between different realms of reality, such as mind and matter, existence and non-existence, and potentiality and actuality. Recursive Realism suggests that time's role as a metaphysical connector involves a feedback loop, where each exploration of time reveals new layers of how reality is structured. For example, the concept of time as a dimension that allows for the unfolding of events invites a recursive reflection on whether time is itself a property of the universe or a framework imposed by consciousness. This perspective highlights how metaphysical inquiry uses self-referential thinking to explore time's role in the deeper connections between existence and the universe's structure.

Time is a complex concept that spans philosophical reflection and scientific theory, involving self-referential processes that deepen our understanding of its nature. Recursive Realism provides a framework for exploring how the perception of time, scientific discoveries, and metaphysical ideas interact in a feedback loop, revealing how each cycle of reflection adds to the understanding of time as both a mental construct and a physical reality. By recognising the recursive nature of temporal inquiry, Recursive Realism offers a vision of time that is dynamic, interconnected, and endlessly unfolding in the dialogue between mind and universe.

In Chapter 42, we have explored how Recursive Realism provides insights into the nature of time, revealing how self-referential processes shape our understanding of temporal experience through memory, anticipation, scientific theories, and philosophical reflection. By examining the flow of time from both subjective and cosmic perspectives, Recursive Realism offers a framework for understanding how time is not simply linear but involves complex feedback loops that shape our perception and the structure of reality itself. This perspective allows us to see time as a multi-dimensional phenomenon, where each reflection on time's nature contributes to a deeper understanding of existence.

References

Aristotle (1984). *The Complete Works of Aristotle: The Revised Oxford Translation.* Edited by Jonathan Barnes. Princeton University Press.

Einstein, A. (1920). *Relativity: The Special and General Theory.* Translated by Robert W. Lawson. Methuen & Co.

Gell-Mann, M. (1995). *The Quark and the Jaguar: Adventures in the Simple and the Complex.* W.H. Freeman and Company.

Hawking, S. (1988). *A Brief History of Time: From the Big Bang to Black Holes.* Bantam Books.

Heidegger, M. (1962). *Being and Time.* Translated by John Macquarrie and Edward Robinson. Harper & Row.

Kant, I. (1998). *Critique of Pure Reason.* Translated by Paul Guyer and Allen W. Wood. Cambridge University Press.

Penrose, R. (2004). *The Road to Reality: A Complete Guide to the Laws of the Universe.* Jonathan Cape.

Prigogine, I. (1997). *The End of Certainty: Time, Chaos, and the New Laws of Nature.* Free Press.

Rovelli, C. (2018). *The Order of Time.* Translated by Erica Segre and Simon Carnell. Allen Lane.

Smolin, L. (2013). *Time Reborn: From the Crisis in Physics to the Future of the Universe.* Houghton Mifflin Harcourt.

Stapp, H. P. (2007). *Mindful Universe: Quantum Mechanics and the Participating Observer.* Springer.

Wheeler, J. A. (1990). "Information, Physics, Quantum: The Search for Links." In *Complexity, Entropy, and the Physics of Information,* edited by W. H. Zurek, 3–28. Addison-Wesley.

Chapter 43: Recursive Realism and the Intersection of Science and Spirituality

The Complementary Nature of Science and Spirituality Through Recursive Reflection

Science and spirituality are often seen as opposing forces, one grounded in empirical evidence and logical reasoning, the other rooted in personal experience and transcendental insights. However, Recursive Realism suggests that science and spirituality can be understood as complementary perspectives on the nature of reality, each offering unique insights through self-referential processes of inquiry and reflection. This section explores how Recursive Realism enables a dialogue between these realms, focusing on the recursive nature of scientific reasoning, the reflective aspects of spiritual experience, and the ways in which these paths converge in the pursuit of meaning.

Scientific Inquiry and the Recursion of Empirical Exploration

Science is built on the systematic investigation of the natural world, using observation, experimentation, and the formulation of theories to uncover patterns in physical phenomena. Recursive Realism suggests that scientific inquiry involves self-referential loops, where each hypothesis is tested, refined, and revisited, creating a feedback loop that deepens our understanding of the universe.

The Scientific Method as a Recursive Process: The scientific method, involving hypothesis, experimentation, data collection, and the revision of theories, is inherently recursive, as it relies on the ability to reexamine assumptions in light of new evidence. Recursive Realism suggests that scientific progress involves a loop, where theories are revisited and reinterpreted as knowledge expands. For example, the shift from Newtonian mechanics to Einstein's theory of relativity involved a re-examination of the nature of space and time, creating a feedback loop where new discoveries prompted a rethinking of fundamental concepts. This perspective highlights how science is not static but dynamic, using recursively structured inquiry to refine our understanding of natural laws.

The Role of Doubt and the Recursion of Sceptical Reflection: Scepticism is a key component of scientific inquiry, allowing for the questioning of assumptions and the critical evaluation of existing theories. Recursive Realism suggests that scepticism involves a self-referential loop, where each doubt leads to a deeper investigation, creating a cycle that refines our understanding of reality. For example, the development of quantum mechanics required questioning classical notions of determinism, leading to new insights into the probabilistic nature of the quantum world. This perspective allows for a deeper appreciation of how scepticism is not an obstacle to knowledge but a recursive process that allows for the continuous refinement of scientific understanding.

Science's Pursuit of Unity and the Recursion of Theoretical Models: Scientific inquiry often seeks a unified understanding of diverse phenomena, aiming to create theories that encompass the full scope of reality's complexity. Recursive Realism suggests that the pursuit of unity involves a feedback loop, where the search for overarching principles is balanced by the need to revisit specific details, creating a cycle that aims for coherence without oversimplification. For example, the quest for a Theory of Everything, a framework that would unify general relativity and quantum mechanics, involves repeatedly revisiting concepts like gravity, space-time, and particle interactions to create a coherent model. This perspective highlights how science's search for unity is not a linear path but a recursive process of synthesis and analysis, where each iteration brings us closer to a holistic understanding of the universe.

Spiritual Exploration and the Recursion of Inner Reflection

Spirituality involves the search for meaning, the exploration of inner experiences, and the contemplation of the transcendent, offering insights into the mysteries of existence beyond empirical observation. Recursive Realism suggests that spiritual exploration involves self-referential processes, where awareness turns inward to reflect on the nature of being, creating a feedback loop that deepens the sense of connection to the cosmos and the self.

Meditation and the Recursion of Self-Awareness: Meditative practices, such as mindfulness or contemplation, involve a process of turning inward, focusing on the nature of thought, breath, or consciousness itself. Recursive Realism suggests that meditation involves a feedback loop, where each moment of awareness leads to a deeper reflection on the nature of the mind, creating a cycle of self-awareness. For example, mindfulness meditation focuses on observing thoughts without attachment, creating a loop where each observation deepens the understanding of mental processes and the nature of selfhood. This perspective highlights how meditation uses recursive reflection to explore the layers of consciousness, allowing for insights into the nature of existence that are not easily accessible through rational analysis.

Mystical Experiences and the Recursion of Transcendence: Mystical experiences, moments where individuals feel a sense of unity with the universe or a connection to something greater than themselves, are often described in spiritual traditions as transcendent. Recursive Realism suggests that mystical experiences involve a feedback loop, where the boundaries of self are dissolved through deep reflection, creating a sense of merging with the cosmos. For example, Sufi mystics describe the dissolution of the self into the divine, creating a cycle where the individual experiences oneness with the universe through recursively engaging in contemplative practices. This perspective allows for a deeper appreciation of how spiritual experiences use recursive awareness to transcend the limitations of individuality, offering insights that complement the empirical focus of science.

The Recursion of Purpose in Spiritual Inquiry: Spirituality often involves questions about purpose, the nature of good and evil, and the meaning of life, creating a framework for understanding one's place in the cosmos. Recursive Realism suggests that the search for purpose involves a self-referential loop, where each reflection on existence leads to new questions and deeper insights into the nature of reality. For example, reflecting on the nature of suffering might involve exploring how challenges can lead to growth and spiritual insight, creating a cycle where each experience becomes a source of wisdom. This perspective highlights how spiritual inquiry is not about finding final answers but involves a recursive process of exploring deeper layers of meaning through reflection and contemplation.

The Convergence of Science and Spirituality Through Recursive Realism

Science and spirituality both seek to understand reality, yet they often use different methods and approaches. Recursive Realism suggests that the convergence of these domains involves self-referential processes, where scientific insights and spiritual experiences inform each other, creating a feedback loop that allows for a more holistic understanding of the universe.

Integrating Scientific Discoveries with Spiritual Insights: Scientific discoveries, from the vastness of the cosmos to the complexity of life, often evoke a sense of wonder that resonates with spiritual awe. Recursive Realism suggests that integrating these perspectives involves a feedback loop, where the empirical knowledge gained through science deepens spiritual reflections on the nature of existence, creating a cycle where the sense of mystery is enriched rather than diminished. For example, the understanding of the universe's scale through astronomy can inspire a spiritual sense of our place in the cosmos, creating a recursive dialogue between the empirical and the transcendent. This perspective highlights how science and spirituality can mutually enrich each other, using recursively structured inquiry to explore the depths of the unknown.

The Role of Wonder and the Recursion of Curiosity: Wonder is a shared foundation of both scientific curiosity and spiritual reverence, prompting the exploration of the mysteries of life. Recursive Realism suggests that wonder involves a self-referential loop, where the desire to understand leads to new questions, and each discovery opens new dimensions of curiosity. For example, the discovery of the genetic code revealed the intricate machinery of life, but it also inspired deeper questions about the origin of life and the nature of consciousness, creating a cycle where science's discoveries enhance the sense of spiritual wonder. This perspective allows for a deeper appreciation of how the recursive nature of curiosity drives both scientific inquiry and spiritual exploration, leading to a continual unfolding of new mysteries.

A Holistic Perspective: The Recursive Realism Approach: Recursive Realism offers a framework for viewing science and spirituality as complementary approaches to understanding reality, suggesting that each perspective contributes to a deeper awareness of the universe's complexity. By recognising the self-referential nature of both empirical inquiry and spiritual contemplation, Recursive Realism allows for a vision where science's precision and spirituality's insights coexist, creating a loop where understanding is continually refined through reflection and integration. This perspective highlights how the mind's capacity for recursion enables a dialogue between the rational and the mystical, offering a holistic understanding of existence.

Science and spirituality both involve self-referential processes that shape our understanding of reality, each offering unique perspectives that can enrich the other. Recursive Realism provides a framework for exploring the dynamic interplay between empirical inquiry and spiritual experience, revealing how the search for knowledge and the quest for meaning are interconnected through recursive reflection. By recognising the common ground between science's exploration of the natural world and spirituality's exploration of the inner world, Recursive Realism offers a vision of understanding that is holistic, deeply connected, and open to the mysteries of existence.

Recursive Realism and the Dialogue Between Empirical Knowledge and Transcendental Understanding

Empirical knowledge and transcendental understanding represent two ways of approaching reality, each offering insights that are often seen as distinct or even in conflict. Recursive Realism suggests that a deeper understanding of existence can emerge from a dialogue between these approaches, using self-referential processes to reconcile and integrate insights from both domains. This section explores how scientific concepts like cosmic evolution, complexity, and quantum phenomena intersect with spiritual ideas about interconnectedness, the nature of consciousness, and the experience of the divine, revealing how Recursive Realism allows for a unified exploration of the mysteries of the universe.

The Interconnectedness of Life and the Recursion of Cosmic Evolution

Science and spirituality both explore the interconnected nature of the universe, though they often use different languages to describe this reality. Recursive Realism suggests that the concept of interconnectedness involves a self-referential process, where the understanding of unity is revisited and reinterpreted through scientific discoveries and spiritual insights, creating a feedback loop that deepens our appreciation of the web of existence.

The Recursion of Complexity in Cosmic Evolution: Scientific theories about the evolution of the universe, from the Big Bang to the emergence of life, reveal a progression from simplicity to complexity, where fundamental particles eventually form atoms, molecules, stars, planets, and living organisms. Recursive Realism suggests that this process of cosmic evolution involves self-referential loops, where each stage of complexity builds on the structures that came before, creating a cycle of emergent order. For example, the formation of carbon-based molecules, essential for life, in the cores of stars highlights how the universe's complexity increases through recursively structured processes of nuclear fusion and element formation. This perspective allows for a deeper appreciation of how scientific insights into cosmic evolution align with spiritual concepts of interconnectedness, where each stage of existence is interdependent with the whole.

Interconnectedness in Spiritual Traditions and the Recursion of Unity: Spiritual traditions often emphasise the unity of all life, using concepts like the Tao in Taoism, Brahman in Hinduism, or the Oneness of Being in Sufism to describe a fundamental interconnectedness that underlies all existence. Recursive Realism suggests that these concepts of unity involve a self-referential loop, where awareness reflects on its connection to the whole, creating a cycle where the sense of individuality is reintegrated into a larger reality. For example, meditative practices that emphasise oneness with nature create a feedback loop where the individual experiences a merging with the cosmos, reflecting a spiritual sense of interconnectedness. This perspective highlights how spiritual insights into unity and interconnectedness resonate with scientific views of the universe's evolutionary complexity, using recursive awareness to explore the connections between all things.

The Dialogue Between Ecology and Spiritual Ecology: Ecology, as a scientific field, studies the interdependence of organisms and their environments, revealing the intricate web of life's interactions. Spiritual ecology, in turn, emphasises the sacredness of these connections, viewing the earth and its ecosystems as a living whole that demands respect and stewardship. Recursive Realism suggests that the integration of ecological science with spiritual perspectives involves a feedback loop, where each new understanding of ecosystem dynamics deepens the sense of reverence for life's interdependence. For example, understanding the role of mycorrhizal networks in forest ecosystems reveals the interconnectedness of trees and fungi, aligning with spiritual notions of the earth as a living entity. This perspective allows for a richer understanding of how scientific knowledge and spiritual reverence for nature can mutually enrich each other, creating a recursive appreciation of life's interconnected web.

Consciousness and the Recursion of Awareness in Science and Spirituality

Consciousness is a central theme in both scientific inquiry and spiritual exploration, prompting questions about its nature, origin, and role in the universe. Recursive Realism suggests that the study of consciousness involves a self-referential process, where the mind reflects on its own awareness, creating a feedback loop that shapes our understanding of the mind's place in reality.

Neuroscience and the Recursion of Cognitive Processes: Neuroscience has made significant strides in understanding the brain, revealing the intricate networks that give rise to thought, emotion, and perception. Recursive Realism suggests that neuroscientific inquiry into consciousness involves a feedback loop, where the study of brain activity informs our understanding of mental processes, and this understanding reshapes the questions we ask about the mind. For example, the concept of neural feedback loops, where sensory information is processed and reprocessed through different brain regions, aligns with recursive models of how awareness emerges. This perspective highlights how neuroscience offers empirical insights into the recursive nature of consciousness, using self-referential thinking to explore the mind's complexity.

Mystical Experiences and the Recursion of Expanded Awareness: Mystical experiences, moments where individuals feel a profound sense of unity with the universe, are often described in spiritual traditions as expansions of consciousness. Recursive Realism suggests that these experiences involve a self-referential loop, where the dissolution of boundaries between self and other leads to a deeper reflection on the nature of awareness. For example, in Buddhist meditation, the experience of non-duality, where the distinction between self and world disappears, creates a cycle of reflection that challenges the ordinary sense of individuality. This perspective allows for a deeper appreciation of how mystical experiences offer insights into the recursive nature of consciousness, revealing how awareness can expand beyond the confines of the self.

The Convergence of Quantum Physics and Consciousness Studies: Quantum physics has raised philosophical questions about the role of the observer in the collapse of wave functions, suggesting a possible connection between consciousness and the nature of reality. Recursive Realism suggests that exploring this connection involves a feedback loop, where the study of quantum phenomena prompts reflections on the role of the mind, creating a cycle where scientific models and metaphysical ideas intersect. For example, the concept of quantum entanglement challenges classical notions of separateness, leading to questions about the role of consciousness in connecting distant particles. This perspective highlights how Recursive Realism provides a framework for engaging with the deep questions at the intersection of quantum theory and consciousness studies, using recursive reflection to explore the mysteries of awareness.

Bridging Rationality and Transcendence Through Recursive Realism

Rational analysis and transcendental insights represent two aspects of the human quest for understanding, often seen as distinct but interrelated. Recursive Realism suggests that the integration of these perspectives involves self-referential thinking, where each reflection on the mysteries of existence invites a synthesis of rational thought and spiritual insight, creating a feedback loop that expands our understanding of reality.

The Rationality of Spiritual Insights and the Recursion of Meaning: Spiritual philosophies often provide deep insights into the nature of life, suffering, and existence, offering perspectives that can complement rational analysis. Recursive Realism suggests that understanding these insights involves a feedback loop, where rational inquiry revisits spiritual teachings to find logical coherence, creating a cycle that reveals new dimensions of meaning. For example, the concept of karma in Eastern philosophy, where actions have spiritual consequences, can be re-examined through psychological theories of cause and effect, creating a dialogue between spiritual wisdom and empirical evidence. This perspective allows for a deeper appreciation of how rationality can be used to refine spiritual insights, using recursive thinking to explore the nature of moral and existential truths.

Transcendence as a Recursion Beyond Conceptual Boundaries: Transcendence involves moving beyond the limitations of ordinary thought, reaching a state of awareness that surpasses conceptual distinctions. Recursive Realism suggests that the experience of transcendence involves a feedback loop, where concepts are deconstructed through deep reflection, allowing for a direct experience of the ineffable. For example, mystical experiences often involve a dissolution of dualities, such as self and other or mind and matter, creating a cycle where each step of reflection leads to a more expansive awareness. This perspective highlights how Recursive Realism allows for a synthesis of rationality and transcendence, using recursive processes to explore the boundaries of thought and the possibilities beyond.

The Unity of Science, Philosophy, and Spirituality Through Recursive Inquiry: Recursive Realism suggests that the unity of science, philosophy, and spirituality involves a self-referential process, where each domain revisits the questions posed by the others, creating a feedback loop that deepens the understanding of existence. For example, the study of the cosmos through astronomy can inspire philosophical reflections on the nature of the infinite, which in turn can lead to spiritual contemplations about the vastness of the universe. This perspective highlights how Recursive Realism provides a framework for bringing together diverse ways of knowing, using recursively structured inquiry to explore the most profound questions about life, consciousness, and the cosmos.

Empirical knowledge and transcendental understanding both offer insights into the nature of existence, and Recursive Realism provides a framework for integrating these perspectives through self-referential processes. By recognising the feedback loops that shape scientific inquiry and spiritual exploration, Recursive Realism allows for a holistic approach to the mysteries of reality, revealing how rationality and wonder can mutually enrich each other. This perspective offers a vision of understanding that is open to the complexities of existence, using recursive thinking to explore the intersections between knowledge and meaning.

In Chapter 43, we have explored how Recursive Realism provides insights into the intersection of science and spirituality, revealing how self-referential processes create a space for dialogue between empirical knowledge and transcendental insights. By examining how concepts of interconnectedness, consciousness, and the mysteries of the cosmos are understood through both rational analysis and spiritual reflection, Recursive Realism offers a framework for integrating diverse perspectives, creating a more comprehensive vision of the universe. This perspective allows us to see science and spirituality as complementary approaches to understanding existence, each contributing to a deeper engagement with the mysteries of life.

References

Capra, F. (1996). *The Web of Life: A New Scientific Understanding of Living Systems.* Anchor Books.

Davies, P. (2006). *The Goldilocks Enigma: Why is the Universe Just Right for Life?* Penguin Books.

Einstein, A. (1936). "Physics and Reality." *Journal of the Franklin Institute,* 221(3), pp. 349–382.

Heisenberg, W. (1958). *Physics and Philosophy: The Revolution in Modern Science.* Harper & Row.

Kabat-Zinn, J. (1990). *Full Catastrophe Living: Using the Wisdom of Your Body and Mind to Face Stress, Pain, and Illness.* Dell Publishing.

Kauffman, S. (1995). *At Home in the Universe: The Search for the Laws of Self-Organization and Complexity.* Oxford University Press.

Penrose, R. (1989). *The Emperor's New Mind: Concerning Computers, Minds and the Laws of Physics.* Oxford University Press.

Prigogine, I. (1997). *The End of Certainty: Time, Chaos, and the New Laws of Nature.* Free Press.

Rovelli, C. (2018). *The Order of Time.* Translated by Erica Segre and Simon Carnell. Allen Lane.

Schrödinger, E. (1944). *What Is Life? The Physical Aspect of the Living Cell.* Cambridge University Press.

Sheldrake, R. (2009). *A New Science of Life: The Hypothesis of Morphic Resonance.* Icon Books.

Sufi, I. (2014). *The Heart of Sufism: Essential Writings of Hazrat Inayat Khan.* Shambhala Publications.

Wilber, K. (2000). *Integral Psychology: Consciousness, Spirit, Psychology, Therapy.* Shambhala Publications.

Chapter 44: Recursive Realism and the Limits of Human Knowledge

The Boundaries of Understanding and the Recursion of Complexity

Human knowledge is shaped by our capacity to perceive patterns, formulate theories, and explore the universe, yet it is also constrained by the limits of our cognitive abilities and the complexity of reality itself. Recursive Realism suggests that understanding these limits involves self-referential processes, where each attempt to grasp complexity leads to a deeper reflection on our epistemological boundaries, creating a feedback loop that continually redefines the scope of what can be known. This section explores how the nature of complexity, the problem of infinity, and the inherent uncertainties of scientific theories shape our understanding of knowledge's limits.

Complexity and the Recursion of Systems Thinking

Complexity arises in natural phenomena, biological systems, and social dynamics, revealing patterns that cannot be fully understood through linear thinking. Recursive Realism suggests that grappling with complexity involves a self-referential loop, where each new understanding of a system's behaviour invites a re-examination of its components, creating a cycle that expands our comprehension while revealing new layers of intricacy.

The Challenge of Complexity in Scientific Models: Scientific models often attempt to simplify reality to make it comprehensible, yet complex systems, such as climate dynamics, ecosystems, or the human brain, defy simple reduction. Recursive Realism suggests that modelling complexity involves a feedback loop, where each simplified model is revised as new data reveals unanticipated interactions, creating a cycle where understanding evolves. For example, climate models must account for the interactions between ocean currents, atmospheric conditions, and biological feedbacks, creating a recursive process of refinement. This perspective highlights how scientific inquiry into complexity reveals the limits of reductionism, showing that knowledge must continually revisit and adjust to capture the dynamic nature of complex systems.

Emergence and the Recursion of New Properties: Emergence refers to the appearance of new properties in complex systems that cannot be predicted from the behaviour of individual components. Recursive Realism suggests that the study of emergence involves a self-referential process, where each new property that emerges invites a reassessment of the system's nature, creating a cycle of revisiting and refining our models. For example, the emergence of consciousness from neural networks or the emergence of social behaviour in animal populations challenges linear explanations, requiring a recursive exploration of how collective dynamics shape new forms of order. This perspective allows for a deeper appreciation of how the study of emergence reveals the limits of prediction, showing that knowledge must be adaptable to new layers of complexity as they unfold.

Chaos Theory and the Recursion of Sensitivity to Initial Conditions: Chaos theory explores how small variations in initial conditions can lead to vastly different outcomes, revealing the unpredictable nature of certain systems. Recursive Realism suggests that chaos involves a feedback loop, where each attempt to understand a chaotic system requires a revisiting of initial assumptions, creating a cycle of adjusting models to account for sensitivity. For example, in meteorology, the challenge of predicting weather patterns is influenced by chaotic dynamics, where tiny changes in temperature or pressure can amplify into large-scale shifts. This perspective highlights how chaos theory reveals the boundaries of predictability, showing that knowledge is inherently limited by the recursive complexity of natural systems.

Infinity and the Recursion of the Infinite

Infinity poses conceptual challenges in both mathematics and philosophy, representing quantities or scales that are beyond comprehension. Recursive Realism suggests that engaging with infinity involves self-referential processes, where each reflection on the nature of the infinite leads to a deeper paradox, creating a feedback loop that challenges our understanding of reality's boundaries.

Mathematical Infinity and the Recursion of Paradoxes: Mathematics has developed concepts like infinite sets, limits, and transfinite numbers to explore the nature of infinity, yet these concepts often lead to paradoxes that challenge intuitive understanding. Recursive Realism suggests that mathematical inquiry into infinity involves a feedback loop, where each solution to a paradox, such as Zeno's paradoxes or Cantor's diagonal argument, leads to new questions, creating a cycle of revisiting the nature of the infinite. For example, Cantor's discovery that infinity comes in different sizes through the concept of cardinality reveals how each exploration of infinity leads to deeper complexities. This perspective allows for a deeper appreciation of how mathematics grapples with the concept of infinity, using recursive reasoning to navigate the limits of abstraction.

Cosmological Infinity and the Recursion of the Universe's Boundaries: Cosmology explores the scale of the universe, raising questions about whether space-time is finite or infinite. Recursive Realism suggests that cosmological reflections on infinity involve a feedback loop, where each model of the universe's structure is revisited in light of new observations, creating a cycle of expanding our understanding of the cosmos. For example, the question of whether the universe will continue expanding forever or will eventually contract leads to a recursive exploration of the nature of cosmic time and space's boundaries. This perspective highlights how cosmological theories use recursive thinking to explore the concept of infinity, revealing how the vastness of the universe stretches the limits of human comprehension.

Philosophical Reflections on the Infinite and the Recursion of Metaphysical Thought: Philosophers have long grappled with the concept of infinity, using it to explore the nature of existence, God, and the structure of reality. Recursive Realism suggests that philosophical reflections on infinity involve a self-referential process, where each attempt to define the infinite leads to a deeper reflection on the limits of human thought, creating a cycle of conceptual expansion. For example, theological concepts like the infinite nature of God challenge finite human understanding, creating a loop where each description of the divine reveals new dimensions of mystery. This perspective highlights how philosophical inquiry into the infinite uses recursive reflection to navigate the boundaries of comprehension, revealing how certain aspects of reality remain beyond human grasp.

Uncertainty, Paradox, and the Recursion of Knowledge's Boundaries

Uncertainty and paradox are inherent in the quest for knowledge, revealing the limitations of our ability to fully grasp the nature of reality. Recursive Realism suggests that engaging with uncertainty involves a feedback loop, where each paradox or uncertainty leads to a deeper inquiry, creating a cycle that allows us to expand our understanding while acknowledging the mysteries that persist.

Quantum Uncertainty and the Recursion of Indeterminacy: Quantum mechanics has introduced the concept of uncertainty as a fundamental aspect of physical reality, where particles do not have precise positions and momenta until measured. Recursive Realism suggests that grappling with quantum uncertainty involves a self-referential loop, where each experiment with particles prompts a reevaluation of the nature of measurement and reality, creating a cycle of exploring the boundaries of certainty. For example, Heisenberg's uncertainty principle reveals that the more precisely we know one property of a particle, the less precisely we can know another, leading to a recursive understanding of the limits of knowledge. This perspective highlights how quantum physics uses self-referential thinking to explore the nature of reality, revealing how uncertainty is an intrinsic aspect of the physical world.

Epistemological Paradoxes and the Recursion of Limits: Philosophical paradoxes, such as the liar paradox or Gödel's incompleteness theorems, reveal the limitations of formal systems and logical structures, showing that no system can fully explain itself. Recursive Realism suggests that engaging with these paradoxes involves a feedback loop, where each paradox challenges the assumptions of a system, creating a cycle that reveals the inherent limitations of any attempt to encompass all truth. For example, Gödel's theorems show that within any consistent mathematical system, there are statements that are true but cannot be proven, creating a recursive boundary between what can be known and what remains outside formal proof. This perspective allows for a deeper understanding of how philosophical inquiry into paradox uses recursive thinking to explore the edges of rationality, revealing how certain truths lie beyond formal explanation.

The Role of Mystery and the Recursion of Knowledge's Humility: Mystery is often seen as an obstacle to understanding, yet it can also be a source of wonder that motivates further inquiry. Recursive Realism suggests that embracing mystery involves a feedback loop, where each encounter with the unknown invites a deeper reflection on the nature of knowledge, creating a cycle that balances the pursuit of answers with the acceptance of uncertainty. For example, the question of consciousness's nature, whether it is emergent or fundamental, remains a profound mystery, creating a loop where each new theory leads to further questions. This perspective highlights how Recursive Realism allows for a balanced approach to the limits of knowledge, using recursive inquiry to explore the unknown while respecting the mysteries that remain beyond human reach.

Knowledge is shaped by the interplay between discovery and the recognition of limits, where the complexity of reality and the mysteries of the universe reveal the boundaries of human understanding. Recursive Realism provides a framework for exploring how self-referential processes shape our engagement with complexity, infinity, and uncertainty, revealing how each reflection on the limits of knowledge invites a deeper appreciation of the mysteries that remain. By recognising the recursive nature of our pursuit of understanding, Recursive Realism offers a vision of knowledge that is both humble and expansive, embracing the unknown as a catalyst for ongoing inquiry.

Recursive Realism and the Paradox of Knowing the Unknowable

Paradoxes challenge our understanding of truth, revealing the limitations of logical reasoning and empirical inquiry. Recursive Realism suggests that engaging with paradoxes involves a self-referential process, where each encounter with a contradiction or an unsolvable mystery invites a deeper reflection, creating a feedback loop that deepens our understanding of the nature of knowledge. This section explores how Recursive Realism allows us to navigate the paradoxes that arise when attempting to know the unknowable, focusing on the recursive nature of doubt, the challenges of conceptualising the infinite, and the ways in which paradox serves as a bridge between rationality and mystery.

The Role of Paradox in Deepening Understanding

Paradoxes, statements or situations that appear self-contradictory yet reveal a deeper truth, have been central to philosophical inquiry, challenging our assumptions and expanding the boundaries of rational thought. Recursive Realism suggests that paradox plays a crucial role in deepening understanding, serving as a point of reflection that creates a feedback loop, where each attempt to resolve a contradiction leads to a more nuanced comprehension of the underlying issue.

The Liar Paradox and the Recursion of Self-Reference: The liar paradox, where a statement declares itself to be false, challenges the nature of truth by creating a loop of self-reference that cannot be resolved within classical logic. Recursive Realism suggests that the liar paradox exemplifies how self-reference can create infinite regressions, where each attempt to determine the truth value of the statement leads back to the original contradiction. This paradox highlights the limits of binary thinking, revealing how reality may involve layers of truth that cannot be fully captured by simplistic categories. Recursive Realism uses this insight to suggest that paradoxes serve as a reminder of the complexity of truth, inviting a recursive reflection on the nature of meaning.

Gödel's Incompleteness Theorems and the Recursion of Formal Systems: Gödel's incompleteness theorems demonstrate that within any consistent formal system, there are true statements that cannot be proven within that system, creating a paradoxical boundary between truth and proof. Recursive Realism suggests that Gödel's findings reveal the recursive nature of logical systems, where the act of proving inherently involves a loop that leaves certain truths outside the system's reach. For example, Gödel's theorems suggest that mathematics, while powerful, cannot be complete, as it must always refer to something beyond itself to account for all truths. This perspective allows for a deeper appreciation of how formal systems are inherently limited by their recursive structure, revealing how the act of knowing involves encountering boundaries that cannot be fully transcended.

Paradox as a Tool for Conceptual Expansion: Philosophers and mystics have often used paradox as a tool to challenge conventional thinking and to expand awareness beyond the ordinary limits of language and logic. Recursive Realism suggests that engaging with paradoxes involves a self-referential process, where each reflection on the contradictory nature of a statement leads to a deeper engagement with the mysteries of existence. For example, the Zen koan, a paradoxical riddle used in Zen practice, is designed to break down rational thinking, creating a loop where the mind's attempts to resolve the riddle lead to a state of insight that transcends logical analysis. This perspective highlights how paradox can serve as a gateway to deeper levels of understanding, using recursive reflection to move beyond the limits of rational thought and to engage with the ineffable aspects of reality.

The Recursion of Doubt and the Search for Certainty

Doubt is often seen as an obstacle to certainty, yet it also serves as a catalyst for deeper inquiry, prompting questions that lead to new insights. Recursive Realism suggests that doubt involves a feedback loop, where each question creates a cycle of reflection, leading to a more refined understanding of what can be known and what must remain uncertain.

Scepticism and the Recursion of Critical Inquiry: Philosophical scepticism, the practice of questioning the possibility of knowledge, has played a central role in the history of philosophy, challenging assumptions and forcing deeper reflection. Recursive Realism suggests that scepticism involves a self-referential loop, where each doubt about a claim leads to further questions, creating a cycle that allows for the refinement of beliefs. For example, Descartes' method of doubt, which sought to question all beliefs until only indubitable truths remained, ultimately led to the discovery of self-awareness as a foundational certainty. This perspective highlights how scepticism is not merely destructive but serves as a recursive process of uncovering deeper truths by challenging surface-level assumptions.

The Paradox of Certainty and the Recursion of Knowledge: The search for certainty often leads to paradoxes, where the very attempt to find absolute knowledge reveals the limits of what can be known. Recursive Realism suggests that the paradox of certainty involves a feedback loop, where the pursuit of unshakable truths leads to a confrontation with the inherent uncertainties of reality, creating a cycle of adjustment and reassessment. For example, the scientific quest to define the fundamental nature of matter has led from the discovery of atoms to the exploration of subatomic particles, and further into the mysteries of quantum fields, revealing how each level of certainty gives way to new layers of uncertainty. This perspective allows for a deeper appreciation of how the desire for certainty is balanced by the recursive process of navigating doubt, showing that true understanding involves an openness to uncertainty.

Mystical Doubt and the Recursion of the Ineffable: Mystical traditions often embrace doubt as a path to spiritual insight, using questions about the nature of reality to deepen awareness of the mystery of existence. Recursive Realism suggests that mystical doubt involves a self-referential process, where each encounter with the ineffable, that which cannot be fully articulated, leads to a cycle of reflection that opens the mind to deeper dimensions of experience. For example, in Sufi mysticism, the concept of divine unknowability invites a continual reflection on the limits of human understanding, creating a loop where the mind is led beyond concepts into a direct encounter with the divine mystery. This perspective highlights how mystical traditions use recursive doubt to explore the boundaries of knowledge, revealing how the unknown serves as a source of wonder and spiritual insight.

Beyond Language: The Recursion of Articulating the Ineffable

Language is a powerful tool for communicating knowledge, yet it is also limited in its ability to capture the full scope of reality. Recursive Realism suggests that the attempt to articulate the ineffable involves a feedback loop, where each effort to put experiences into words leads to a deeper reflection on the limits of expression, creating a cycle that both reveals and obscures the nature of the mystery.

The Limits of Language in Describing Reality: Philosophers and linguists have long explored the limitations of language, recognising that words often fail to capture the richness of experience. Recursive Realism suggests that the limits of language involve a self-referential loop, where each attempt to describe the ineffable creates a deeper awareness of language's boundaries, revealing the gap between words and reality. For example, the experience of love, awe, or transcendence often defies simple description, creating a cycle where language points to something beyond itself. This perspective highlights how language is both a bridge and a barrier, using recursive reflection to approach the mysteries that lie beyond words.

Art and the Recursion of Expressing the Inexpressible: Art, whether visual, musical, or poetic, often seeks to express aspects of reality that cannot be fully articulated through rational discourse. Recursive Realism suggests that artistic expression involves a feedback loop, where each creation attempts to capture the ineffable, creating a cycle of interpretation and reinterpretation. For example, abstract art or avant-garde music often seeks to evoke emotions or experiences that transcend literal meaning, inviting the viewer or listener into a recursive engagement with the artwork's deeper message. This perspective allows for a richer understanding of how art can serve as a recursive exploration of the mysteries of existence, revealing how the ineffable can be approached through creative means.

Mystical Language and the Paradox of Saying the Unsayable: Mystics often use paradoxical language to describe their experiences, recognising that words are inadequate to convey the full reality of transcendent states. Recursive Realism suggests that mystical language involves a feedback loop, where each attempt to articulate the divine leads to a recognition of language's failure, creating a cycle that invites the listener into the mystery. For example, Rumi's poetry often uses paradox and imagery to point beyond words to the nature of the divine. This perspective highlights how mystical traditions use recursive reflection to approach the ineffable, revealing how language can both hint at deeper truths and acknowledge its own limitations.

Paradox, doubt, and the limits of language reveal the boundaries of human understanding, challenging us to confront the mysteries that lie beyond rationality. Recursive Realism provides a framework for engaging with these challenges, using self-referential processes to explore how the search for knowledge inevitably leads to encounters with the unknown. By embracing the recursive nature of our inquiry, Recursive Realism offers a vision of understanding that recognises the value of uncertainty and the richness of mystery, showing that the journey of knowledge is both a quest for truth and an acceptance of the mysteries that remain.

In Chapter 44, we have explored how Recursive Realism provides insights into the limits of human knowledge, revealing how self-referential processes shape our engagement with complexity, infinity, and paradox. By examining how the act of knowing involves confronting boundaries and navigating uncertainties, Recursive Realism offers a framework for understanding how the mysteries of existence can inspire both inquiry and wonder. This perspective allows us to see knowledge as an ongoing dialogue between certainty and the unknown, where the recursive nature of thought enables a deeper engagement with the profound questions that lie at the heart of reality.

References

Aristotle (1984) *Physics*. Translated by R.P. Hardie and R.K. Gaye. Princeton: Princeton University Press.

Cantor, G. (1955) *Contributions to the Founding of the Theory of Transfinite Numbers*. New York: Dover Publications.

Descartes, R. (1998) *Discourse on Method and Meditations on First Philosophy*. Translated by D.A. Cress. 4th edn. Indianapolis: Hackett Publishing.

Gödel, K. (1931) 'On Formally Undecidable Propositions of Principia Mathematica and Related Systems', *Monatshefte für Mathematik und Physik*, 38, pp. 173–198.

Heisenberg, W. (1958) *Physics and Philosophy: The Revolution in Modern Science*. New York: Harper & Row.

Lorenz, E.N. (1963) 'Deterministic Nonperiodic Flow', *Journal of the Atmospheric Sciences*, 20(2), pp. 130–141. doi:10.1175/1520-0469(1963)020<0130:DNF>2.0.CO;2.

Prigogine, I. (1984) *Order Out of Chaos: Man's New Dialogue with Nature*. New York: Bantam Books.

Rumi, J. (2004) *The Essential Rumi*. Translated by C. Barks. San Francisco: HarperOne.

Tao, T. (2008) *Structure and Randomness: Pages from Year One of a Mathematical Blog*. Providence: American Mathematical Society.

Tegmark, M. (2014) *Our Mathematical Universe: My Quest for the Ultimate Nature of Reality*. New York: Alfred A. Knopf.

Wheeler, J.A. (1990) 'Information, Physics, Quantum: The Search for Links', in Zurek, W.H. (ed.) *Complexity, Entropy, and the Physics of Information*. Redwood City: Addison-Wesley, pp. 309–336.

Zenkei, S. (1971) *Zen Comments on the Mumonkan*. New York: Harper & Row.

Zeno of Elea (2010) *Fragments and Paradoxes*. Edited by S. Hargreaves. London: Routledge.

Chapter 45: Recursive Realism and the Role of Imagination in Shaping Reality

The Nature of Imagination and the Recursion of Creative Thought

Imagination is often considered a uniquely human ability, allowing us to envision possibilities, create new worlds, and redefine reality. Recursive Realism suggests that imagination involves self-referential processes, where the mind engages in a feedback loop between memory, perception, and creative synthesis, allowing for the generation of new ideas and novel insights. This section explores how Recursive Realism provides insights into the nature of imagination, focusing on the recursive interplay between experience and creation, the role of imagination in shaping perception, and the ways in which imagination contributes to the evolution of culture.

Imagination as a Recursive Process of Reflection and Synthesis

Imagination allows individuals to combine elements of past experiences with novel ideas, creating mental representations that are not directly tied to the present moment. Recursive Realism suggests that imaginative thought involves a self-referential loop, where the mind revisits memories, modifies details, and projects new possibilities, creating a cycle that expands the boundaries of thought.

Mental Imagery and the Recursion of Perceptual Elements: Mental imagery, the ability to visualise scenes, objects, or concepts in the mind's eye, involves a recursive process, where sensory experiences are reassembled and reimagined in new configurations. Recursive Realism suggests that creating mental images involves a feedback loop, where perceptual details are recombined in novel ways, allowing for a synthesis of imagination and memory. For example, when imagining a landscape, the mind might draw on memories of past environments, modifying them to create a new scene that has never been directly perceived. This perspective highlights how imagination uses recursive reflection to generate new possibilities, expanding the realm of experience beyond the confines of immediate perception.

Daydreaming and the Recursion of Thought Patterns: Daydreaming, the act of letting the mind wander through varied scenarios, demonstrates how imagination engages in a recursive loop, where thoughts shift between current concerns, future aspirations, and fantastical ideas. Recursive Realism suggests that daydreaming involves a cycle where the mind explores different possibilities, revisiting previous thoughts and modifying scenarios based on new desires or emerging themes. For example, someone might imagine a different outcome to a past conversation, adjusting the details each time they revisit the scene. This perspective highlights how daydreaming serves as a form of recursive creativity, allowing the mind to experiment with variations of reality and to explore different outcomes through imaginative reflection.

Creative Problem-Solving and the Recursion of Idea Generation: Creativity often involves the ability to approach problems from new angles, using imagination to generate novel solutions. Recursive Realism suggests that creative problem-solving involves a self-referential loop, where the mind revisits existing knowledge, modifies approaches, and tests new ideas, creating a cycle that allows for breakthroughs in thinking. For example, scientists working on a complex problem might cycle through different hypotheses, adjusting their models until a new insight emerges that resolves the issue. This perspective allows for a deeper appreciation of how imagination plays a crucial role in expanding the limits of what is possible, using recursive reflection to transform old problems into new opportunities for discovery.

The Role of Imagination in Shaping Perception and Worldviews

Imagination not only allows for the creation of new ideas but also shapes how we perceive reality, influencing our understanding of the world and our place within it. Recursive Realism suggests that imagination involves a feedback loop between our mental models and our sensory experiences, where new interpretations of reality are tested, refined, and integrated into a coherent worldview.

Imagination as a Lens for Perception: Imagination can alter the way we perceive reality by providing a framework through which new experiences are interpreted. Recursive Realism suggests that imaginative frameworks involve a feedback loop, where the mind's expectations shape how sensory information is processed, creating a cycle where imagination influences perception. For example, reading a work of fiction can lead to imagining the world differently, influencing how we perceive social dynamics, nature, or even ourselves. This perspective highlights how imagination serves as a lens through which reality is filtered, using recursive reflection to transform the raw data of perception into a meaningful narrative.

Myths, Metaphors, and the Recursion of Cultural Imagination: Myths and metaphors are products of collective imagination that shape cultural worldviews and influence how societies understand their place in the universe. Recursive Realism suggests that myths and metaphors involve a feedback loop, where stories are retold, reinterpreted, and reimagined, creating a cycle that continually reshapes cultural understanding. For example, creation myths often reflect a society's understanding of the cosmos, evolving over time as new interpretations emerge. This perspective allows for a richer appreciation of how cultural imagination shapes worldviews, using recursive storytelling to explore the mysteries of existence and to create a shared sense of meaning.

Imagination's Role in Scientific Paradigms: Science is often seen as grounded in empirical evidence, yet the development of new theories often relies on the creative power of imagination. Recursive Realism suggests that scientific paradigms are shaped by a feedback loop between observations and the imaginative models that scientists create to explain phenomena, creating a cycle where theories evolve as imaginative insights are tested and refined. For example, Einstein's theory of relativity began as a thought experiment, imagining what it would be like to ride a beam of light, and later developed into a rigorous scientific model that transformed our understanding of space-time. This perspective highlights how imagination is integral to scientific advancement, using recursive thought experiments to explore the boundaries of reality.

Imagination and the Evolution of Culture Through Recursive Creativity

Imagination is a driving force behind cultural evolution, allowing societies to reinterpret traditions, invent new forms of expression, and adapt to changing circumstances. Recursive Realism suggests that cultural imagination involves a self-referential process, where creative expressions, such as art, literature, and philosophy, are revisited and recontextualised, creating a feedback loop that shapes the direction of cultural change.

Artistic Movements and the Recursion of Aesthetic Innovation: Artistic movements, from Renaissance art to Modernism, often represent a reimagining of aesthetics, where artists revisit existing forms and transform them through new creative insights. Recursive Realism suggests that artistic innovation involves a feedback loop, where each new style emerges through a dialogue with previous traditions, creating a cycle that allows for continuous evolution. For example, Impressionism challenged classical artistic norms by reimagining the use of light and colour, leading to a recursive transformation of artistic perception that influenced the development of subsequent movements like Post-Impressionism and Abstract Expressionism. This perspective highlights how cultural imagination uses recursive creativity to expand the boundaries of expression, creating new ways of seeing and understanding the world.

Philosophical Ideas and the Recursion of Conceptual Revolutions: Philosophy is often driven by imaginative insights that challenge existing paradigms, leading to new ways of thinking about reality and human experience. Recursive Realism suggests that philosophical revolutions involve a self-referential loop, where new ideas revisit previous concepts, creating a cycle of reinterpretation that leads to conceptual shifts. For example, the shift from Medieval Scholasticism to Enlightenment thinking involved a reimagining of knowledge, where thinkers like Descartes and Kant challenged traditional views of reason and existence, creating a recursive dialogue between past and present. This perspective allows for a deeper appreciation of how philosophy evolves through recursive reflection, using imaginative thought to push the boundaries of intellectual inquiry.

Cultural Adaptation and the Recursion of Tradition and Innovation: Cultures adapt to changing environments and new challenges by reimagining traditions and inventing new practices. Recursive Realism suggests that cultural adaptation involves a feedback loop, where traditions are revisited, modified, and reapplied to new contexts, creating a cycle of innovation and continuity. For example, oral storytelling traditions adapt to new cultural narratives, incorporating modern themes while preserving core elements of mythology. This perspective highlights how imagination plays a vital role in cultural resilience, using recursive creativity to navigate the balance between honouring the past and embracing the future.

Imagination is a powerful force that allows for the creation of new possibilities, the reinterpretation of reality, and the evolution of culture. Recursive Realism provides a framework for exploring how imaginative thought involves self-referential processes, where ideas are revisited, modified, and synthesised into new visions of reality. By recognising the recursive nature of imagination, Recursive Realism offers a vision of creativity that is dynamic, interconnected, and central to the process of shaping the world.

Recursive Realism and the Power of Imagination in Science, Art, and Culture

Imagination plays a critical role in driving progress in science, art, and culture, allowing individuals and societies to conceive possibilities beyond the limitations of current knowledge and experience. Recursive Realism suggests that the power of imagination lies in its recursive nature, where each creative insight generates new perspectives, which are then refined through reflection and interaction with the world. This section explores how Recursive Realism reveals the interplay between imagination and reality, focusing on the role of imagination in scientific breakthroughs, artistic innovation, and the shaping of cultural identities.

Imagination's Role in Scientific Discovery and the Recursion of Inquiry

Scientific discovery often begins with imaginative thinking, where new hypotheses and theories are conceived through thought experiments, speculative models, and creative interpretations of data. Recursive Realism suggests that scientific imagination involves a feedback loop, where ideas are proposed, tested, and refined through empirical validation, creating a cycle that leads to new understandings of the natural world.

Thought Experiments and the Recursion of Hypothetical Thinking: Thought experiments allow scientists to imagine scenarios that may not be practically testable, using imagination to explore the implications of scientific principles. Recursive Realism suggests that thought experiments involve a self-referential loop, where hypothetical scenarios are reflected upon and adjusted, creating a cycle of exploration and refinement. For example, Einstein's thought experiments on relativity, such as imagining riding alongside a beam of light, led to the conceptual breakthroughs that reshaped our understanding of space-time. This perspective highlights how thought experiments use imaginative reflection to probe the limits of current theories, allowing for breakthroughs that transform the scientific paradigm.

Model-Building and the Recursion of Theoretical Constructs: Scientific models, such as the structure of DNA, quantum field theories, or climate models, often begin as imaginative constructs that seek to capture complex phenomena in a coherent framework. Recursive Realism suggests that model-building involves a feedback loop, where models are imagined, tested against data, and refined based on empirical observations, creating a cycle that gradually aligns theory with reality. For example, the development of the double-helix model of DNA by Watson and Crick involved imagining the possible arrangements of nucleotides and examining X-ray diffraction patterns, leading to a recursive refinement of their initial ideas. This perspective allows for a deeper appreciation of how theoretical constructs are shaped through imaginative iteration, revealing how scientific models evolve through a recursive dialogue between creativity and empirical data.

Exploring the Unseen and the Recursion of Speculative Physics: Speculative branches of physics, such as string theory, multiverse theories, or the search for dark matter, rely heavily on imagination to conceptualise phenomena that are beyond the reach of current technology. Recursive Realism suggests that speculative physics involves a feedback loop, where imaginative concepts, like extra dimensions or parallel universes, are explored through mathematics and simulations, creating a cycle of hypothesis and re-evaluation. For example, string theory imagines that fundamental particles are vibrating strings in higher-dimensional space, leading to a recursive exploration of the mathematical structures that might underpin the universe. This perspective highlights how speculative imagination drives the boundaries of scientific inquiry, using recursive thinking to probe the limits of our understanding of the cosmos.

Imagination's Role in Art and the Recursion of Aesthetic Innovation

Artistic expression is often seen as the purest form of imagination, where new forms, styles, and meanings are created through the interaction between the artist's vision and the cultural context. Recursive Realism suggests that artistic creativity involves a feedback loop, where each work of art is reinterpreted and reimagined through dialogue with the artist's influences, the audience's perceptions, and the evolving art movements, creating a cycle that continually reshapes the landscape of aesthetic expression.

The Evolution of Styles and the Recursion of Artistic Movements: Artistic movements often emerge through a process of reflecting on past styles, challenging conventions, and imagining new ways of expressing the human experience. Recursive Realism suggests that the evolution of artistic styles involves a self-referential loop, where each generation of artists revisits the work of their predecessors, creating a cycle of reinterpretation and innovation. For example, Cubism, pioneered by Picasso and Braque, involved a reimagining of perspective that challenged traditional representations of space and form, leading to a recursive transformation of the visual arts. This perspective highlights how artistic movements use imagination to expand the boundaries of aesthetic perception, creating new ways of seeing the world through the interplay between tradition and novelty.

Abstract Art and the Recursion of Non-Representational Imagination: Abstract art, which moves away from literal representation to focus on form, colour, and texture, demonstrates how imagination can reshape the concept of art itself. Recursive Realism suggests that abstract art involves a feedback loop, where the artist's imaginative vision interacts with the viewer's interpretation, creating a cycle of meaning-making that is open-ended and subjective. For example, Jackson Pollock's drip paintings invite the viewer to see patterns in seemingly random splatters, engaging in a recursive process of finding form within chaos. This perspective allows for a richer appreciation of how abstract art uses imagination to explore the boundaries of perception, revealing how art can transform reality through the act of creative reimagining.

Literature, Storytelling, and the Recursion of Narrative Imagination: Literature and storytelling allow for the creation of worlds and characters that engage the reader's imagination, offering new perspectives on life and human nature. Recursive Realism suggests that narrative imagination involves a feedback loop, where stories are interpreted and reinterpreted through the reader's experiences, creating a cycle that shapes cultural understanding and personal insight. For example, science fiction often imagines future societies and technological possibilities, inviting readers to reflect on the implications of current trends. This perspective highlights how literature uses recursive storytelling to expand the scope of imagination, offering new ways of understanding the human condition and the nature of reality.

Imagination's Role in Culture and the Recursion of Identity Formation

Cultural identities are shaped by imaginative visions of who we are, where we come from, and what we aspire to become. Recursive Realism suggests that cultural imagination involves a feedback loop, where myths, symbols, and narratives are revisited and reimagined through the lens of history, social change, and personal interpretation, creating a cycle that continually reshapes collective identity.

National Myths and the Recursion of Collective Identity: National myths, stories about a people's origins, values, and destinies, play a central role in shaping cultural identity. Recursive Realism suggests that national myths involve a feedback loop, where each retelling of a nation's story adapts the narrative to new historical contexts, creating a cycle of reimagining the past to inform the present. For example, the myth of the American frontier has been reinterpreted through different eras, influencing ideas about individualism, progress, and freedom. This perspective highlights how cultural imagination uses recursive storytelling to adapt myths to new social realities, shaping collective identity through the interaction between past ideals and present challenges.

Symbolic Imagination and the Recursion of Cultural Symbols: Symbols, such as national flags, religious icons, or cultural artefacts, carry imaginative meanings that are continuously reinterpreted. Recursive Realism suggests that cultural symbols involve a self-referential loop, where each generation attaches new significance to old symbols, creating a cycle of cultural reinterpretation. For example, the yin-yang symbol in Chinese philosophy has been used to represent the balance of opposites, yet its meaning has evolved as it has been reinterpreted in new philosophical and cultural contexts. This perspective allows for a deeper appreciation of how cultural symbols use imagination to adapt to changing values, revealing how the process of cultural evolution is driven by recursive reflection on shared symbols.

Utopian Imagination and the Recursion of Social Vision: Utopian visions, imaginary societies where ideals of justice, equality, or harmony are realised, have inspired social movements and reform efforts throughout history. Recursive Realism suggests that utopian imagination involves a feedback loop, where visions of a better world are conceived, challenged, and refined through the interaction with social realities, creating a cycle that shapes the aspirations of future generations. For example, Thomas More's Utopia provided a blueprint for imagining a society without private property, influencing socialist thought and modern critiques of capitalism. This perspective highlights how imagination is central to the evolution of social ideals, using recursive thinking to explore the possibilities of human progress.

Imagination is a powerful force that drives innovation, cultural evolution, and the creation of new realities. Recursive Realism provides a framework for understanding how imaginative thought involves self-referential processes, where ideas are continually reimagined, refined, and tested through interaction with the world. By recognising the recursive nature of imagination, Recursive Realism offers a vision of creativity that is dynamic, interconnected, and central to our ability to transform reality. This perspective allows us to see imagination as not only a reflection of the world but also a force that shapes and redefines what is possible.

In Chapter 45, we have explored how Recursive Realism provides insights into the role of imagination in shaping reality, revealing how self-referential processes enable the creative power of imaginative thought. By examining how imagination shapes scientific discovery, artistic innovation, and cultural identities, Recursive Realism offers a framework for understanding how the mind's capacity for creativity allows us to envision new worlds, redefine boundaries, and transform the possibilities of human experience. This perspective allows us to see imagination as a recursive process that continually reimagines reality, using creativity to expand the horizons of knowledge and the depth of human understanding.

References

Baars, B.J. (1988) *A Cognitive Theory of Consciousness*. Cambridge: Cambridge University Press.

Bohm, D. (1980) *Wholeness and the Implicate Order*. London: Routledge.

Einstein, A. (1938) *The Evolution of Physics*. Co-authored with L. Infeld. New York: Simon and Schuster.

Fauconnier, G. and Turner, M. (2002) *The Way We Think: Conceptual Blending and the Mind's Hidden Complexities*. New York: Basic Books.

Freud, S. (1908) 'Creative Writers and Day-Dreaming', *The Standard Edition of the Complete Psychological Works of Sigmund Freud*, Volume IX (1906-1908). Translated and edited by J. Strachey. London: Hogarth Press.

Jung, C.G. (1964) *Man and His Symbols*. London: Aldus Books.

Kearney, R. (1988) *The Wake of Imagination: Toward a Postmodern Culture*. Minneapolis: University of Minnesota Press.

Koestler, A. (1964) *The Act of Creation*. London: Hutchinson & Co.

Lakoff, G. and Johnson, M. (1980) *Metaphors We Live By*. Chicago: University of Chicago Press.

More, T. (1516) *Utopia*. Translated and edited by R. M. Adams (1992). Cambridge: Cambridge University Press.

Picasso, P. and Braque, G. (2001) *Cubism*. Edited by E. Cowling. London: Thames & Hudson.

Pollock, J. (1956) *Jackson Pollock: Paintings and Drawings*. Edited by F.V. O'Connor. New York: Museum of Modern Art.

Popper, K.R. (1963) *Conjectures and Refutations: The Growth of Scientific Knowledge*. London: Routledge.

Turner, M. (1996) *The Literary Mind: The Origins of Thought and Language*. New York: Oxford University Press.

Wheeler, J.A. (1990) 'Information, Physics, Quantum: The Search for Links', in Zurek, W.H. (ed.) *Complexity, Entropy, and the Physics of Information*. Redwood City: Addison-Wesley, pp. 309–336.

Chapter 46: Recursive Realism and the Dynamics of Memory and Future Thinking

The Nature of Memory and the Recursion of Reflective Recall

Memory plays a central role in how we understand ourselves, our history, and the continuity of our experiences. Recursive Realism suggests that memory is not a static record of past events but involves self-referential processes, where each act of remembering reshapes the memory itself, creating a feedback loop that influences our sense of identity and our understanding of the world. This section explores how Recursive Realism provides insights into the nature of memory, focusing on the recursive dynamics of autobiographical memory, the role of memory in shaping selfhood, and the interplay between memory and meaning.

Autobiographical Memory and the Recursion of Personal Narratives

Autobiographical memory refers to the recollection of events and experiences that are central to the formation of personal identity. Recursive Realism suggests that autobiographical memory involves a feedback loop, where each act of recall revisits and reinterprets past experiences, creating a cycle that continuously reshapes the narrative of the self.

The Reconstructive Nature of Memory: Memory is often described as reconstructive, meaning that recalling an event involves reassembling fragments of sensory information and emotions, rather than simply retrieving a stored record. Recursive Realism suggests that this reconstructive process involves a self-referential loop, where each act of remembering alters the memory based on current emotions, new insights, or the passage of time. For example, revisiting a childhood memory as an adult might involve reinterpreting the experience in light of new understanding, creating a recursive process where the past is continuously reshaped by the present. This perspective highlights how autobiographical memory is not fixed but dynamic, using recursive reflection to maintain a sense of continuity while allowing for the evolution of personal narratives.

Memory as a Narrative Process and the Recursion of Selfhood: Autobiographical memory allows individuals to construct a narrative of their lives, connecting disparate events into a cohesive story that defines who they are. Recursive Realism suggests that the narrative nature of memory involves a feedback loop, where each new experience prompts a reconfiguration of the life story, creating a cycle that maintains a sense of self while adapting to new contexts. For example, a person might reinterpret a past failure as a turning point in their personal growth, allowing for a narrative shift that changes how the past is understood. This perspective highlights how the self is not static but continuously constructed through a recursive dialogue between memory and new experiences, revealing how identity evolves through the ongoing interplay of past and present.

The Recursion of Meaning in Memory: Memory is not only about recalling facts but also about assigning meaning to past events, interpreting their significance within the broader context of one's life. Recursive Realism suggests that the search for meaning in memory involves a feedback loop, where each reflection on a past event leads to a deeper understanding of its significance, creating a cycle where the meaning of experiences evolves over time. For example, the memory of a difficult relationship might initially be associated with pain but later be seen as a source of resilience and wisdom, reshaping its meaning as the individual revisits and reinterprets it. This perspective allows for a richer appreciation of how the process of remembering involves recursively reflecting on the meaning of events, revealing how our understanding of the past is malleable and responsive to our ongoing search for purpose.

Memory's Role in Time Perception and the Recursion of Temporal Continuity

Memory shapes our perception of time, providing a sense of continuity that connects past experiences with present awareness. Recursive Realism suggests that the perception of time involves self-referential processes, where memories are revisited and reintegrated, creating a feedback loop that allows for the construction of temporal continuity.

The Recursion of Time and the Mental Timeline: Human cognition creates a mental timeline that allows us to organise events into a coherent sequence, maintaining a sense of past, present, and future. Recursive Realism suggests that this mental timeline involves a feedback loop, where each memory is recontextualised within the larger framework of personal history, creating a cycle that shapes how we perceive the flow of time. For example, when recalling a significant life event, an individual might place it in relation to other key moments, creating a temporal structure that defines the trajectory of their life. This perspective highlights how the mind's ability to perceive time is not linear but involves a recursive organisation of experiences, allowing for a flexible understanding of how past events relate to the present.

Temporal Anchors and the Recursion of Emotional Memory: Certain memories serve as temporal anchors, emotional experiences that shape our sense of time by creating points of reference in our mental timeline. Recursive Realism suggests that emotional memory involves a feedback loop, where strong emotions associated with particular events are revisited and reinterpreted, creating a cycle that influences how we understand the passage of time. For example, the memory of a joyful celebration or a traumatic loss might stand out as a significant marker in the narrative of one's life, influencing how time is perceived before and after the event. This perspective allows for a deeper understanding of how emotions shape memory, using recursive reflection to anchor our sense of continuity and temporal flow.

The Recursion of Collective Memory and Cultural Time: Memory is not only individual but also collective, with shared memories shaping how cultures perceive time and understand their history. Recursive Realism suggests that collective memory involves a self-referential process, where societies revisit and reinterpret key events in their history, creating a feedback loop that shapes cultural identity. For example, national holidays that commemorate historical events involve a recursive reflection on the past, allowing societies to redefine the meaning of these events in light of current values. This perspective highlights how collective memory is dynamic, using recursive storytelling to create a shared sense of time that connects the past with the present and future aspirations.

Memory is a dynamic process that involves self-referential loops, allowing for the continuous reimagining of the past and the construction of temporal continuity. Recursive Realism provides a framework for exploring how the mind's ability to reflect on memories shapes our understanding of time, identity, and meaning. By recognising the recursive nature of memory, Recursive Realism reveals how our sense of self is continually redefined through the ongoing interaction between past experiences and present reflection, allowing for a deeper understanding of how we perceive time and our place within it.

Recursive Realism and the Dynamics of Future Thinking

Future thinking is a fundamental aspect of human cognition, allowing us to anticipate possibilities, plan for outcomes, and adapt to change. Recursive Realism suggests that the mind's capacity for envisioning the future involves self-referential processes, where imagined scenarios are revisited and revised in light of new experiences and reflective insights, creating a feedback loop that shapes how we perceive potential futures. This section explores how Recursive Realism provides a framework for understanding the nature of anticipation, focusing on the interplay between imagination and prediction, the role of future thinking in decision-making, and the ways in which anticipation shapes resilience and adaptability.

Imagination and the Recursion of Anticipation

Anticipation involves imagining possible futures based on current knowledge and past experiences, allowing individuals to prepare for opportunities and avoid potential risks. Recursive Realism suggests that anticipation involves a self-referential loop, where the mind envisions scenarios, tests their plausibility, and adjusts them based on new information, creating a cycle that shapes our understanding of what the future might hold.

Scenario Building and the Recursion of Predictive Models: Scenario building is a cognitive process where individuals imagine different potential outcomes and evaluate their likelihood. Recursive Realism suggests that scenario building involves a feedback loop, where each imagined outcome is revisited and refined based on new data or changing circumstances, creating a cycle that allows for a nuanced understanding of possible futures. For example, when planning for a business venture, an entrepreneur might consider different market conditions, adjust their strategies, and reimagine scenarios as they gather more information. This perspective highlights how imagination plays a central role in future thinking, using recursive reflection to balance creativity with practicality.

Imagining Best-Case and Worst-Case Scenarios: Future thinking often involves imagining both the best-case and worst-case scenarios, allowing individuals to prepare for a range of possibilities. Recursive Realism suggests that the process of envisioning extremes involves a self-referential cycle, where each imagined scenario prompts a reconsideration of the assumptions underlying it, creating a feedback loop that refines our expectations of the future. For example, a scientist considering the implications of climate change might imagine worst-case outcomes such as extreme weather events, and best-case scenarios like successful mitigation efforts, using each vision to guide policy recommendations. This perspective highlights how recursive thinking allows for a balanced approach to future uncertainties, integrating imagination with strategic foresight.

Future Thinking as a Bridge Between Memory and Possibility: Anticipation is inherently linked to memory, as the recollection of past experiences informs our expectations of what might happen next. Recursive Realism suggests that future thinking involves a feedback loop between memory and imagination, where each new vision of the future is shaped by lessons from the past, creating a cycle that grounds imaginative projections in real-world experiences. For example, a person who has experienced financial difficulties might anticipate future risks more conservatively, adjusting their plans based on past challenges. This perspective allows for a deeper appreciation of how memory and imagination work together in the recursive process of envisioning the future, creating a continuum between what has been and what might be.

Decision-Making and the Recursion of Planning

Decision-making is a critical aspect of future thinking, involving the evaluation of options and the selection of actions that will shape future outcomes. Recursive Realism suggests that decision-making involves a feedback loop, where choices are revisited and adjusted based on ongoing reflection, creating a cycle that allows for adaptation and refinement of strategies.

Strategic Planning and the Recursion of Adjustments: Strategic planning involves setting long-term goals and creating plans of action to achieve them. Recursive Realism suggests that effective planning involves a feedback loop, where initial plans are revisited and revised as new information becomes available, creating a cycle that allows for flexibility and resilience. For example, a company's strategic plan might involve adjusting marketing strategies in response to changing market trends, creating a recursive process where goals are continuously refined. This perspective highlights how decision-making is not a one-time event but involves recursive cycles of evaluation, allowing for a dynamic approach to complex challenges.

Risk Assessment and the Recursion of Uncertainty: Assessing risks is a key component of decision-making, where individuals weigh the potential costs and benefits of different actions. Recursive Realism suggests that risk assessment involves a feedback loop, where each evaluation of potential risks leads to adjustments in strategy, creating a cycle that balances caution with opportunity. For example, investors in financial markets continually reevaluate risks based on market fluctuations, adjusting their portfolios to align with new predictions. This perspective highlights how risk assessment uses recursive thinking to navigate uncertainty, allowing for adaptation to changing conditions through a process of continual reassessment.

The Role of Reflection in Decision-Making: Reflection allows individuals to reconsider past decisions and adjust future plans based on what they have learned. Recursive Realism suggests that reflection involves a self-referential loop, where each new insight prompts a reassessment of previous choices, creating a cycle of learning that shapes future decisions. For example, after failing to achieve a goal, a person might reflect on their approach, identify mistakes, and adjust their strategy for future efforts. This perspective highlights how reflection is a crucial part of effective decision-making, using recursive processes to integrate past experiences into future-oriented actions.

Resilience, Adaptability, and the Recursion of Future Thinking

Resilience and adaptability are critical for navigating uncertainty and responding to change, allowing individuals and societies to adjust to new circumstances and maintain continuity in the face of challenges. Recursive Realism suggests that resilience involves self-referential processes, where future thinking allows for the continuous revision of plans and expectations, creating a feedback loop that enhances the ability to adapt.

Adapting to Change Through Recursive Learning: Adaptability involves the ability to learn from new experiences and adjust behaviour in response to changing environments. Recursive

Realism suggests that adaptability involves a feedback loop, where each new experience prompts a reassessment of strategies, creating a cycle that allows for ongoing adjustment. For example, during a global crisis, such as a pandemic, individuals and organisations must adapt their behaviours to align with new safety protocols, using recursive reflection to update their responses as conditions change. This perspective highlights how adaptability relies on future thinking that is dynamic and responsive, allowing for a continuous alignment between plans and real-world challenges.

Resilience as a Cycle of Recovery and Growth: Resilience is often defined as the ability to recover from setbacks and emerge stronger. Recursive Realism suggests that resilience involves a self-referential loop, where each recovery involves revisiting past challenges, learning from adversity, and using those lessons to shape future efforts, creating a cycle of growth and adaptation. For example, communities affected by natural disasters often build greater resilience by learning from past events, creating recovery plans that are better suited for future threats. This perspective allows for a deeper appreciation of how resilience is not merely a static trait but involves an ongoing process of recursive learning, allowing for the integration of past experiences into future preparation.

Building a Vision of the Future Through Recursive Ideals: Future thinking is often guided by visions and ideals that provide a sense of direction and purpose. Recursive Realism suggests that the creation of a vision involves a feedback loop, where ideals are imagined, tested against reality, and revised to align with changing circumstances, creating a cycle that shapes long-term goals. For example, social movements might create a vision of a more just society, using recursively updated strategies to align their goals with political realities. This perspective highlights how visions of the future use recursive thinking to remain flexible yet purpose-driven, allowing for a dynamic approach to building a better world.

Future thinking is a dynamic process that involves self-referential loops, allowing for the continual adjustment of plans and visions in light of new information and reflective insights. Recursive Realism provides a framework for exploring how anticipation, decision-making, and resilience are shaped by recursive reflection, revealing how the mind's capacity for imagining possibilities allows for adaptation to the uncertainties of life. By recognising the recursive nature of future thinking, Recursive Realism offers a vision of cognition that is responsive, adaptive, and oriented towards growth, allowing for a deeper understanding of how we shape the future through the iterative process of reflection and imagination.

In Chapter 46, we have explored how Recursive Realism provides insights into the dynamic interplay between memory and future thinking, revealing how self-referential processes shape our understanding of time, continuity, and possibility. By examining how memory involves the recursive reimagining of the past, and how future thinking shapes anticipation and adaptation, Recursive Realism offers a framework for understanding how the mind navigates the temporal flow of experience. This perspective allows us to see time as a continuum shaped by the recursive interplay between what has been and what might be, using reflection and imagination to build a coherent narrative of the self and the world.

References

Addis, D. R., Wong, A. T., & Schacter, D. L., 2007. Remembering the past and imagining the future: Common and distinct neural substrates during event construction and elaboration. *Neuropsychologia*, 45(7), pp.1363–1377.

Bar, M., 2007. The proactive brain: Using analogies and associations to generate predictions. *Trends in Cognitive Sciences*, 11(7), pp.280–289.

Boyd, R. & Richerson, P. J., 1985. *Culture and the Evolutionary Process*. Chicago: University of Chicago Press.

Conway, M. A. & Pleydell-Pearce, C. W., 2000. The construction of autobiographical memories in the self-memory system. *Psychological Review*, 107(2), pp.261–288.

Gilbert, D. T. & Wilson, T. D., 2007. Prospection: Experiencing the future. *Science*, 317(5843), pp.1351–1354.

Hassabis, D. & Maguire, E. A., 2007. Deconstructing episodic memory with construction. *Trends in Cognitive Sciences*, 11(7), pp.299–306.

Klein, S. B., 2013. The temporal orientation of memory: It's time for a change of direction. *Journal of Applied Research in Memory and Cognition*, 2(4), pp.222–234.

McAdams, D. P., 2001. The psychology of life stories. *Review of General Psychology*, 5(2), pp.100–122.

Suddendorf, T. & Corballis, M. C., 2007. The evolution of foresight: What is mental time travel, and is it unique to humans? *Behavioral and Brain Sciences*, 30(3), pp.299–313.

Tulving, E., 2002. Episodic memory: From mind to brain. *Annual Review of Psychology*, 53(1), pp.1–25.

Chapter 47: Recursive Realism and the Relationship Between Language, Thought, and Reality

The Nature of Language and the Recursion of Meaning

Language is fundamental to human thought, allowing us to encode ideas, share experiences, and construct models of reality. Recursive Realism suggests that language is not merely a tool for communication but involves self-referential processes, where the meaning of words evolves through a cycle of interpretation and recontextualisation. This section explores how Recursive Realism provides insights into the nature of language, focusing on the recursive dynamics of meaning-making, the role of metaphor in shaping thought, and the interplay between language and perception.

The Dynamics of Meaning-Making and the Recursion of Interpretation

Meaning-making is a process through which words and symbols are interpreted to convey ideas and experiences. Recursive Realism suggests that meaning-making involves a feedback loop, where each use of language revisits the context and cultural assumptions that shape its interpretation, creating a cycle that allows for the evolution of meaning over time.

Semantic Networks and the Recursion of Language Context: Words derive their meaning through their relationships to other words within a language system, creating semantic networks that form the basis of communication. Recursive Realism suggests that understanding a word involves a self-referential process, where each word's meaning is revisited through its connections to other terms, creating a cycle that allows for nuanced interpretations. For example, the word freedom carries different connotations depending on cultural context and philosophical perspective, and its meaning evolves as it is re-examined in new social and political settings. This perspective highlights how language is not static but involves recursive networks of meaning, allowing for rich interpretations that adapt to changing cultural dynamics.

The Evolution of Language Through Recursive Use: Languages change over time as new words are introduced, old words take on new meanings, and dialects merge or diverge. Recursive Realism suggests that language evolution involves a feedback loop, where each generation of speakers revisits the meanings of words and phrases, creating a cycle that adapts language to new realities. For example, words related to technology, such as cloud, stream, or tweet, have taken on new meanings as they have been recontextualised in the digital age. This perspective highlights how language evolves through recursive cycles of redefinition, allowing for a dynamic adaptation to new concepts and cultural shifts.

Interpretation as a Recursive Process of Understanding: Interpreting language involves a self-referential loop, where the listener or reader integrates new information with existing knowledge, creating a feedback cycle that shapes understanding. Recursive Realism suggests that interpretation is not linear but involves a recursive dialogue between the text and the interpreter's perspective, allowing for multiple layers of meaning. For example, reading a poem involves interpreting each line in the context of the whole poem, and revisiting previous lines as new meanings emerge. This perspective allows for a deeper appreciation of how language creates recursive opportunities for insight, revealing how meaning is constructed through an ongoing cycle of reflection and reinterpretation.

Metaphor and the Recursion of Conceptual Frameworks

Metaphor is a fundamental aspect of language, allowing us to understand abstract concepts through concrete imagery. Recursive Realism suggests that metaphors involve self-referential processes, where conceptual frameworks are revisited and extended, creating a cycle that shapes how we think about reality.

Metaphorical Thinking and the Recursion of Abstract Concepts: Metaphors allow us to connect unfamiliar ideas to known experiences, creating bridges between abstract concepts and tangible imagery. Recursive Realism suggests that metaphorical thinking involves a feedback loop, where each metaphor reshapes our understanding of both the source and the target concept, creating a cycle that deepens conceptual insights. For example, the metaphor "time is a river" allows us to imagine time as flowing, but also reframes our understanding of rivers as symbolic of change and continuity. This perspective highlights how metaphors use recursive reflection to reorganise concepts, allowing us to perceive reality through new lenses.

Extending Metaphors and the Recursion of Analogies: Analogies and extended metaphors play a key role in scientific theories, allowing complex ideas to be explained through more familiar concepts. Recursive Realism suggests that extending a metaphor involves a feedback loop, where the analogy is tested, refined, and expanded, creating a cycle that shapes our understanding of the subject matter. For example, the metaphor of the mind as a computer has been extended through analogies with information processing, influencing cognitive science and artificial intelligence research. This perspective allows for a richer appreciation of how metaphors can transform disciplines by revisiting familiar ideas to reimagine complex phenomena, revealing how metaphorical thinking is central to conceptual innovation.

The Role of Metaphor in Shaping Worldviews: Metaphors are not only linguistic tools but also shape worldviews by influencing how we interpret reality. Recursive Realism suggests that the metaphors we use create a feedback loop, where each metaphor shapes our worldview, which in turn influences the metaphors we create, forming a cycle that shapes cultural understanding. For example, viewing the earth as a mother has shaped environmental ethics, creating a recursive relationship between language and our attitudes toward nature. This perspective highlights how metaphors can act as catalysts for cultural change, using recursive thinking to reshape our understanding of the world and our place within it.

Language, Perception, and the Recursion of Reality

Language shapes how we perceive reality, providing a framework for interpreting sensory experiences and understanding the world. Recursive Realism suggests that the relationship between language and perception involves self-referential processes, where each word or concept influences how we perceive the world, creating a feedback loop that shapes the boundaries of reality.

The Sapir-Whorf Hypothesis and the Recursion of Linguistic Relativity: The Sapir-Whorf hypothesis, which suggests that language influences thought, explores how different languages create different cognitive frameworks for understanding reality. Recursive Realism suggests that linguistic relativity involves a feedback loop, where the structure of a language shapes perception, and new experiences reshape the use of language, creating a cycle that influences cognitive development. For example, speakers of languages that distinguish between many shades of a colour might be more sensitive to visual differences, creating a recursive interaction between language and perception. This perspective allows for a deeper appreciation of how language is not merely descriptive but actively shapes how we experience the world, using recursive thought to explore the nuances of reality.

Naming and the Recursion of Object Identity: The act of naming involves assigning words to objects and phenomena, giving them distinct identities within our cognitive framework. Recursive Realism suggests that naming involves a feedback loop, where each name creates a category that shapes how we perceive similarities and differences, creating a cycle that structures our understanding of the environment. For example, naming a particular species of plant influences how we classify and study its characteristics, shaping our understanding of biodiversity. This perspective highlights how language creates boundaries within the natural world, using recursive categorisation to define the contours of knowledge.

Words as Tools for Constructing Reality: Language not only describes the world but also constructs realities through the creation of social norms, legal definitions, and scientific concepts. Recursive Realism suggests that the constructive power of language involves a self-referential process, where words create frameworks that shape our actions and beliefs, creating a cycle that influences how society evolves. For example, terms like "freedom of speech" or "human rights" shape social expectations and legal frameworks, creating a recursive dialogue between language and cultural values. This perspective allows for a richer understanding of how language is not passive but actively participates in the construction of reality, using recursive reflection to shape our collective vision of what is possible.

Language is a dynamic system that involves self-referential loops, allowing for the continuous reinterpretation of words, metaphors, and conceptual frameworks. Recursive Realism provides a framework for exploring how language shapes our understanding of reality, revealing how meaning-making is an evolving process that balances stability with innovation. By recognising the recursive nature of language, Recursive Realism offers a vision of communication that is interconnected, adaptive, and deeply influential in shaping our perception of the world.

Recursive Realism and the Role of Language in Shaping Thought and Culture

Language is not only a means of communication but also a tool for shaping thought and influencing culture. Recursive Realism suggests that the relationship between language and thought involves self-referential processes, where each new expression shapes how we think, and new ways of thinking influence how we use language, creating a feedback loop that drives intellectual development and cultural evolution. This section explores how Recursive Realism reveals the role of language in shaping social norms, philosophical concepts, and the narratives that define cultural identity.

Language as a Medium for Recursive Reflection and Social Norms

Social norms, the shared expectations of behaviour within a community, are shaped by the language we use to describe our values, rules, and ideals. Recursive Realism suggests that the establishment of social norms involves a feedback loop, where language both reflects and reinforces societal values, creating a cycle that allows norms to evolve in response to changing social conditions.

The Recursion of Norms Through Everyday Language: Everyday language serves as a vehicle for the transmission of norms, where repeated phrases and expressions shape expectations about acceptable behaviour. Recursive Realism suggests that norms are reinforced through a feedback loop, where each use of language reflects social standards, and social standards influence the words and phrases that become culturally accepted. For example, terms like "politically correct" or "woke" reflect current social expectations about inclusivity, and their usage shapes how individuals perceive their responsibilities toward marginalised groups. This perspective highlights how language acts as a recursive mechanism for norm reinforcement, creating a dynamic interplay between how we speak and how we behave.

Changing Norms and the Recursion of Language Reform: Language reform, efforts to change terms and phrases to align with new social values, demonstrates how language can drive social change through recursive processes. Recursive Realism suggests that language reform involves a feedback loop, where new terms challenge old norms, creating a cycle of linguistic adaptation that reshapes cultural standards. For example, shifts in gender pronouns aim to reflect a broader understanding of gender identity, prompting revisions in how society speaks about individuals. This perspective allows for a deeper appreciation of how language is a dynamic force in cultural evolution, using recursive reflection to align words with shifting values.

Social Narratives and the Recursion of Collective Beliefs: Social narratives, the stories that communities tell about themselves, are shaped by the language used to describe events, values, and aspirations. Recursive Realism suggests that social narratives involve a feedback loop, where stories shape beliefs, and beliefs in turn influence the stories that are told, creating a cycle that shapes cultural identity. For example, narratives about national resilience or struggles for justice shape how societies view themselves and their role in the world, influencing political discourse and public policy. This perspective highlights how language is integral to the construction of cultural narratives, using recursive storytelling to shape the collective understanding of history and values.

Philosophical Concepts and the Recursion of Abstract Thought

Philosophical thinking often explores abstract concepts like existence, freedom, and truth, using language to articulate ideas that are difficult to define. Recursive Realism suggests that philosophical inquiry involves a feedback loop, where language shapes conceptual frameworks, and conceptual shifts influence how language is used, creating a cycle that deepens our understanding of complex ideas.

Defining Abstract Concepts Through Recursive Debate: Philosophers often engage in debates over the definitions of abstract concepts, using language to explore the nuances of meaning. Recursive Realism suggests that these debates involve a feedback loop, where each definition is revisited and refined through critique and counter-argument, creating a cycle that leads to more precise understanding. For example, debates over the nature of free will have evolved through a recursive dialogue between deterministic perspectives and arguments for autonomy, with each side redefining terms to address new philosophical challenges. This perspective highlights how philosophical concepts are not fixed but involve recursive refinement, allowing for a deeper exploration of the nature of reality.

Language and the Recursion of Epistemological Inquiry: Epistemology, the study of knowledge and belief, relies on language to articulate the conditions of knowing. Recursive Realism suggests that epistemological inquiry involves a feedback loop, where each attempt to define knowledge is revisited as new perspectives emerge, creating a cycle of reflection and adjustment. For example, the shift from empiricism to postmodern critiques of objectivity has involved a recursive questioning of what it means to know, with language adapting to new models of understanding. This perspective allows for a richer appreciation of how language shapes our understanding of knowledge, using recursive processes to explore the nature of certainty and the limits of understanding.

Philosophical Language as a Medium for Conceptual Expansion: Philosophical language often seeks to push the boundaries of what can be articulated, using new terms and metaphors to explore the edges of thought. Recursive Realism suggests that the expansion of philosophical language involves a feedback loop, where each new concept invites further reflection and redefinition, creating a cycle that allows for the continuous development of ideas. For example, the introduction of terms like "being" (Heidegger) or "the Other" (Levinas) has led to recursive explorations of existence and ethical responsibility, shaping the discourse around ontology and ethics. This perspective highlights how philosophical language serves as a tool for recursive thought, using words to explore the profound questions that lie at the heart of human experience.

Cultural Narratives and the Recursion of Identity Formation

Cultural narratives provide a framework for understanding who we are as individuals and as members of a community, using stories and symbols to create a sense of belonging. Recursive Realism suggests that cultural narratives involve self-referential processes, where each generation reinterprets the stories of the past, creating a feedback loop that allows for the evolution of identity and shared meaning.

Myths, Legends, and the Recursion of Cultural Memory: Myths and legends serve as anchors for cultural memory, providing narratives that explain origins, moral values, and cultural achievements. Recursive Realism suggests that myths involve a feedback loop, where each retelling adapts the story to new contexts, creating a cycle that allows for the preservation and evolution of cultural memory. For example, Greek myths like Prometheus have been reinterpreted through philosophical and artistic lenses, influencing concepts of rebellion, human progress, and the pursuit of knowledge. This perspective highlights how myths use recursive storytelling to maintain cultural continuity while allowing for new interpretations that reflect the changing values of society.

Literature and the Recursion of Cultural Self-Understanding: Literature allows societies to explore their values, critique their norms, and imagine alternative realities, using stories to reflect the complexities of human life. Recursive Realism suggests that literature involves a feedback loop, where each story is revisited by new generations, creating a cycle of reinterpretation that shapes cultural self-understanding. For example, Shakespeare's plays have been adapted and reinterpreted countless times, with each adaptation reflecting the concerns and perspectives of a new cultural context. This perspective allows for a deeper appreciation of how literature serves as a recursive dialogue between the past and the present, using storytelling to explore the evolving nature of human identity.

National Narratives and the Recursion of Identity Politics: National narratives shape how societies understand themselves, defining their place in history and their aspirations for the future. Recursive Realism suggests that national identity involves a feedback loop, where each generation reexamines the stories of national struggle, triumph, or injustice, creating a cycle that adapts identity to current challenges. For example, the Civil Rights Movement in the United States redefined American identity by reinterpreting the principles of freedom and equality, shaping national narratives about justice and inclusion. This perspective highlights how cultural identity is not fixed but involves a dynamic process of revisiting and reinterpreting stories, using recursive reflection to align national values with the realities of the present.

Language is a powerful tool that shapes our understanding of ourselves, our values, and our world. Recursive Realism provides a framework for exploring how self-referential processes shape the interaction between language and thought, revealing how words and concepts create feedback loops that influence social norms, philosophical ideas, and cultural identities. By recognising the recursive nature of language, Recursive Realism offers a vision of communication that is interconnected, dynamic, and capable of evolving with changing social realities. This perspective allows us to see language as not merely descriptive but a creative force that actively participates in shaping the contours of human experience.

In Chapter 47, we have explored how Recursive Realism provides insights into the relationship between language and thought, revealing how self-referential processes shape our understanding of reality, our engagement with abstract concepts, and our collective sense of identity. By examining how language creates feedback loops that influence meaning, social norms, and cultural narratives, Recursive Realism offers a framework for understanding how the words we use shape our perception of the world and our interactions with each other. This perspective allows us to see language as a recursive process that continually reshapes reality, using reflection and imagination to navigate the complexities of existence.

References

Atance, C.M. and O'Neill, D.K. (2001) 'Episodic future thinking', *Trends in Cognitive Sciences*, 5(12), pp. 533-539. https://doi.org/10.1016/S1364-6613(00)01804-0.

Addis, D.R., Wong, A.T. and Schacter, D.L. (2007) 'Remembering the past and imagining the future: Common and distinct neural substrates during event construction and elaboration', *Neuropsychologia*, 45(7), pp. 1363-1377. https://doi.org/10.1016/j.neuropsychologia.2006.10.016.

Conway, M.A. and Pleydell-Pearce, C.W. (2000) 'The construction of autobiographical memories in the self-memory system', *Psychological Review*, 107(2), pp. 261-288. https://doi.org/10.1037/0033-295X.107.2.261.

Tulving, E. (2002) 'Episodic memory: From mind to brain', *Annual Review of Psychology*, 53(1), pp. 1-25. https://doi.org/10.1146/annurev.psych.53.100901.135114.

Schacter, D.L., Addis, D.R. and Buckner, R.L. (2008) 'Episodic simulation of future events: Concepts, data, and applications', *Annals of the New York Academy of Sciences*, 1124(1), pp. 39-60. https://doi.org/10.1196/annals.1440.001.

Boyer, P. (2008) 'Evolutionary economics of mental time travel?', *Trends in Cognitive Sciences*, 12(6), pp. 219-224. https://doi.org/10.1016/j.tics.2008.03.003.

Suddendorf, T. and Corballis, M.C. (2007) 'The evolution of foresight: What is mental time travel, and is it unique to humans?', *Behavioral and Brain Sciences*, 30(3), pp. 299-351. https://doi.org/10.1017/S0140525X07001975.

Buckner, R.L. and Carroll, D.C. (2007) 'Self-projection and the brain', *Trends in Cognitive Sciences*, 11(2), pp. 49-57. https://doi.org/10.1016/j.tics.2006.11.004.

Gilbert, D.T. and Wilson, T.D. (2007) 'Prospection: Experiencing the future', *Science*, 317(5843), pp. 1351-1354. https://doi.org/10.1126/science.1144161.

Gopnik, A. (2010) 'How we know our minds: The illusion of first-person knowledge of intentionality', *Behavioral and Brain Sciences*, 16(1), pp. 1-14. https://doi.org/10.1017/S0140525X0003944X.

Levine, B., Svoboda, E., Hay, J.F. and Winocur, G. (2002) 'Aging and autobiographical memory: Dissociating episodic from semantic retrieval', *Psychology and Aging*, 17(4), pp. 677-689. https://doi.org/10.1037/0882-7974.17.4.677.

Smallwood, J. and Schooler, J.W. (2006) 'The restless mind', *Psychological Bulletin*, 132(6), pp. 946-958. https://doi.org/10.1037/0033-2909.132.6.946.

Baumeister, R.F. and Vohs, K.D. (2002) 'The pursuit of meaningfulness in life', *Psychological Inquiry*, 13(1), pp. 1-15. https://doi.org/10.1207/S15327965PLI1301_01.

Mischel, W., Shoda, Y. and Rodriguez, M.L. (1989) 'Delay of gratification in children', *Science*, 244(4907), pp. 933-938. https://doi.org/10.1126/science.2658056.

Rubin, D.C., Wetzler, S.E. and Nebes, R.D. (1986) 'Autobiographical memory across the lifespan', *Advances in Psychology*, 25, pp. 202-221. https://doi.org/10.1016/S0166-4115(08)62029-4.

Spreng, R.N., Mar, R.A. and Kim, A.S. (2009) 'The common neural basis of autobiographical memory, prospection, navigation, theory of mind, and the default mode: A quantitative meta-analysis', *Journal of Cognitive Neuroscience*, 21(3), pp. 489-510. https://doi.org/10.1162/jocn.2008.21029.

Chapter 48: Recursive Realism and the Interplay Between Consciousness and the Unconscious

The Nature of Consciousness and the Recursion of Self-Awareness

Consciousness is the foundation of human experience, allowing us to perceive, reflect, and engage with the world. Recursive Realism suggests that consciousness involves self-referential processes, where awareness loops back on itself, creating a cycle that allows for reflection, introspection, and the formation of selfhood. This section explores how Recursive Realism provides insights into the nature of consciousness, focusing on the recursive dynamics of self-awareness, the relationship between attention and perception, and the role of conscious thought in shaping reality.

The Dynamics of Self-Awareness and the Recursion of the Reflective Mind

Self-awareness, the ability to reflect on one's own thoughts and experiences, is a defining feature of human consciousness. Recursive Realism suggests that self-awareness involves a feedback loop, where conscious thought reflects on itself, creating a cycle that allows for the formation of a cohesive sense of self.

The Mirror of the Mind: Recursion in Reflective Thought: Reflective thought allows us to think about our own thinking, creating a self-referential loop that shapes our understanding of our own mental processes. Recursive Realism suggests that this process involves a cycle, where each act of reflection deepens our awareness of our thoughts, creating a layered understanding of the self. For example, when contemplating a decision, an individual might reflect on their motivations, and then reflect on why they hold those motivations, creating a recursive examination of their desires and values. This perspective highlights how self-awareness is not a static state but involves continuous cycles of introspection, allowing for a richer understanding of the mind.

Recursive Feedback and the Development of Self-Identity: Self-identity, the sense of who we are as individuals, develops through a recursive process of reflecting on experiences, values, and memories. Recursive Realism suggests that the formation of self-identity involves a feedback loop, where each new experience is integrated into the narrative of the self, creating a cycle that allows for growth and adaptation. For example, a person might revisit past memories in light of current beliefs, leading to a reinterpretation of their past that reshapes their understanding of themselves. This perspective allows for a deeper appreciation of how self-identity is a dynamic construct, using recursive reflection to maintain a sense of continuity while adapting to new experiences.

Introspection and the Recursion of Conscious Awareness: Introspection, the process of examining one's inner thoughts and feelings, allows for a deeper engagement with the contents of the mind. Recursive Realism suggests that introspection involves a feedback loop, where each act of self-observation leads to further insights, creating a cycle that allows for the refinement of self-knowledge. For example, during meditation, individuals might observe their thoughts and emotions as they arise, creating a recursive process of awareness that deepens their understanding of mental patterns. This perspective highlights how introspection is not merely passive but involves a dynamic engagement with the mind, using recursive reflection to explore the depths of consciousness.

The Role of Attention in Shaping Perception and the Recursion of Focus

Attention is the mechanism through which consciousness selects what to focus on, allowing us to prioritise certain aspects of our experience over others. Recursive Realism suggests that attention involves self-referential processes, where the act of focusing on one aspect of experience influences how we perceive others, creating a feedback loop that shapes our awareness of the world.

Attention as a Recursive Process of Filtering Experience: Attention functions as a filter, determining which sensory inputs and thoughts become the focus of conscious awareness. Recursive Realism suggests that this filtering process involves a feedback loop, where each shift in attention changes the context of perception, creating a cycle that allows for adaptive focus. For example, when listening to music, focusing on the melody might change how we hear the rhythm, creating a recursive adjustment of how different elements of the music are perceived. This perspective highlights how attention is not fixed but involves a dynamic process of shifting focus, using recursive engagement to shape our experience of the present moment.

The Recursion of Focus and the Creation of Flow States: Flow states, moments of intense focus and engagement with an activity, demonstrate how attention can shape the quality of consciousness. Recursive Realism suggests that flow states involve a feedback loop, where sustained attention creates a cycle of engagement, allowing for a seamless integration of action and awareness. For example, during creative writing, a person might become so absorbed in their thoughts that time seems to disappear, creating a recursive process where each thought seamlessly leads to the next. This perspective highlights how attention can create profoundly immersive experiences, using recursive focus to align conscious thought with the flow of creativity.

Mindfulness and the Recursion of Present Awareness: Mindfulness, the practice of bringing one's full attention to the present moment, involves a recursive process of observing thoughts and letting them go, allowing for a clear perception of experience. Recursive Realism suggests that mindfulness involves a feedback loop, where the practice of attention creates a cycle of self-awareness that clarifies the contents of consciousness. For example, focusing on the breath during meditation allows for a recursive awareness of thoughts, emotions, and bodily sensations, creating a deeper understanding of the mind's patterns. This perspective allows for a richer appreciation of how mindfulness uses recursive reflection to enhance clarity and presence, shaping the nature of conscious experience.

The Role of Conscious Thought in Shaping Reality and the Recursion of Intentions

Conscious thought allows us to form intentions, make decisions, and shape our reality through deliberate action. Recursive Realism suggests that intentional thought involves a feedback loop, where plans and goals are revisited and adjusted, creating a cycle that shapes how we engage with the world.

Intentions as a Recursive Process of Goal Formation: Setting intentions involves imagining future outcomes and organising actions to achieve them. Recursive Realism suggests that goal formation involves a self-referential loop, where each new intention is revisited in light of new experiences, creating a cycle that allows for flexibility and persistence. For example, a person working on a long-term project might revise their goals as new challenges arise, creating a recursive adjustment of their intentions. This perspective highlights how intentions shape our reality through an ongoing process of reflection and adaptation, using recursive thought to align plans with changing circumstances.

Decision-Making and the Recursion of Reflective Choice: Decision-making involves choosing between options based on anticipated outcomes and personal values. Recursive Realism suggests that decision-making involves a feedback loop, where each choice is reconsidered through reflective thought, creating a cycle that refines our understanding of what is best. For example, choosing a career path might involve considering different possibilities, reflecting on past experiences, and revising decisions as new opportunities arise. This perspective allows for a deeper appreciation of how decisions are not isolated events but involve a recursive dialogue between goals, reflection, and experience.

Conscious Thought as a Tool for Shaping Reality: Conscious thought allows us to imagine possibilities, create plans, and manifest change through deliberate action. Recursive Realism suggests that the power of thought to shape reality involves a self-referential loop, where thoughts influence actions, and the outcomes of actions influence future thoughts, creating a cycle of world-building. For example, envisioning a new invention involves imagining how it might function, testing prototypes, and adjusting the design based on feedback, creating a recursive process that brings ideas into reality. This perspective highlights how the mind's ability to reflect on possibilities is central to the creative process, using recursive thought to transform imagination into action.

Consciousness is a dynamic process that involves self-referential loops, allowing for the continuous refinement of self-awareness, attention, and intentional thought. Recursive Realism provides a framework for exploring how the mind's ability to reflect on its own processes shapes our understanding of the self and the world. By recognising the recursive nature of conscious thought, Recursive Realism reveals how our sense of self, our focus, and our goals are continually redefined through the interplay between reflection and action, allowing for a deeper engagement with the complexities of experience.

Recursive Realism and the Depths of the Unconscious Mind

The unconscious mind is a domain that remains largely hidden from direct awareness, yet it profoundly shapes our thoughts, behaviours, and emotional responses. Recursive Realism suggests that the unconscious is not separate from conscious thought but is intertwined with it through self-referential processes, where patterns of thought and emotion are revisited and transformed in non-conscious ways. This section explores how Recursive Realism provides insights into the dynamic interplay between conscious awareness and unconscious processes, focusing on the role of dreams, the nature of creative intuition, and the hidden patterns that shape our behaviour.

The Role of Dreams in the Recursion of the Mind

Dreams are one of the most mysterious manifestations of the unconscious mind, offering glimpses into hidden desires, fears, and symbolic representations of our inner world. Recursive Realism suggests that dreaming involves self-referential processes, where the mind revisits experiences, recombines images, and constructs narratives, creating a feedback loop that allows for psychological integration and emotional processing.

Dreams as a Recursion of Symbolic Thought: Dreams often contain symbols and metaphors that represent deeper emotional truths, using imagery to explore aspects of the self that may be difficult to confront directly. Recursive Realism suggests that the creation of dream symbols involves a feedback loop, where the unconscious mind revisits emotional experiences and reframes them through symbolic representation, creating a cycle that allows for psychological exploration. For example, a recurring dream about falling might represent feelings of insecurity, with each iteration of the dream modifying the context to reveal new layers of meaning. This perspective highlights how dreams use recursive symbolism to provide insight into unconscious processes, offering a deeper understanding of the mind's hidden dynamics.

The Recursion of Memory in Dreaming: Dreams often draw on fragments of memory, recombining elements of past experiences into new scenarios. Recursive Realism suggests that this recombination process involves a feedback loop, where memories are revisited and recontextualised in light of current emotional states, creating a cycle that shapes how past events are integrated into the present self. For example, dreaming about a childhood home while experiencing major life changes might reflect a recursive reflection on feelings of safety or nostalgia. This perspective highlights how dreams serve as a space where the mind's recursive processes allow for the reimagining of past experiences, using the symbolic language of the unconscious to integrate memory and emotion.

Dreams as a Recursive Dialogue Between Conscious and Unconscious: Lucid dreaming, the ability to be aware of and control one's dreams, demonstrates how the boundary between conscious and unconscious thought can become fluid. Recursive Realism suggests that lucid dreaming involves a feedback loop, where conscious awareness interacts with the imagery of the unconscious mind, creating a cycle that allows for direct engagement with the dream narrative. For example, in a lucid dream, a person might change the outcome of a recurring nightmare, creating a recursive transformation of the dream's emotional impact. This perspective allows for a richer appreciation of how the conscious mind can interact with the unconscious, using recursive reflection to explore the hidden realms of the psyche.

Creative Intuition and the Recursion of Insight

Creativity often draws on intuitive processes that emerge from the unconscious, allowing for novel ideas and inspirations that seem to arise spontaneously. Recursive Realism suggests that creative intuition involves self-referential processes, where patterns of thought are revisited and reassembled in ways that transcend logical reasoning, creating a feedback loop that allows for the emergence of new insights.

The Recursion of Unconscious Associations: Creative intuition often involves the ability to see connections between seemingly unrelated ideas, drawing on the mind's capacity for forming associations below the level of conscious thought. Recursive Realism suggests that these associative processes involve a feedback loop, where the unconscious mind revisits stored memories and concepts, creating new combinations that rise into consciousness as creative insights. For example, a scientist might suddenly realise a solution to a problem after dreaming about a metaphorical scenario that mirrors the scientific challenge. This perspective highlights how creativity is not merely the product of conscious effort but involves recursive interactions with the deep structures of the unconscious mind.

Eureka Moments and the Recursion of Cognitive Breakthroughs: Eureka moments, sudden flashes of insight, demonstrate how the mind can solve complex problems in a seemingly instant leap. Recursive Realism suggests that these moments of insight involve a self-referential loop, where the unconscious mind processes patterns and connections until a solution becomes conscious, creating a cycle that allows for breakthroughs. For example, mathematician Henri Poincaré described how solutions to complex equations would suddenly appear to him after a period of relaxation, suggesting that the unconscious mind was working recursively through the problem. This perspective allows for a deeper appreciation of how creative breakthroughs emerge from the interplay between unconscious processing and conscious reflection, using recursive thought to transcend logical barriers.

Artistic Inspiration and the Recursion of Imaginary Worlds: Artists often describe feeling guided by a muse or inner voice, suggesting that the creative process involves a dialogue with the unconscious. Recursive Realism suggests that artistic inspiration involves a feedback loop, where imaginary worlds and visions emerge from unconscious processes and are shaped through conscious refinement, creating a cycle that allows for the creation of new forms. For example, a painter might visualise a scene in a dream and later develop it into a detailed artwork, creating a recursive process where unconscious imagery is transformed through conscious artistry. This perspective highlights how art uses recursive interaction with the unconscious mind to explore the boundaries of imagination.

Hidden Patterns of the Unconscious and the Recursion of Behaviour

The unconscious mind shapes our behaviour in ways that are often outside of conscious awareness, influencing our habits, reactions, and emotional responses. Recursive Realism suggests that the patterns of the unconscious involve self-referential processes, where repeated experiences create deep-seated patterns that influence our actions, creating a feedback loop that shapes how we navigate the world.

The Recursion of Habits and the Unconscious Mind: Habits, automatic patterns of behaviour, are shaped by repeated experiences and reinforcement over time. Recursive Realism suggests that habit formation involves a feedback loop, where each repetition of a behaviour reinforces neural pathways, creating a cycle that makes the habit more automatic. For example, a person who regularly practices a musical instrument might find that certain finger movements become second nature, reflecting the unconscious mastery of a skill through recursive practice. This perspective highlights how the unconscious mind uses recursive processes to create stable patterns of behaviour, allowing for the development of expertise and automaticity.

Emotional Conditioning and the Recursion of Subconscious Responses: Emotional responses are often conditioned by past experiences, with certain stimuli triggering automatic feelings or reactions. Recursive Realism suggests that emotional conditioning involves a feedback loop, where the mind revisits past emotional associations, creating a cycle that shapes how we respond to new situations. For example, someone who has experienced trauma might have automatic fear responses to specific triggers, even if they are not consciously aware of the association. This perspective allows for a richer appreciation of how the unconscious mind shapes emotional responses through recursive associations, revealing how past experiences continue to influence the present.

The Unconscious as a Source of Intuitive Wisdom: Intuition, the ability to understand something instinctively, without conscious reasoning, often draws on the hidden depths of the unconscious mind. Recursive Realism suggests that intuition involves a feedback loop, where the unconscious mind integrates patterns from past experiences, creating a cycle that allows for fast, adaptive decisions. For example, a seasoned chess player might intuitively recognise the best move in a complex position, reflecting a recursive refinement of pattern recognition over years of practice. This perspective highlights how the unconscious mind uses recursive processes to provide insights that can guide conscious thought, allowing for a seamless interaction between intuition and analysis.

The unconscious mind is a dynamic realm that interacts with conscious thought through self-referential loops, shaping our dreams, creative insights, and intuitive responses. Recursive Realism provides a framework for understanding how the mind's hidden processes influence our awareness and actions, revealing how patterns and associations are continually refined through recursive cycles of reflection and reinterpretation. By recognising the recursive nature of the unconscious, Recursive Realism offers a vision of the mind that is interconnected, adaptive, and deeply shaped by the interplay between consciousness and the unknown depths of the psyche.

In Chapter 48, we have explored how Recursive Realism provides insights into the dynamic interplay between conscious awareness and the unconscious mind, revealing how self-referential processes shape our thoughts, dreams, and intuitive insights. By examining how the unconscious interacts with conscious thought through feedback loops, Recursive Realism offers a framework for understanding how the mind's hidden depths shape our experiences and creativity, allowing us to navigate the complexities of human psychology. This perspective allows us to see the mind as a recursive system that continually reimagines itself, using reflection and intuitive understanding to explore the mysteries of existence.

References

Baars, B. J. (1988) *A Cognitive Theory of Consciousness*. Cambridge: Cambridge University Press.

Brown, J. W. (2014) *The Self-Embodying Mind: A Neuropsychological Perspective*. 2nd edn. New York: Psychology Press.

Chalmers, D. J. (1996) *The Conscious Mind: In Search of a Fundamental Theory*. Oxford: Oxford University Press.

Freud, S. (1915) *The Unconscious*. Translated by J. Riviere. London: Hogarth Press.

Gazzaniga, M. S. (2018) *The Consciousness Instinct: Unraveling the Mystery of How the Brain Makes the Mind.* New York: Farrar, Straus and Giroux.

Hobson, J. A. (2002) *Dreaming: An Introduction to the Science of Sleep.* Oxford: Oxford University Press.

James, W. (1890) *The Principles of Psychology.* Vol. 1. New York: Henry Holt and Company.

Kahneman, D. (2011) *Thinking, Fast and Slow.* New York: Farrar, Straus and Giroux.

Lakoff, G. and Johnson, M. (1980) *Metaphors We Live By.* Chicago: University of Chicago Press.

Penrose, R. (1994) *Shadows of the Mind: A Search for the Missing Science of Consciousness.* Oxford: Oxford University Press.

Poincaré, H. (1952) *Science and Method.* Translated by F. Maitland. New York: Dover Publications.

Reber, A. S. (1993) *Implicit Learning and Tacit Knowledge: An Essay on the Cognitive Unconscious.* Oxford: Oxford University Press.

Schneider, S. and Velmans, M. (eds.) (2007) *The Blackwell Companion to Consciousness.* Malden: Wiley-Blackwell.

Thompson, E. (2007) *Mind in Life: Biology, Phenomenology, and the Sciences of Mind.* Cambridge: Belknap Press.

Zeman, A. (2001) *Consciousness: A User's Guide.* New Haven: Yale University Press.

Chapter 49: Recursive Realism and the Integration of Science, Philosophy, and Spirituality

The Intersection of Science and Recursive Realism

Science provides a systematic approach to understanding the natural world, relying on empirical evidence and testable hypotheses. Recursive Realism suggests that scientific inquiry is inherently self-referential, involving a feedback loop where theories are revisited, refined, and adjusted in response to new data, creating a dynamic cycle that drives the evolution of knowledge. This section explores how Recursive Realism provides insights into the nature of scientific inquiry, focusing on the role of recursion in theory formation, the convergence of disciplines, and the boundaries of empirical knowledge.

Recursion in Theory Formation and the Evolution of Scientific Ideas

Scientific theories often evolve through cycles of hypothesis, experimentation, and revision, allowing for the progressive refinement of our understanding of the universe. Recursive Realism suggests that this process involves a feedback loop, where each new finding prompts a reassessment of existing models, creating a cycle that allows for the continuous evolution of theoretical frameworks.

The Recursive Nature of Scientific Models: Scientific models, such as theories of gravity, quantum mechanics, or evolutionary biology, are often refined through a recursive process, where each experiment challenges the assumptions of existing models, leading to adjustments and improvements. Recursive Realism suggests that the strength of scientific models lies in their ability to adapt through feedback, allowing for a dynamic alignment with empirical reality. For example, the shift from Newtonian mechanics to Einstein's theory of relativity involved a recursive refinement of our understanding of space-time, incorporating new insights into the nature of gravity. This perspective highlights how scientific inquiry uses recursive processes to deepen our understanding of the universe, allowing for a flexible and adaptive approach to knowledge.

Hypothesis Testing as a Recursive Dialogue with Nature: Hypothesis testing involves formulating predictions and testing them through controlled experiments, creating a recursive loop where each result informs the next hypothesis. Recursive Realism suggests that the process of testing is not linear but involves cycles of iteration, where scientists revisit their assumptions and adjust their models to align with observed phenomena. For example, the discovery of subatomic particles required a reimagining of atomic models, creating a recursive dialogue between theoretical predictions and empirical observations. This perspective allows for a deeper appreciation of how scientific knowledge is not static but evolves through a continuous interaction with the natural world, using recursive thinking to explore the unknown.

The Role of Recursion in Scientific Paradigm Shifts: Paradigm shifts, fundamental changes in scientific perspective, often arise through recursive processes, where anomalies challenge existing frameworks, leading to the emergence of new theories. Recursive Realism suggests that paradigm shifts involve a feedback loop, where each new insight reconfigures the conceptual structure of a discipline, creating a cycle that transforms the way we understand reality. For example, the shift from geocentric models to heliocentric astronomy involved a recursive questioning of the assumptions underlying cosmology, leading to a redefinition of our place in the universe. This perspective highlights how recursion is integral to scientific revolutions, allowing for the transformation of knowledge systems through a dynamic process of revaluation.

The Convergence of Disciplines and the Role of Interdisciplinary Inquiry

Interdisciplinary research, which combines insights from multiple fields, is increasingly important in addressing complex questions about the universe and human experience. Recursive Realism suggests that the convergence of disciplines involves self-referential processes, where each field revisits its concepts in light of new perspectives, creating a feedback loop that enriches our understanding of reality.

The Recursion of Concepts Across Scientific Disciplines: Many scientific concepts, such as entropy, symmetry, or emergence, appear in multiple disciplines, allowing for a recursive dialogue between fields. Recursive Realism suggests that these shared concepts create a feedback loop, where each discipline contributes new insights that shape how other fields understand the same phenomena. For example, the concept of entropy plays a role in thermodynamics, information theory, and evolutionary biology, creating a recursive exchange that enriches the interpretation of disorder and complexity. This perspective highlights how interdisciplinary inquiry uses recursive reflection to connect different domains, allowing for a more holistic understanding of the universe.

Systems Thinking and the Recursion of Complexity: Systems thinking, which focuses on the interactions between components of complex systems, relies on a recursive approach to understanding dynamics across scales and levels. Recursive Realism suggests that systems thinking involves a feedback loop, where models of complexity are revised as new connections are discovered, creating a cycle that allows for the integration of diverse insights. For example, understanding ecosystems requires a recursive examination of biotic and abiotic factors, feedback mechanisms, and emergent properties that shape the interactions within the natural world. This perspective allows for a richer appreciation of how systems thinking aligns with the principles of Recursive Realism, offering a way to explore the interconnectedness of life.

The Role of Recursion in the Boundaries of Empirical Knowledge: Science has boundaries defined by the limits of empirical observation, yet it often approaches questions that border on philosophical or metaphysical considerations. Recursive Realism suggests that the exploration of these boundaries involves a feedback loop, where scientific inquiry informs philosophical reflection, creating a cycle that pushes the limits of what can be known. For example, the study of quantum mechanics raises questions about the nature of reality, causality, and consciousness, prompting a recursive dialogue between physicists and philosophers. This perspective highlights how the interplay between science and philosophy can extend our understanding beyond the empirical, using recursive thinking to explore the frontiers of knowledge.

The Integration of Spirituality and the Recursion of Meaning

Spirituality often addresses the questions of purpose, existence, and the nature of reality that go beyond the scope of empirical science. Recursive Realism suggests that spiritual understanding involves self-referential processes, where beliefs and experiences are revisited in light of personal insight and cultural context, creating a feedback loop that allows for the deepening of spiritual awareness.

The Role of Recursion in Spiritual Reflection: Spiritual practices, such as meditation, prayer, or contemplation, often involve a recursive process of reflecting on the nature of existence, creating a dialogue between the self and a higher reality. Recursive Realism suggests that these practices involve a feedback loop, where each reflection deepens the understanding of spiritual truths, creating a cycle that shapes personal growth. For example, meditative practices in Buddhism involve revisiting thoughts and letting them go, creating a recursive process that allows for insight into the nature of mind and interconnectedness. This perspective highlights how spiritual reflection aligns with the principles of recursion, using self-referential thought to explore the mysteries of existence.

Mythology and the Recursion of Transcendent Narratives: Mythological stories often explore the relationship between humanity and the divine, providing narratives that address the origins and purpose of life. Recursive Realism suggests that mythological narratives involve a feedback loop, where each retelling adapts the story to align with new cultural values, creating a cycle that allows for the evolution of spiritual meaning. For example, the myth of the hero's journey has been reinterpreted in different religious traditions and literary works, offering a recursive framework for understanding the challenges of spiritual transformation. This perspective allows for a deeper appreciation of how mythology uses recursive storytelling to engage with the profound questions of life and death, revealing how cultural narratives shape spiritual understanding.

The Convergence of Spirituality and Scientific Inquiry Through Recursive Reflection: Some spiritual traditions seek to integrate scientific discoveries into their understanding of the universe, creating a dialogue between empirical knowledge and spiritual insight. Recursive Realism suggests that this convergence involves a feedback loop, where scientific concepts, such as the nature of the cosmos or the emergence of life, are reinterpreted within spiritual frameworks, creating a cycle that enriches both perspectives. For example, the exploration of cosmology has led to spiritual reflections on the nature of time and the origins of the universe, prompting a recursive exchange between astrophysics and theological reflection. This perspective highlights how Recursive Realism can serve as a bridge between science and spirituality, allowing for a holistic vision that respects both the empirical and the transcendent.

Science is a dynamic process that involves self-referential loops, allowing for the continuous refinement of theories and models through a feedback loop between empirical evidence and conceptual frameworks. Recursive Realism provides a framework for understanding how scientific inquiry evolves through a recursive dialogue with nature, revealing how the convergence of disciplines and the interplay with philosophy can push the boundaries of knowledge. By recognising the recursive nature of scientific thought, Recursive Realism offers a vision of knowledge that is adaptive, interconnected, and open to the mysteries that lie beyond empirical observation.

Recursive Realism and the Role of Philosophy and Spirituality

Philosophy and spirituality have long addressed the fundamental questions of existence, the nature of reality, and the search for meaning. Recursive Realism suggests that the inquiries of philosophy and spirituality involve self-referential processes, where each reflection deepens the understanding of fundamental concepts, creating a feedback loop that allows for the continuous evolution of thought and belief. This section explores how Recursive Realism provides insights into the nature of philosophical inquiry, the role of spirituality in understanding the self, and the integration of different ways of knowing into a cohesive worldview.

The Role of Philosophy in Recursive Reflection and the Nature of Existence

Philosophy seeks to explore the foundational principles that underlie our understanding of reality, using reason and conceptual analysis to address the nature of existence, consciousness, and the universe. Recursive Realism suggests that philosophical reflection involves self-referential processes, where each argument is revisited in light of new perspectives, creating a cycle that allows for the deepening of understanding.

Metaphysics and the Recursion of Ontological Inquiry: Metaphysics, the branch of philosophy concerned with the nature of being and the structure of reality, often involves recursive reflection on the fundamental nature of existence. Recursive Realism suggests that metaphysical inquiry involves a feedback loop, where each exploration of ontological questions, such as the nature of time, space, and causality, leads to further reflection, creating a cycle that deepens our understanding of what it means to exist. For example, philosophical debates about the nature of consciousness have evolved through recursive dialogues between materialist, dualistic, and panpsychist perspectives, with each position revisiting fundamental assumptions about the mind. This perspective highlights how Recursive Realism allows for a dynamic approach to metaphysical questions, using self-referential thought to explore the deepest aspects of reality.

Epistemology and the Recursion of Knowing: Epistemology, the study of knowledge and belief, relies on recursive processes of self-reflection, where the nature of knowing is examined and revised based on new insights. Recursive Realism suggests that epistemological inquiry involves a feedback loop, where each exploration of the conditions for knowledge leads to reconsiderations of what it means to understand and to believe, creating a cycle that refines our concepts of truth. For example, the debate between empiricism and rationalism involves a recursive examination of whether knowledge comes from sense experience or reason, with each side revisiting their assumptions in light of philosophical critiques. This perspective allows for a richer appreciation of how Recursive Realism provides a framework for understanding the nature of knowledge through recursive reflection, revealing how the search for truth is an ongoing process.

Ethics and the Recursion of Moral Reflection: Ethics, the study of moral values and principles, often involves a recursive process of examining actions and revisiting moral judgments, creating a cycle that allows for the refinement of moral understanding. Recursive Realism suggests that moral reflection involves a feedback loop, where ethical dilemmas prompt reconsiderations of what is right, leading to the evolution of moral principles. For example, debates over justice and equality have evolved through recursive dialogues that revisit the principles of fairness in light of new social contexts and historical experiences. This perspective highlights how Recursive Realism provides a way to explore the dynamic nature of ethics, allowing for a continuous engagement with the question of how we should live.

Spirituality, Self-Understanding, and the Recursion of Inner Wisdom

Spirituality often involves a search for deeper meaning, seeking to understand the nature of the self and its connection to the universe through inner reflection and contemplative practices. Recursive Realism suggests that spiritual understanding involves self-referential processes, where the individual revisits beliefs, experiences mystical insights, and deepens their sense of connection, creating a feedback loop that allows for the integration of spiritual wisdom.

Mystical Experience and the Recursion of the Self: Mystical experiences, moments of profound connection to the universe or a sense of oneness, often involve a deep engagement with the unconscious mind and a transcendence of ordinary perception. Recursive Realism suggests that mystical experiences involve a feedback loop, where the boundaries of the self are revisited and redefined through experiences of unity, creating a cycle that shapes spiritual awareness. For example, experiences of ego dissolution during meditation or altered states of consciousness allow for a recursive exploration of the nature of selfhood, offering insights into the interconnectedness of life. This perspective highlights how spiritual experiences use recursive processes to explore the depths of consciousness, revealing how the mind's interaction with the divine can reshape our understanding of existence.

Contemplative Practices and the Recursion of Inner Reflection: Meditation, prayer, and contemplation are central to many spiritual traditions, providing a framework for self-reflection and connection to a greater reality. Recursive Realism suggests that these practices involve a feedback loop, where each session of contemplation deepens the understanding of spiritual truths, creating a cycle that refines the experience of transcendence. For example, the practice of mindfulness involves a recursive focus on the present moment, allowing for a deeper awareness of thoughts and sensations that leads to a more profound sense of inner peace. This perspective allows for a richer appreciation of how spirituality uses recursive thought to explore the inner dimensions of the self, creating a space for spiritual growth and self-transcendence.

The Recursion of Belief and the Evolution of Spiritual Traditions: Religious and spiritual traditions often evolve through the reinterpretation of sacred texts, doctrines, and rituals. Recursive Realism suggests that the evolution of spiritual beliefs involves a feedback loop, where each generation revisits the teachings of the past to align them with new insights, creating a cycle that allows for the renewal of spiritual meaning. For example, the reformation movements within Christianity, Buddhism, or Islam have involved a recursive return to foundational texts, seeking a deeper understanding that resonates with contemporary experiences. This perspective highlights how Recursive Realism provides a framework for understanding the evolution of spiritual traditions, using recursive reflection to balance continuity with adaptation.

Integrating Science, Philosophy, and Spirituality: A Recursive Vision of Reality

Recursive Realism seeks to provide a unifying framework that allows for the integration of scientific discovery, philosophical inquiry, and spiritual insight, creating a cohesive vision of reality that embraces different ways of knowing. This integration is grounded in self-referential processes, where the principles of recursion allow for a dialogue between disciplines, creating a cycle that enriches our understanding of the universe.

Bridging the Gap Between Empirical and Metaphysical Inquiry: Science and spirituality often appear to address different realms, one focusing on the material world and the other on the transcendent. Recursive Realism suggests that the dialogue between empirical and metaphysical inquiry involves a feedback loop, where scientific insights inform spiritual perspectives, and spiritual questions challenge scientific assumptions, creating a cycle that expands the scope of understanding. For example, the exploration of consciousness through neuroscience raises questions about the nature of subjective experience, prompting a recursive reflection on the limits of materialist explanations. This perspective highlights how Recursive Realism allows for a holistic approach that respects both the empirical and the spiritual, offering a vision of reality that is grounded yet open to mystery.

The Role of Philosophy as a Mediator in the Recursive Dialogue: Philosophy often serves as a bridge between science and spirituality, providing conceptual tools to explore questions of meaning, existence, and the nature of knowledge. Recursive Realism suggests that philosophy plays a key role in the recursive dialogue between different modes of understanding, creating a feedback loop that allows for the integration of insights from diverse fields. For example, philosophical concepts like emergence and non-duality provide a framework for exploring how consciousness arises from physical processes, while also addressing the spiritual notion of interconnectedness. This perspective highlights how Recursive Realism allows for a synthesis of philosophical reflection, scientific inquiry, and spiritual wisdom, creating a cohesive vision of reality.

A Recursive Vision of Reality as a Continuum of Experience: Recursive Realism suggests that reality can be understood as a continuum, where the material, the conceptual, and the spiritual are not separate domains but interconnected aspects of a larger whole. This perspective involves a feedback loop, where each level of experience, from physical phenomena to abstract thought and spiritual insight, reflects and reframes the others, creating a cycle that allows for a deeper engagement with the universe. For example, understanding the nature of time involves exploring its physical aspects through relativity, its conceptual dimensions through philosophical thought, and its experiential qualities through spiritual contemplation. This perspective allows for a richer appreciation of how Recursive Realism offers a holistic framework that can encompass the full range of human experience, providing a unified approach to understanding reality.

Philosophy and spirituality are dynamic processes that involve self-referential loops, allowing for the continuous exploration of the nature of existence, the self, and the universe. Recursive Realism provides a framework for understanding how the inquiries of philosophy and spirituality evolve through a recursive dialogue with inner experiences and external knowledge, revealing how reflection and contemplation can deepen our understanding of the mysteries of life. By recognising the recursive nature of spiritual and philosophical thought, Recursive Realism offers a vision of reality that is integrated, adaptive, and capable of embracing both the known and the unknowable.

In Chapter 49, we have explored how Recursive Realism provides insights into the integration of science, philosophy, and spirituality, revealing how self-referential processes shape our understanding of reality, the search for meaning, and the ultimate questions of existence. By examining how the recursive dynamics of thought and reflection allow for a dialogue between disciplines, Recursive Realism offers a framework for understanding how the empirical, the conceptual, and the spiritual can come together to create a cohesive vision of the universe. This perspective allows us to see the universe as a continuum of interconnected experiences, using recursive reflection to navigate the complexities of knowledge and the mysteries of being.

References

Barrow, J. D. and Tipler, F. J. (1986) *The Anthropic Cosmological Principle*. Oxford: Oxford University Press.

Capra, F. (1996) *The Web of Life: A New Scientific Understanding of Living Systems*. New York: Anchor Books.

Chalmers, D. J. (1996) *The Conscious Mind: In Search of a Fundamental Theory*. Oxford: Oxford University Press.

Einstein, A. (1920) *Relativity: The Special and General Theory*. Translated by R. W. Lawson. New York: Henry Holt and Company.

Hawking, S. (1988) *A Brief History of Time: From the Big Bang to Black Holes*. London: Bantam Books.

Heidegger, M. (1962) *Being and Time*. Translated by J. Macquarrie and E. Robinson. Oxford: Blackwell.

Kuhn, T. S. (1970) *The Structure of Scientific Revolutions*. 2nd edn. Chicago: University of Chicago Press.

Lakoff, G. and Johnson, M. (1980) *Metaphors We Live By*. Chicago: University of Chicago Press.

Laszlo, E. (2004) *Science and the Akashic Field: An Integral Theory of Everything*. Rochester: Inner Traditions.

Poincaré, H. (1952) *Science and Method*. Translated by F. Maitland. New York: Dover Publications.

Rovelli, C. (2018) *The Order of Time*. London: Allen Lane.

Schneider, S. and Velmans, M. (eds.) (2007) *The Blackwell Companion to Consciousness*. Malden: Wiley-Blackwell.

Sheldrake, R. (2009) *A New Science of Life: The Hypothesis of Morphic Resonance*. 3rd edn. Rochester: Inner Traditions.

Thompson, E. (2007) *Mind in Life: Biology, Phenomenology, and the Sciences of Mind*. Cambridge: Belknap Press.

Zohar, D. and Marshall, I. (2000) *Spiritual Intelligence: The Ultimate Intelligence*. London: Bloomsbury.

Chapter 50: Recursive Realism and the Future of Human Thought

Recursive Realism and the Advancement of Scientific Knowledge

The future of scientific inquiry will be defined by its ability to integrate complexity, explore new paradigms, and address the fundamental questions of existence and reality. Recursive Realism suggests that advancing scientific knowledge involves self-referential processes, where each new discovery invites a deeper reflection on the nature of the universe, creating a feedback loop that allows for the expansion of scientific understanding. This section explores how Recursive Realism can shape the future of science, focusing on the potential for interdisciplinary breakthroughs, the exploration of consciousness, and the reimagining of fundamental concepts.

Interdisciplinary Breakthroughs and the Recursion of Scientific Paradigms

Scientific progress often arises through the convergence of insights from multiple disciplines, allowing for the synthesis of new ideas and the creation of comprehensive models. Recursive Realism suggests that the future of science will involve a feedback loop where each discipline revisits its foundations in light of advances in other fields, creating a cycle that allows for the emergence of new paradigms.

The Role of Recursion in Quantum Biology and Consciousness Studies: Quantum biology, which explores the role of quantum phenomena in biological processes, and consciousness studies represent fields where the convergence of physics and life sciences can lead to new paradigms. Recursive Realism suggests that understanding the mind-body relationship involves a feedback loop, where biological models are revised in light of quantum principles, creating a cycle that could reshape our understanding of the nature of life and awareness. For example, exploring the role of quantum coherence in photosynthesis or neural processes could reveal recursive patterns that link the fundamental nature of matter to the emergence of life. This perspective highlights how Recursive Realism can guide interdisciplinary research, using recursion to bridge the gap between the physical and the biological.

Artificial Intelligence and the Recursion of Cognitive Models: Artificial intelligence (AI) and machine learning rely on recursive algorithms that allow systems to learn from data, identify patterns, and adapt to new information. Recursive Realism suggests that advances in AI could benefit from a deeper understanding of the recursive dynamics of human thought, allowing for the development of models that better emulate the complexity of consciousness. For example, understanding how the brain uses recursive loops to process sensory data and construct a cohesive experience of the world could inform the design of AI systems that more closely mimic human cognition. This perspective allows for a richer appreciation of how recursive thought can shape the future of technology, offering a vision of machines that are capable of more nuanced interactions with the world.

Recursion in Astrobiology and the Search for Life Beyond Earth: Astrobiology, the study of life's potential beyond Earth, involves revisiting the principles of biology and planetary science in light of new discoveries about the conditions for life. Recursive Realism suggests that the search for extraterrestrial life involves a feedback loop, where each new finding about extreme environments on Earth or exoplanetary atmospheres prompts a reconsideration of what it means to be alive, creating a cycle that could reshape our understanding of biology. For example, discovering microbial life in subsurface oceans on icy moons like Europa or Enceladus could prompt a recursive reflection on the potential for life in extreme conditions, leading to new theories about the origins of life. This perspective highlights how Recursive Realism can guide astrobiological exploration, using recursion to expand our definition of life itself.

The Exploration of Consciousness and the Recursion of Subjective Experience

Consciousness remains one of the greatest mysteries of human inquiry, challenging scientific models with its subjective nature and complexity. Recursive Realism suggests that advancing the study of consciousness involves a feedback loop, where the exploration of subjective experience informs new models of the brain, creating a cycle that allows for a deeper understanding of awareness and the mind's place in the universe.

Neuroscience and the Recursion of Self-Models: Neuroscientific models of the brain often involve recursive processes, where self-referential neural circuits contribute to the experience of selfhood and awareness. Recursive Realism suggests that understanding consciousness involves a feedback loop, where the study of neural recursion informs our understanding of how the brain constructs a sense of identity and continuity. For example, exploring how the brain integrates sensory data through recursive pathways could reveal how awareness emerges from neural activity, offering a bridge between physical processes and subjective experience. This perspective highlights how Recursive Realism can shape neuroscientific research, using the principles of recursion to explore the nature of the mind.

Phenomenology and the Recursion of Inner Experience: Phenomenology, the study of the structures of experience, offers insights into how consciousness constructs reality through the intentional act of perceiving. Recursive Realism suggests that phenomenological inquiry involves a feedback loop, where reflective awareness of one's own perceptions allows for a deeper engagement with the nature of consciousness, creating a cycle that shapes our understanding of subjective reality. For example, reflecting on the experience of time or the sense of self involves a recursive examination of how the mind organises experience, revealing the processes that construct our sense of presence. This perspective allows for a richer appreciation of how Recursive Realism can guide phenomenological research, using recursive thought to explore the hidden dimensions of consciousness.

Psychedelics and the Recursion of Altered States: Psychedelic research explores how substances like psilocybin, LSD, and DMT can alter consciousness, revealing the malleability of perception and the potential for transcendent experiences. Recursive Realism suggests that the study of altered states involves a feedback loop, where each exploration of the boundaries of consciousness informs new models of the mind, creating a cycle that expands our understanding of the nature of reality. For example, experiences of ego dissolution during psychedelic journeys challenge the boundaries of self-perception, prompting a recursive reflection on what it means to be conscious. This perspective highlights how Recursive Realism can shape psychedelic research, using recursion to explore the depths of the mind and the potential for cognitive transformation.

Reimagining Fundamental Concepts and the Recursion of Scientific Paradigms

The future of scientific inquiry involves reimagining fundamental concepts like time, space, energy, and information. Recursive Realism suggests that revisiting these concepts through self-referential processes allows for a feedback loop, where each new interpretation deepens our understanding of the universe's fundamental nature, creating a cycle that drives scientific evolution.

Time and the Recursion of Temporal Understanding: Time is a concept that challenges scientific models due to its complex nature and relationship to change. Recursive Realism suggests that the study of time involves a feedback loop, where physical theories of time, such as relativity, interact with philosophical reflections on the experience of time, creating a cycle that shapes our understanding of temporal reality. For example, exploring how time's flow is experienced subjectively and how it can be relative in different frames of reference could reveal recursive connections between consciousness and cosmic processes. This perspective allows for a deeper appreciation of how Recursive Realism can shape the study of time, using recursion to bridge the empirical and the experiential.

Space and the Recursion of Spatial Models: The nature of space is central to our understanding of the universe, yet the concept of space itself evolves through new discoveries and theories. Recursive Realism suggests that understanding space involves a feedback loop, where models of physical space, such as string theory or multiverse hypotheses, interact with philosophical reflections on the nature of dimensionality, creating a cycle that shapes our understanding of the cosmos. For example, exploring how higher-dimensional spaces might intersect with our three-dimensional reality could reveal recursive patterns that connect the macroscopic and the quantum realms. This perspective highlights how Recursive Realism can guide cosmological inquiry, using recursion to explore the structure of space itself.

Information and the Recursion of Knowledge Systems: Information theory has become central to understanding the nature of reality, from the behaviour of subatomic particles to the complexity of life. Recursive Realism suggests that the study of information involves a feedback loop, where theories of data encoding and transmission interact with philosophical questions about meaning and consciousness, creating a cycle that expands our conception of reality. For example, the idea of the universe as a computational process or holographic principles suggests recursive reflections on how information structures can generate physical reality. This perspective allows for a richer appreciation of how Recursive Realism can shape our understanding of information, using recursion to explore the foundations of reality.

The future of scientific inquiry will be shaped by its ability to revisit fundamental concepts through self-referential processes, allowing for the continuous refinement of models and the integration of diverse insights. Recursive Realism provides a framework for understanding how scientific paradigms evolve through a recursive dialogue with new discoveries, revealing how interdisciplinary breakthroughs, the study of consciousness, and the reimagining of time and space can push the boundaries of knowledge. By recognising the recursive nature of scientific thought, Recursive Realism offers a vision of the future that is expansive, adaptive, and capable of embracing complexity.

Recursive Realism and the Transformation of Human Thought and Culture

Human culture is a complex web of ideas, beliefs, and practices that evolves through the reinterpretation of values, the integration of new knowledge, and the adaptation to changing conditions. Recursive Realism suggests that cultural evolution involves self-referential processes, where each generation revisits the narratives and philosophical concepts of the past, creating a feedback loop that shapes the future direction of societies. This section explores how Recursive Realism provides insights into the transformation of cultural narratives, the evolution of philosophical perspectives, and the deepening of spiritual understanding, offering a vision of human thought that is adaptive, reflective, and aligned with the complexities of modernity.

The Transformation of Cultural Narratives Through Recursive Reflection

Cultural narratives, the stories that define a society's sense of identity, purpose, and values, evolve through a process of reinterpretation and adaptation. Recursive Realism suggests that the transformation of cultural narratives involves a feedback loop, where each generation revisits the stories of the past in light of current experiences, creating a cycle that allows for the renewal of cultural meaning.

The Recursion of Myth and the Reimagining of Identity: Myths and legends serve as cultural anchors, providing stories that define a community's understanding of its origins and its values. Recursive Realism suggests that the reinterpretation of myth involves a feedback loop, where each retelling adapts the story to new contexts, creating a cycle that allows for the evolution of identity. For example, the myth of the American Dream has evolved over time, being reinterpreted in light of social movements, economic changes, and cultural critiques, shaping the way Americans perceive success and individualism. This perspective highlights how Recursive Realism can guide the transformation of cultural myths, using recursion to maintain continuity while allowing for adaptation.

Cultural Memory and the Recursion of Collective Trauma: Collective trauma, such as wars, colonialism, or social upheaval, can shape the narratives that define a society's sense of self. Recursive Realism suggests that the process of remembering involves a feedback loop, where each generation revisits the memory of trauma through new cultural lenses, creating a cycle that shapes how societies confront their past. For example, the legacy of the Holocaust has been revisited in literature, cinema, and political discourse, allowing for a recursive reflection on the nature of human rights, memory, and forgiveness. This perspective allows for a deeper appreciation of how Recursive Realism can guide the healing and reimagining of collective memory, using recursion to explore the impact of the past on the present.

The Role of Media in the Recursion of Cultural Values: Media, from literature and film to digital platforms, plays a key role in shaping cultural narratives by revisiting themes and introducing new perspectives. Recursive Realism suggests that the influence of media involves a feedback loop, where cultural values are reflected and reshaped through stories, creating a cycle that allows for the evolution of social norms. For example, science fiction often explores the impact of technology on society, allowing for a recursive reflection on the ethical implications of scientific progress. This perspective highlights how Recursive Realism can guide the role of media in shaping cultural evolution, using recursion to explore the changing boundaries of societal values.

The Evolution of Philosophical Thought Through Recursive Analysis

Philosophy evolves by revisiting fundamental questions about existence, knowledge, and values, allowing for the continuous refinement of conceptual frameworks. Recursive Realism suggests that philosophical thought involves self-referential processes, where each new perspective invites a reconsideration of old ideas, creating a feedback loop that shapes the trajectory of philosophical inquiry.

Philosophical Paradigms and the Recursion of Metaphysical Questions: Philosophical paradigms, such as existentialism, idealism, or postmodernism, often arise through a process of revisiting previous concepts, challenging existing ideas, and proposing new interpretations. Recursive Realism suggests that the evolution of metaphysical inquiry involves a feedback loop, where each paradigm shift prompts a re-evaluation of core assumptions, creating a cycle that allows for the deepening of metaphysical insight. For example, existentialist thought emerged through a recursive critique of traditional metaphysics, focusing on the experience of the individual in an uncertain universe. This perspective highlights how Recursive Realism can guide the evolution of philosophy, using recursion to explore the complexities of being and knowing.

Ethical Reflection and the Recursion of Moral Concepts: Ethics evolves as societies confront new moral dilemmas, revisiting the principles of right and wrong in light of new social realities. Recursive Realism suggests that ethical reflection involves a feedback loop, where each generation revisits moral concepts like justice, autonomy, and human rights, creating a cycle that allows for the adaptation of ethical frameworks. For example, the evolution of bioethics has required a recursive examination of the moral implications of biotechnology, genetic engineering, and artificial intelligence, shaping new ethical principles for the modern era. This perspective allows for a richer appreciation of how Recursive Realism can guide ethical inquiry, using recursion to navigate the evolving nature of moral questions.

The Role of Dialectic in the Recursion of Philosophical Debate: Dialectic, the method of argument and counter-argument, is central to philosophical inquiry, allowing for a recursive process of thesis, antithesis, and synthesis. Recursive Realism suggests that philosophical debate involves a feedback loop, where each position is revisited in light of new critiques, creating a cycle that allows for the emergence of more nuanced perspectives. For example, the dialectic between determinism and free will has evolved through a recursive exploration of how individual agency relates to causal laws, shaping the philosophy of mind. This perspective highlights how Recursive Realism can guide philosophical dialogue, using recursion to deepen our understanding of the nature of thought.

Deepening Spiritual Insights Through Recursive Contemplation

Spirituality offers a means of engaging with the mysteries of existence, allowing for the exploration of the self and its connection to the universe. Recursive Realism suggests that spiritual contemplation involves self-referential processes, where the individual revisits beliefs, experiences states of transcendence, and redefines spiritual understanding, creating a feedback loop that allows for the deepening of spiritual wisdom.

Meditative Practices and the Recursion of Inner Awareness: Meditation and mindfulness involve a recursive process of observing the mind, letting go of thoughts, and returning to a point of focus, creating a cycle that deepens inner awareness. Recursive Realism suggests that these practices involve a feedback loop, where each session of contemplation allows for a deeper understanding of the nature of consciousness, leading to spiritual insights. For example, the practice of Zen meditation involves a recursive focus on the breath, allowing practitioners to explore the mind's patterns and achieve a state of inner calm. This perspective highlights how Recursive Realism can guide meditative practices, using recursion to explore the mind's depths and cultivate spiritual clarity.

Mystical Experiences and the Recursion of Transcendence: Mystical experiences, such as feelings of oneness with the universe, often involve a transcendence of ordinary perception and a connection to a deeper reality. Recursive Realism suggests that mystical states involve a feedback loop, where the boundaries of selfhood are dissolved and reconstructed through experiences of unity, creating a cycle that shapes spiritual understanding. For example, experiences of ego dissolution during profound meditative states allow for a recursive reimagining of the nature of selfhood, offering insights into the interconnectedness of life. This perspective allows for a richer appreciation of how Recursive Realism can guide the exploration of mystical states, using recursion to delve into the nature of spiritual experience.

The Evolution of Spiritual Traditions and the Recursion of Belief Systems: Spiritual traditions evolve through the reinterpretation of teachings and the adaptation of rituals, allowing for the continuous renewal of spiritual meaning. Recursive Realism suggests that the evolution of spiritual traditions involves a feedback loop, where each generation revisits the doctrines of the past, creating a cycle that aligns ancient wisdom with modern experiences. For example, the integration of contemplative practices into secular mindfulness movements reflects a recursive engagement with Buddhist teachings, adapting ancient techniques for contemporary life. This perspective highlights how Recursive Realism can guide the evolution of spirituality, using recursion to balance continuity with adaptation, allowing for spiritual traditions to remain relevant in an ever-changing world.

Human culture is a dynamic process that involves self-referential loops, allowing for the continuous reinterpretation of cultural narratives, philosophical concepts, and spiritual insights. Recursive Realism provides a framework for understanding how societies adapt to new ideas through a recursive dialogue with the past, revealing how cultural values evolve through a cycle of reflection and adaptation. By recognising the recursive nature of thought and belief, Recursive Realism offers a vision of cultural transformation that is dynamic, resilient, and capable of navigating the complexities of the modern world.

In Chapter 50, we have explored how Recursive Realism provides insights into the future of scientific discovery, the evolution of philosophical thought, and the transformation of cultural and spiritual narratives, revealing how self-referential processes shape the next phase of human understanding. By examining how recursive dynamics can guide the exploration of new frontiers, the reinterpretation of beliefs, and the deepening of spiritual insights, Recursive Realism offers a roadmap for the future of human thought that is expansive, integrative, and capable of embracing complexity. This perspective allows us to see the universe and our place within it through a lens of interconnectedness, using recursive reflection to navigate the mysteries of existence and the possibilities of the future.

References

Barrow, J. D. and Tipler, F. J. (1986) *The Anthropic Cosmological Principle*. Oxford: Oxford University Press.

Chalmers, D. J. (1996) *The Conscious Mind: In Search of a Fundamental Theory*. Oxford: Oxford University Press.

Dennett, D. C. (1991) *Consciousness Explained*. Boston: Little, Brown and Company.

Einstein, A. (1920) *Relativity: The Special and General Theory*. Translated by R. W. Lawson. New York: Henry Holt and Company.

Grof, S. (2009) *The Ultimate Journey: Consciousness and the Mystery of Death*. Santa Cruz: Multidisciplinary Association for Psychedelic Studies.

Hawking, S. (1988) *A Brief History of Time: From the Big Bang to Black Holes*. London: Bantam Books.

Heidegger, M. (1962) *Being and Time*. Translated by J. Macquarrie and E. Robinson. Oxford: Blackwell.

Kuhn, T. S. (1970) *The Structure of Scientific Revolutions*. 2nd edn. Chicago: University of Chicago Press.

Lakoff, G. and Johnson, M. (1980) *Metaphors We Live By*. Chicago: University of Chicago Press.

Laszlo, E. (2004) *Science and the Akashic Field: An Integral Theory of Everything*. Rochester: Inner Traditions.

Penrose, R. (1994) *Shadows of the Mind: A Search for the Missing Science of Consciousness*. Oxford: Oxford University Press.

Poincaré, H. (1952) *Science and Method*. Translated by F. Maitland. New York: Dover Publications.

Rovelli, C. (2018) *The Order of Time*. London: Allen Lane.

Schneider, S. and Velmans, M. (eds.) (2007) *The Blackwell Companion to Consciousness*. Malden: Wiley-Blackwell.

Sheldrake, R. (2009) *A New Science of Life: The Hypothesis of Morphic Resonance*. 3rd edn. Rochester: Inner Traditions.

Varela, F. J., Thompson, E. and Rosch, E. (1992) *The Embodied Mind: Cognitive Science and Human Experience*. Cambridge: MIT Press.

Zohar, D. and Marshall, I. (2000) *Spiritual Intelligence: The Ultimate Intelligence*. London: Bloomsbury.

Chapter 51: Modelling and Predicting the Future with Recursive Realism

The Necessity of Recursive Models in Understanding Human Thought

Binary thinking, characterized by dividing ideas into oppositional categories like good vs. evil, true vs. false, or nature vs. nurture, has historically provided simplistic frameworks for understanding complex issues. However, recursive models reveal that human thought and societal dynamics cannot be fully captured by such dichotomous thinking. Recursive Realism argues that predicting human thought and cultural evolution requires a model that recognizes the complexity and feedback inherent in mental processes and social interactions. This section examines why recursive thinking is critical for understanding the adaptive nature of human cognition and how it offers a more comprehensive model than binary frameworks.

The Limitations of Binary Extremism and the Need for Recursive Understanding

Binary models provide clarity by simplifying complex phenomena into either/or frameworks, yet they often overlook the interconnected processes that drive adaptation and change in thought and society. Recursive Realism suggests that this simplification fails to capture the continuous interplay of self-reflection, adaptation, and adjustment that defines human behaviour.

Nature vs. Nurture: Revisiting an Outdated Dichotomy: The nature vs. nurture debate is a classic example of binary thinking that fails to recognize the recursive interplay between genes and environment. Recursive Realism proposes that genetic expression and environmental influences form a feedback loop, where each modifies the other in a dynamic process. For instance, epigenetic changes, where environmental factors influence gene expression, demonstrate that genes and environment are not oppositional forces but part of a recursive system that shapes development. This perspective emphasizes that predicting developmental outcomes requires a model that incorporates the recursive nature of these interactions, rather than framing them as oppositional.

Political Polarization and the Need for Complexity: Political discourse is often polarized into binary categories, such as liberal vs. conservative, which oversimplify the spectrum of beliefs and ignore the feedback processes that shape political identity. Recursive Realism suggests that individual political beliefs are shaped by a continuous loop of experience, reflection, and social feedback, where exposure to diverse perspectives can lead to evolving views. For example, individuals might shift their political stance as they encounter new information and revisit past beliefs in light of current contexts. This model reveals how political identities are not static categories but dynamic systems that adapt through recursive interactions.

The Challenge of Extremism in Social Movements: Social movements often emerge as reactions to perceived injustices, but they can become rigid in their ideologies, leading to extremist positions that resist adaptation. Recursive Realism posits that successful movements involve a recursive process of reflecting on goals, adapting strategies, and responding to feedback from wider society. For instance, movements that adjust their tactics to address new social dynamics are more likely to achieve sustainable change than those that maintain fixed ideologies. This perspective highlights how recursive models can guide social change, offering a path forward that balances principled action with adaptability.

Recursive Models as a Tool for Predicting Human Thought

Recursive models are essential for predicting human behaviour and thought because they account for the adaptability and complexity of cognitive processes. Recursive Realism emphasizes that thought patterns involve self-referential feedback, where each reflection can lead to new insights and adjustments in beliefs.

The Role of Recursion in Cognitive Processes: Human cognition is characterized by the ability to reflect on thoughts, anticipate outcomes, and adjust beliefs based on past experiences. Recursive Realism suggests that cognitive processes operate through a feedback loop, where each decision or thought is revised as new information is integrated, allowing for flexibility in adaptation. For example, learning involves revisiting concepts, adjusting mental models, and incorporating feedback, creating a dynamic process that fosters growth. This model provides a more accurate representation of human cognition than linear models, offering insights into how thought patterns adapt to changing contexts.

Modelling Social Dynamics Through Recursive Reflection: Societal changes often arise from the interplay of individual actions, cultural values, and collective experiences. Recursive Realism suggests that modelling social evolution involves a feedback loop, where cultural norms are revisited and redefined through dialogue and reflection, allowing for continuous adaptation. For example, the rise of digital communication has transformed how cultural narratives are shared, leading to recursive adaptations in social behaviour and norms. This perspective emphasizes that anticipating cultural trends requires a model that can capture the cyclical nature of cultural evolution.

The Principle of Moderation as a Test for Recursive Models: Moderation, the approach of avoiding extremes and seeking balance, serves as a key measure for evaluating the practicality of Recursive Realism. Moderate approaches are often more adaptable because they avoid the rigidity of extreme positions, allowing for continuous adjustment. Recursive Realism suggests that models that incorporate moderation can better respond to complex realities, as they remain flexible in their interpretations and applications. For example, moderate political strategies often succeed because they can adjust to diverse viewpoints, creating a more inclusive dialogue. This perspective underscores how moderation provides a pragmatic test for the resilience of recursive models.

Recursive Realism as a Framework for Modelling the Future

The future of human thought will be shaped by technological advancements, cultural evolution, and shifts in social values. Recursive Realism offers a framework for modelling these changes, emphasizing how self-referential feedback loops can be used to predict emergent trends and guide adaptation. This section explores how Recursive Realism can help anticipate future developments, focusing on the role of technology, the evolution of social structures, and the potential for a world beyond binary thinking.

Technology and the Recursion of Human Cognition

Technological innovations, such as artificial intelligence, virtual reality, and biotechnology, are transforming the nature of human thought. Recursive Realism suggests that technology introduces new feedback loops into cognitive processes, reshaping how individuals interact with information and each other.

Artificial Intelligence and the Recursive Nature of Learning: AI systems that use recursive algorithms for pattern recognition and adaptive learning mirror human cognitive processes, offering new insights into the evolution of intelligence. Recursive Realism posits that the interaction between humans and AI creates a feedback loop, where each interaction influences the cognitive capabilities of both machines and humans. For example, AI assistants that learn from user behaviour can adapt to individual preferences, while users adjust their engagement based on AI suggestions, creating a dynamic cycle of mutual adaptation. This model highlights how recursive interactions can shape the evolution of intelligent systems and human thought.

Digital Media and the Recursion of Cultural Expression: The rise of digital media has transformed how cultural narratives are shared and evolve, creating a recursive space for dialogue, innovation, and cultural exchange. Recursive Realism suggests that digital platforms enable new forms of recursive engagement, where individuals and communities constantly revisit ideas, challenge norms, and reinterpret cultural values. For example, online communities that engage in collaborative content creation, such as open-source projects or digital storytelling, reflect a recursive culture that adapts through feedback and iteration. This perspective emphasizes how Recursive Realism can guide our understanding of the cultural impacts of digital technology, offering a framework for predicting the evolution of social interaction.

The Evolution of Society Beyond Binary Extremism

Societies that move beyond binary thinking can better navigate complexity, creating a culture that embraces diversity and moderation. Recursive Realism suggests that the future of social evolution involves a shift toward models that recognize the interplay between different values and perspectives, creating a feedback loop that supports adaptive change.

Moderation as a Tool for Social Stability: Moderation involves balancing diverse viewpoints, avoiding extremes, and seeking common ground. Recursive Realism posits that moderation is essential for stable social systems, as it allows for a dynamic exchange of ideas that encourages adaptation and cooperation. For example, societies that embrace inclusive policies and adapt to cultural differences are often more resilient to social unrest. This perspective highlights how Recursive Realism can guide social strategies toward moderation, using feedback loops to refine social structures in response to changing needs.

Recursive Realism offers a model that captures the complexity of human thought and social dynamics through self-referential feedback loops, providing a framework for predicting the future that moves beyond binary extremism. By embracing moderation and recognizing the dynamic interplay between thought processes and cultural evolution, Recursive Realism presents a path for understanding the nuances of cognitive adaptation and social change. This chapter highlights how Recursive Realism allows for a more comprehensive understanding of human evolution, using recursion to model the intricate relationships that shape our future.

References

Bateson, G. (1972) *Steps to an Ecology of Mind: Collected Essays in Anthropology, Psychiatry, Evolution, and Epistemology*. Chicago: University of Chicago Press.

Dennett, D. C. (1991) *Consciousness Explained*. Boston: Little, Brown and Company.

Giddens, A. (1984) *The Constitution of Society: Outline of the Theory of Structuration*. Cambridge: Polity Press.

Kahneman, D. (2011) *Thinking, Fast and Slow*. London: Penguin Books.

Kuhn, T. S. (1970) *The Structure of Scientific Revolutions*. 2nd edn. Chicago: University of Chicago Press.

Lakoff, G. and Johnson, M. (1980) *Metaphors We Live By*. Chicago: University of Chicago Press.

Lewin, K. (1951) *Field Theory in Social Science: Selected Theoretical Papers*. Edited by D. Cartwright. New York: Harper & Row.

Luhmann, N. (1995) *Social Systems*. Translated by J. Bednarz Jr. and D. Baecker. Stanford: Stanford University Press.

Prigogine, I. and Stengers, I. (1984) *Order Out of Chaos: Man's New Dialogue with Nature*. New York: Bantam Books.

Tversky, A. and Kahneman, D. (1974) 'Judgment Under Uncertainty: Heuristics and Biases', *Science*, 185(4157), pp. 1124–1131.

Varela, F. J., Thompson, E. and Rosch, E. (1992) *The Embodied Mind: Cognitive Science and Human Experience*. Cambridge: MIT Press.

von Bertalanffy, L. (1968) *General System Theory: Foundations, Development, Applications*. New York: George Braziller.

Wiener, N. (1948) *Cybernetics: Or Control and Communication in the Animal and the Machine*. Cambridge: MIT Press.

General References

Dennett, D. C. (1991). *Consciousness Explained*. Boston: Little, Brown and Company.

This foundational work explores the nature of consciousness, providing insights into how self-referential processes and cognitive loops play a role in human perception, aligning with the recursive dynamics discussed in Recursive Realism.

Hofstadter, D. R. (1979). *Gödel, Escher, Bach: An Eternal Golden Braid*. New York: Basic Books.

A classic exploration of recursion, self-reference, and emergent complexity, Hofstadter's work is instrumental in understanding how recursive processes can shape thought, consciousness, and the structure of reality.

Kauffman, S. A. (1993). *The Origins of Order: Self-Organization and Selection in Evolution*. New York: Oxford University Press.

Kauffman's work on self-organization and complexity provides a biological perspective that complements the discussion on the interaction between genes and environment, and the role of recursive processes in evolution.

Penrose, R. (1989). *The Emperor's New Mind: Concerning Computers, Minds, and the Laws of Physics*. Oxford: Oxford University Press.

Penrose examines the connections between physics, consciousness, and computation, offering perspectives on the nature of intelligence that are relevant to the themes of Recursive Realism and its approach to modeling the mind.

Varela, F. J., Thompson, E., & Rosch, E. (1991). *The Embodied Mind: Cognitive Science and Human Experience*. Cambridge: MIT Press.

This book emphasizes the role of the body and environment in shaping cognition, aligning with Recursive Realism's focus on the dynamic interplay between internal processes and external influences.

Bateson, G. (1972). *Steps to an Ecology of Mind: Collected Essays in Anthropology, Psychiatry, Evolution, and Epistemology*. San Francisco: Chandler Publishing Company.

Bateson's exploration of systems theory and recursive feedback in understanding mind and nature is foundational for discussions on how recursive models can integrate biological, psychological, and cultural dimensions.

Prigogine, I., & Stengers, I. (1984). *Order Out of Chaos: Man's New Dialogue with Nature*. New York: Bantam Books.

Prigogine's work on non-equilibrium systems and the emergence of order is relevant to understanding how recursive processes can create adaptive change in complex systems, such as human societies.

Gazzaniga, M. S. (2008). *Human: The Science Behind What Makes Us Unique*. New York: Ecco/HarperCollins.

Gazzaniga's work on neuroscience and cognitive science provides empirical insights into the functioning of the human brain, supporting the discussion on how Recursive Realism is grounded in the actual dynamics of brain processes.

Tomasello, M. (1999). *The Cultural Origins of Human Cognition*. Cambridge: Harvard University Press.

This work explores how cultural practices shape human cognition, aligning with Recursive Realism's emphasis on the feedback loop between culture and individual thought in shaping the evolution of human intelligence.

Barabási, A. L. (2002). *Linked: The New Science of Networks*. Cambridge: Perseus Publishing.

Barabási's insights into the nature of networks and complex connections provide a valuable perspective on how recursive interactions shape complex systems, including social networks and cognitive processes.

Tononi, G. (2012). *Phi: A Voyage from the Brain to the Soul*. New York: Pantheon.

Tononi's theory of integrated information (IIT) explores how consciousness arises from integrated networks, providing a theoretical foundation for understanding the recursive nature of perception and awareness.

Morin, E. (2008). *On Complexity*. Cresskill: Hampton Press.

Morin's examination of complexity and the importance of embracing uncertainty offers a philosophical backdrop for Recursive Realism, especially in its application to understanding the future of human thought.

Scharmer, C. O. (2009). *Theory U: Leading from the Future as It Emerges*. San Francisco: Berrett-Koehler Publishers.

Scharmer's exploration of change and the dynamics of learning through iterative reflection provides insights into how recursive processes can guide personal and collective transformation.

Eagleman, D. (2011). *Incognito: The Secret Lives of the Brain*. New York: Pantheon Books.

Eagleman's work on the subconscious mind and the hidden processes of the brain complements the themes of Recursive Realism, particularly the idea of recursive feedback loops in shaping human perception and decision-making.

Dehaene, S. (2014). *Consciousness and the Brain: Deciphering How the Brain Codes Our Thoughts*. New York: Viking.

Dehaene's research on the neural mechanisms of consciousness supports the empirical foundation of Recursive Realism, exploring how recursive neural activity contributes to conscious awareness.

Further Readings

Cognitive Science and Neuroscience

Edelman, G. M. (1992). *Bright Air, Brilliant Fire: On the Matter of the Mind*. New York: Basic Books.

Edelman explores the dynamic nature of consciousness and neural activity, offering insights into how complex brain processes might align with the principles of recursive thinking.

LeDoux, J. E. (2002). *Synaptic Self: How Our Brains Become Who We Are*. New York: Viking.

This book examines the neural mechanisms underlying identity and consciousness, contributing to our understanding of how recursive loops in the brain shape personality and behavior.

Damasio, A. R. (1999). *The Feeling of What Happens: Body and Emotion in the Making of Consciousness*. New York: Harcourt Brace.

Damasio's work on the relationship between emotion and consciousness provides a foundation for exploring the feedback loops between bodily states and self-awareness.

Churchland, P. S. (2002). *Brain-Wise: Studies in Neurophilosophy*. Cambridge: MIT Press.

Churchland's exploration of neurophilosophy offers a detailed analysis of how brain function shapes concepts of self, free will, and consciousness, relevant to the themes of Recursive Realism.

Metzinger, T. (2009). *The Ego Tunnel: The Science of the Mind and the Myth of the Self*. New York: Basic Books.

Metzinger challenges traditional notions of selfhood, emphasizing the role of brain-generated models in shaping our sense of reality.

Complex Systems and Emergence

Holland, J. H. (1998). *Emergence: From Chaos to Order*. Reading: Addison-Wesley.

Holland provides a thorough examination of how complex systems emerge from simple rules, offering insights into how recursive processes contribute to the organization of life and thought.

Capra, F. (1996). *The Web of Life: A New Scientific Understanding of Living Systems*. New York: Anchor Books.

Capra's holistic approach to understanding the interconnectedness of life aligns with the recursive principles discussed in Recursive Realism, particularly in the context of biological and ecological systems.

Heylighen, F., & Joslyn, C. (2001). *Cybernetics and Second-Order Cybernetics*. In R. A. Meyers (Ed.), *Encyclopedia of Physical Science & Technology* (3rd ed., pp. 155-170). New York: Academic Press.

This work explores how cybernetic principles can be applied to understanding self-referential systems, complementing the discussions on recursion in thought and perception.

Mitchell, M. (2009). *Complexity: A Guided Tour*. Oxford: Oxford University Press.

Mitchell's accessible introduction to complexity science provides insights into the role of feedback and adaptation in complex systems, offering a foundation for understanding recursive dynamics.

Philosophy of Mind and Metaphysics

Nagel, T. (1986). *The View from Nowhere*. New York: Oxford University Press.

Nagel's exploration of objectivity and subjectivity provides a philosophical backdrop for discussions on the nature of reality and the role of self-reference in shaping knowledge.

Chalmers, D. J. (1996). *The Conscious Mind: In Search of a Fundamental Theory*. Oxford: Oxford University Press.

Chalmers's work on the hard problem of consciousness offers a deep inquiry into the nature of experience, contributing to the theoretical foundations of Recursive Realism.

Searle, J. R. (1992). *The Rediscovery of the Mind*. Cambridge: MIT Press.

Searle challenges computational theories of mind, emphasizing the importance of understanding the subjective aspects of consciousness, which aligns with Recursive Realism's emphasis on self-referential processes.

Whitehead, A. N. (1929). *Process and Reality*. New York: Macmillan.

Whitehead's process philosophy, which views reality as a series of interconnected events, offers a theoretical foundation for understanding the fluid and adaptive nature of perception in Recursive Realism.

Clark, A. (2016). *Surfing Uncertainty: Prediction, Action, and the Embodied Mind*. Oxford: Oxford University Press.

Clark's exploration of predictive coding and the brain's constant adjustments to sensory information supports the recursive aspects of how humans perceive and interact with the world.

Evolutionary Theory and Biology

Mayr, E. (2001). *What Evolution Is*. New York: Basic Books.

Mayr's detailed explanation of evolutionary mechanisms provides context for understanding the recursive feedback loops between genetic variation and environmental pressures.

Margulis, L., & Sagan, D. (2002). *Acquiring Genomes: A Theory of the Origins of Species*. New York: Basic Books.

Margulis's work on symbiosis and the role of cooperation in evolution offers a perspective that aligns with Recursive Realism's focus on interconnectedness and adaptation.

Dawkins, R. (2006). *The Selfish Gene* (30th Anniversary ed.). Oxford: Oxford University Press.

Dawkins's concept of genes as units of selection provides a basis for exploring how genetic information evolves through recursive interactions with the environment.

Wilson, E. O. (2012). *The Social Conquest of Earth*. New York: Liveright.

Wilson's exploration of social evolution and cooperation complements the discussion on how recursive processes shape the interplay between individual and group dynamics in human societies.

Psychology and Cognitive Development

Piaget, J. (1934). *The Construction of Reality in the Child*. New York: Basic Books.

Piaget's insights into cognitive development and the role of self-referential thought in learning offer a foundation for understanding the recursive nature of human cognition.

Vygotsky, L. S. (1978). *Mind in Society: The Development of Higher Psychological Processes*. Cambridge: Harvard University Press.

Vygotsky's emphasis on the social context of learning and cognitive development aligns with Recursive Realism's view of the feedback loops between individual thought and cultural influences.

Bandura, A. (1986). *Social Foundations of Thought and Action: A Social Cognitive Theory*. Englewood Cliffs: Prentice-Hall.

Bandura's work on self-efficacy and the role of feedback in shaping behavior provides empirical support for the recursive nature of learning and adaptation.

Spirituality and the Nature of Reality

Wilber, K. (2000). *A Theory of Everything: An Integral Vision for Business, Politics, Science, and Spirituality*. Boston: Shambhala.

Wilber's integral theory offers a framework for integrating science, spirituality, and philosophy, complementing the holistic vision of Recursive Realism.

Tolle, E. (2005). *A New Earth: Awakening to Your Life's Purpose*. New York: Penguin Group.

Tolle's emphasis on self-awareness and the transcendence of ego aligns with Recursive Realism's focus on understanding the deeper layers of consciousness and self-reflection.

Hanh, T. N. (1999). *The Heart of the Buddha's Teaching: Transforming Suffering into Peace, Joy, and Liberation*. New York: Broadway Books.

Thich Nhat Hanh's teachings on mindfulness and the interconnectedness of all life offer a spiritual perspective that resonates with the recursive exploration of self and reality.

McGilchrist, I. (2009). *The Master and His Emissary: The Divided Brain and the Making of the Western World*. New Haven: Yale University Press.

McGilchrist's exploration of the divided nature of brain function provides insights into how different cognitive processes can create a holistic understanding of reality through recursive integration.

385

386

Recursive Realism: The Universe Beyond Binary Extremism

In a world increasingly dominated by polarized thinking and binary extremes, Recursive Realism offers a groundbreaking new model for understanding the complexity of the human mind, the nature of perception, and the vast universe we inhabit. This book challenges conventional frameworks by introducing a theory that explores how our thoughts, experiences, and realities are shaped through recursive processes, where self-reflection, feedback loops, and adaptation lead to deeper understanding.

Recursive Realism guides readers beyond simplistic dichotomies like nature vs. nurture or good vs. evil, instead offering a holistic approach to examining how the brain, society, and the cosmos interact in dynamic, self-referential ways. By engaging with empirical evidence from neuroscience, cognitive science, and evolutionary theory, this book reveals how human intelligence is not static but constantly evolving through recursive adaptation.

In Recursive Realism, the universe is not perceived through rigid, binary distinctions but through an adaptive, evolving lens, where perception itself is fluid and capable of transformation. This model enables us to construct, deconstruct, and reconstruct our understanding of reality, opening new doors to how we might evolve and understand our place in the universe.

Packed with insights into how human intelligence can evolve in a future free from binary extremism, this book is a must-read for anyone interested in the future of thought, philosophy, and the cutting-edge theories that will shape tomorrow's intellectual landscape.

387

388

www.ingramcontent.com/pod-product-compliance
Lightning Source LLC
Chambersburg PA
CBHW052237220526
45471CB00001B/75